Climate Justice Now

Climate Justice Now

Crossing Disciplines to Combat Our Planetary Crisis

Edited by
Rebecca Marwege,
Nikhar Gaikwad,
and Joerg Schaefer

Columbia University Press
New York

Columbia University Press
Publishers Since 1893
New York   Chichester, West Sussex
cup.columbia.edu

Library of Congress Cataloging-in-Publication Data
Names: Marwege, Rebecca editor | Gaikwad, Nikhar editor | Schaefer, Joerg
  editor
Title: Climate justice now : crossing disciplines to combat our planetary
  crisis / edited by Rebecca Marwege, Nikhar Gaikwad, and Joerg Schaefer.
Other titles: Crossing disciplines to combat our planetary crisis
Description: New York : Columbia University Press, [2026] | Includes
  bibliographical references and index.
Identifiers: LCCN 2025032394 (print) | LCCN 2025032395 (ebook) | ISBN
  9780231220231 hardback | ISBN 9780231220248 trade paperback | ISBN
  9780231563130 epub | ISBN 9780231565172 PDF
Subjects: LCSH: Climate justice | Environmental justice | Climate change
  mitigation | Carbon dioxide mitigation | Interdisciplinary approach in
  education
Classification: LCC GE220 .C5594 2026 (print) | LCC GE220 (ebook) | DDC
  363.7--dc23/eng/20251124
LC record available at https://lccn.loc.gov/2025032394
LC ebook record available at https://lccn.loc.gov/2025032395

Cover design: Julia Kushnirsky
Cover image: Shutterstock

GPSR Authorized Representative: Easy Access System Europe, Mustamäe tee 50, 10621 Tallinn,
Estonia, gpsr.requests@easproject.com

Contents

Acknowledgments   *ix*

Part I. Shifting Tides: Uniting Disciplines for Climate Justice

Prelude   3
Sheila R. Foster

1. When Natural Scientists, Social Scientists, and Humanists
   Meet to Discuss Climate Justice   12
   Rebecca S. Marwege, Nikhar Gaikwad, and Joerg M. Schaefer

**Part II. The Intersection of Climate, Environment, and
Justice: Theory, Politics, Science, and Urban Perspectives**

2. Introduction to Theories of Environmental and
   Climate Justice   45
   Mary E. Witlacil

3. The Politics of Climate Justice   66
   Zara Riaz and Page Fortna

4. Transport, Justice, and Climate Crisis in Cities:
   A Paradigm Shift 88
   Jacqueline M. Klopp and Festival G. Boateng

5. The Potential Conflict Between Climate Justice and
   Environmental Justice: Reducing Greenhouse Gas Emissions
   Does Not Improve Air Quality 109
   Róisín Commane

**Part III. Climate Change as a Harm Multiplier:
Exploring Sociology, Migration, and Environmental Health**

6. Sociological Perspectives on Climate Justice 127
   Jennifer E. Givens and Mufti Nadimul Quamar Ahmed

7. Climate Justice and Climate Mobility: International
   Migration from Central America and West Africa 153
   *Alex de Sherbinin, David Wrathall, Susana Adamo,
   Sara Pan-Algarra, and Elena Giacomelli*

8. Climate Justice in the Field: Climate Change and
   the Health of Migrant Agricultural Workers 179
   *Lewis H. Ziska, Jeffrey L. Shaman, Emily Weaver,
   and Ami Zota*

**Part IV. Climate Justice, Capitalism, and Colonialism:
From Literature to History to Science**

9. Justice, the Incommensurable, and the Scale(s) of
   Business as Usual: A Literary Studies Approach 195
   Jennifer Wenzel

10. Climate Justice in the Arctic: Multispecies Approaches in
    Anthropology and History 208
    Emma Gilheany and Julia Lajus

11. Bridging the Gulf: Intersections of Geology, Biology,
    and Climate Justice 226
    Kailani Acosta and Gisela Winckler

12. How Much Positive Influence on the Climate Problem
    Can One Have as an Academic or Industry
    Climate Scientist?   236
    Adam Sobel and Melanie Bieli

    Part V. Rethinking Knowledge at a Time of Crisis:
    Climate Justice, Expertise, and Community

13. Climate (In)justice for Whom? Alternative Theories and
    the Absence of Scientific Language   255
    Sheng Long

14. Climate Justice and Religion   272
    Raffaella Taylor-Seymour and Courtney Bender

15. Building a Better Model for Flood Protection Planning   288
    Paul Gallay

    Part VI. Pursuing Climate Justice in Academia and Beyond:
    Embracing Complexity and Bridging Disciplines

16. The Elusive Challenge of Climate Justice and
    a Call to Action   315
    Rebecca S. Marwege, Nikhar Gaikwad, and Joerg M. Schaefer

    Contributors   337
    Index   343

Acknowledgments

This book emerged from many conversations about climate justice that started at Columbia University during the Covid pandemic and grew over the years. In 2021 we founded an interdisciplinary network, the Decarbonization, Climate Resilience, and Climate Justice Network (DCRJ), under the auspices of the Earth Institute and then the Climate School. This network expanded discussions that began at the Columbia Interdisciplinary Research on Climate (CIRC) workshop and sought to formally establish a network of scholars at Columbia University who were interested in collaborating to further scholarship, pedagogy, and advocacy related to themes of climate justice. The idea of this book was developed in the third year of the DCRJ network as we thought about ways to make the conversations we were having in our network accessible to a wider audience

We thank Page Fortna and Noah Zucker for co-organizing CIRC, Lauren Moseley for co-organizing DCRJ, and the hundreds of faculty, researchers, and graduate students who participated in CIRC and DCRJ workshops, conferences, and events over the five years of their existence. A number of sponsors have generously supported these working groups and the book project. We thank Columbia's Institute for Social and Economic Research and Policy, which sponsored the CIRC workshop. We also thank the Earth Institute and the Climate School for providing us with funding, space, and administrative support over several years. Special thanks to Sandra Goldmark, Andrea Lopez Duarte, and Carly Roberts for their tireless support in developing this book project. We also extend our thanks to the Lamont Climate Center and the Vetlesen Foundation for their generous support.

In addition, we are grateful to the Heyman Center for sponsoring our book workshop and thank Lindsey Schram for providing administrative support in planning it. We also thank the Teachers College at Columbia University for allowing us to relocate our workshop to their spaces at short notice in May 2024.

For believing in this project from the first time we pitched it to them, we extend our deep gratitude to Caelyn Cobb and Emily Elizabeth Simon at Columbia University Press. We are also grateful to three anonymous reviewers who gave invaluable feedback on the initial proposal for this book, as well as two anonymous reviewers who gave detailed and constructive feedback on the final draft of the manuscript. We thank Leel Moka Dias, who helped us to generate the graphs for the first chapter.

Finally, we thank our chapter contributors for their dedication to this project and their permission to use the material. When we started the network, we had no idea of the scope the project would take, and we are grateful to all chapter authors for providing their excellent contributions under a tight timeline. We remain deeply impressed by their enthusiasm and support, which made this edited volume possible.

Climate Justice Now

Shifting Tides

*Uniting Disciplines for
Climate Justice*

Prelude

SHEILA R. FOSTER

The year was 1996, and I was a newly minted assistant professor at Rutgers University Camden when a resident who lived in a heavily polluted area of the city called my office. This resident had heard that I was an expert on environmental justice. The caller explained that a new polluting facility was coming into the neighborhood, Waterfront South, which was predominantly Black and Latino. The neighborhood was host to a regional sewage treatment plant, a trash-to-steam incinerator, and a cogeneration power plant; contained two heavily contaminated Superfund sites, located close to a major highway; and suffered from high rates of respiratory illnesses and cancer. Residents had organized and formed South Camden Citizens in Action (SCCA) not only to push back on the state's proposal to permit an additional facility in their neighborhood, but also to draw attention to the ways that their neighborhood had become a sacrifice zone in the region and state for hazardous land uses that more affluent and suburban communities were able to resist successfully.

Around the same time, I met Zulene Mayfield, an activist in Chester, Pennsylvania, and became involved in the struggle against environmental racism in her community. Zulene was the founder of Chester Residents Concerned for Quality of Life (CRCQL), which grew out of residents' efforts to resist new polluting facilities in their neighborhoods, many of which shared the same social, economic, and pollution profile as the Waterfront South neighborhood in Camden. The huge trucks that roared through their neighborhoods day and night, disturbing the peace in their community and damaging the foundation of their homes and their property values, spurred residents to begin organizing. The residents felt trapped in their neighborhoods. By the late 1990s, CRCQL had become a political force in the region, resisting the further degradation of their neighborhood and taking on powerful real estate interests.

I moved to New York City in 2001 to teach at Fordham University. Soon after, I became involved in the effort by the Harlem-based environmental justice organization, WE ACT, to close down some of the bus depots around the city. All the depots were located in communities of color. It was common for buses to idle near or within those depots, contributing to the poor air quality and increasing rates of asthma in those neighborhoods (one in four children in Harlem had asthma). In addition to the bus depots, northern Manhattan was host to multiple polluting or hazardous sources, including the North River Wastewater Treatment Plant and the (now closed) 135th Street Marine Waste Transfer Station. Founded in 1988, WE ACT has become one of the world's most significant environmental justice organizations over the last thirty-five-plus years, mobilizing more than 600,000 residents locally while shaping federal, state, and local policies that promote environmental justice at various levels.

A core tenet of the environmental justice movement is that environmentally overburdened communities should "speak for themselves" and define the kinds of injustices they face and the type of justice they seek. This is still true. What is also true, and worth highlighting to underscore the importance of a book like this, is that the development and success of the environmental justice movement is as much a product of these communities' organizing and activism as it is the product of a dynamic and multidisciplinary cast of academics, researchers, and practitioners working alongside them.

The period during which I was becoming an academic in the field was an inflection point for the environmental justice movement. It is worth remembering how much learning and contestation there was about the idea that race was a driving factor in the placement or location of environmentally hazardous land uses and in the exposure of the most socially and economically marginalized populations to lead and other toxins in the indoor and outdoor environment. Mainstream environmental organizations, public officials, and agency regulators struggled to address and redress the claims of a movement that labeled itself "environmental" but whose focus—discrimination and racism—went beyond the set of concerns that characterized the traditional environmental movement.

As outsiders to the movement struggled to understand its demands, the movement and its leaders were evolving. They began embracing broader and overlapping issues, including Indigenous land and human rights, in the United States and abroad. At the same time, academic and policy detractors relied heavily on market explanations and economic analysis that normalized the glaringly obvious distributional inequities. These issues became the focus of increasingly robust scholarly literature and policy debate. The central point of contestation in the field at the time was how much race and racism drove the siting decisions and disproportionate pollution exposures that the predominantly Black and lower-income communities were challenging and protesting against.

Dr. Benjamin Chavis coined the term "environmental racism" in 1982 during protests opposing the placement of a waste landfill in Warren County, North Carolina, the birthplace of the environmental justice movement. Chavis defined environmental racism as the intentional siting of polluting and waste facilities in communities primarily populated by African Americans, Latines, Indigenous People, Asian Americans and Pacific Islanders, migrant farmworkers, and low-income workers. He was also the leader of the United Church of Christ (UCC)'s Commission for Racial Justice, which produced the groundbreaking study "Toxic Wastes and Race in the United States" (1987), finding race among the top variables associated with the location of a toxic waste facility. The UCC study was updated in 1994 and found even greater disparities in the demographics of people who lived around hazardous waste sites.

A slew of other studies followed in the 1990s, predominantly conducted by academics and researchers from various disciplines—social scientists, economists, legal scholars, etc. Some of the most prominent academics in the movement published their seminal works during this period, including Robert Bullard's *Dumping in Dixie* (1990) and Bunyan Bryant and Paul Mohai's *Race and the Incidence of Environmental Hazards* (1992). These were among the first works I read on the subject, and they shaped the kinds of questions I pursued in my own research as a young scholar. This empirical work continued over the next two decades, even as the debate over the role of race receded, in part because of the strength of social science proving its predominant role.

Scholars like myself entered the discourse seeking to fill in the more normative dimensions of this emerging academic field. We began to leverage social science and empirical work and to connect that work more deeply with community voices and stories. My early work, for example, articulated and analyzed claims of racial injustice and discrimination from the ground up, by capturing the experiences of some of the communities I was working with at the time. These stories provided an important window into the economic and social factors and dynamics rendering those communities vulnerable to disproportionate hazard exposure and, at times, sacrifice zones for the country's economic progress. The stories also made it possible to observe the legal and administrative processes that were failing these communities and leaving them severely underprotected.

Seeking to ground the empirical debate in the reality of structural exclusion and historical racism, I joined scholars from different disciplines to peel back the deeper forces and drivers behind the distributive patterns the empirical studies were bringing into clear view. These factors included racially biased and exclusionary housing and land use practices and policies, past and present, such as expulsive zoning and redlining. These practices and policies left neighborhoods such as Camden, Chester, and Harlem disinvested and polluted and shaped them into attractive targets for new pollution sources. The historical and social structures lurking behind the distributive patterns were linkages that required researchers trained in history, sociology, and law to work through, often together. These linkages were utilized in policy and legal arenas where practitioners and public officials applied them to new policies directed at undoing and remedying the historical legacies driving the disproportionate exposures that studies revealed.

Additionally, I experienced the benefits of engaging across disciplines when I began to bring legal claims on behalf of communities in Camden, Chester, and Harlem. Legal academics like myself, with practitioners, built on the rich and emerging interdisciplinary field to resuscitate and fortify legal theories and claims in both administrative and judicial arenas. We published articles and books helping to create a new field of law that bridged two legal disciplines that often proved more incompatible than compatible: environmental law and civil rights law. We used quantitative data developed by social scientists to help regulatory agencies like the Environmental Protection Agency realize that pollution emission standards left vulnerable populations overexposed to environmental toxins. Regulators responded by altering those standards to be more protective of those populations. These efforts paid off, initially, for the communities most burdened.

President Clinton signed the first Executive Order on Environmental Justice in 1994, which directed all federal agencies to take actions to ensure that programs did not result in disproportionate impacts on already overburdened low-income communities and communities of color. This order and President Biden's second Executive Order on Environmental Justice, reinforcing the first one, were rescinded by President Trump during his second term in office. In other legal arenas, it is notable that the first environmental justice lawsuits were brought in the early 1980s, and federal courts universally rejected the claims that Black communities that were host to waste facilities and landfills suffered from racial discrimination. It wasn't until the late 1990s and early 2000s, in lawsuits filed on behalf of Camden and Chester residents, that federal courts ruled their racial discrimination claims as meritorious. While these claims were later frustrated by Supreme Court decisions, the initial successes were a testament to the early multidisciplinary academic work in the field.

Climate justice seems to be at a similar inflection point to that faced by environmental justice during this seminal era. Expert bodies like the Intergovernmental Panel on Climate Change (IPCC) and their national and local counterparts, like the New York City Panel on Climate Change (NPCC), have played a significant role in showing that climate impacts such as extreme heat and flooding occur on an unequal landscape across the world, imposing disproportionate harm and losses to socially and economically marginalized populations. Researchers from different disciplines increasingly link these disparate impacts to colonialism, racial capitalism, and the legacy effects of racist and discriminatory land use practices like redlining.

In the United States, for example, studies have demonstrated that historically disinvested neighborhoods are more likely to experience much higher and deadlier temperatures, above their city averages, during increasingly severe heat waves. These areas tend to have older, less energy-efficient housing and higher energy burdens. Increased flooding from heavy rains is particularly consequential for low-income populations as habitable areas shrink and housing prices rise, leaving only affluent residents able to relocate away from flooding risks. Social and economically vulnerable communities tend to lack flood-mitigating infrastructure while most of the benefits of green and resilient infrastructure go to areas with wealthier, whiter, and better-educated residents who are more likely to receive support for seawalls, funding to elevate homes, and drainage infrastructure. These racial disparities have been linked empirically to the legacy of government redlining and disinvestment in Black communities over time, shaping the ability of Black neighborhoods and households to adapt to climate risks and to build household and community wealth.

None of this is news to residents in Camden, Chester, and Harlem, who continue to face disproportionate environmental hazard exposures, such as poor indoor and outdoor air quality, and experience unequal access to environmental amenities such as green space and affordable fresh food. In addition, many of these communities now have to deal with the intersecting risks from legacy environmental hazards and disproportionate climate impacts. They understand that the root causes of environmental racism and injustice are the same causes that render them uniquely vulnerable to climate impacts and stressors. This is why many traditional environmental justice organizations have pivoted seamlessly to focus their organizing and advocacy activities on climate justice.

For US-based grassroots organizations, climate justice, like environmental justice, is about more than pollution, contamination, extreme weather, and natural disasters. It is also about the intersecting challenges—economic, social, and ecological—the affected communities face on a day-to-day basis. Academic researchers from a range of disciplines can and do play a critical role in documenting, predicting, and measuring the intersecting stressors and issues that these communities face. Doing so can help to shape climate mitigation and adaptation policies that address and redress these harms, from the international to the local level. For example, in my role on the NPCC for the last seven years, I have worked with a multidisciplinary group of researchers to create and assess a variety of tools, indices, and metrics to inform city climate policy and to make sure that the most vulnerable and affected populations are centered in that policy. We discovered early on a lack of adequate metrics that are able to capture the intersectionality and complexity of challenges faced by these populations and communities. However, we also discovered nascent research in multiple disciplines that is helping to bridge these gaps.

An emerging literature on "climate gentrification," for instance, is beginning to capture some of the intersectional challenges faced by affected communities. However, this research requires more critical analysis of the factors driving migration and displacement patterns in these geographies. As this literature demonstrates, climate change is inverting the value of elevation in coastal cities like New Orleans and Miami by inducing high- and moderate-income households at risk of sea level rise to move into higher-elevation areas that were once disinvested and historically marginalized, resulting in the displacement of legacy residents. The literature has illuminated that a variety of pathways exist that lead to displacement, some of which are overlapping and poorly understood.

The climate gentrification literature is a great example of the ways that scholars are opening up critical areas of inquiry into the multiple dimensions of climate justice. One aspect of this literature, for example, examines the drivers of climate migration and their relationship to historic injustices. For instance, while risk mitigation explains the out-migration from higher- to lower-risk areas, climate risk alone does not fully explain why certain neighborhoods are targets of and vulnerable to gentrification.

Researchers have struggled to understand the relationships between neighborhood characteristics, such as high elevation, and other factors, such as zoning practices and targeted community redevelopment incentives, as shaping which neighborhoods are likely to become gentrified from climate migration. While still developing, these studies have identified a robust set of questions that will require interdisciplinary contributions to fully explore and answer.

A similar and related critical field of inquiry concerns the challenges that socially vulnerable and marginalized communities face from specific adaptation policies and measures, like managed retreat programs. Federal and state governments have been investing in voluntary buyout programs, taking into account dire predictions of storm surge and flooding risks. Evidence is now emerging that these programs can reinforce and entrench racialized land and property markets, particularly in historically underinvested neighborhoods where property values are low. While a buyout could in theory support housing mobility, scholars are helping policymakers to understand the racially and economically stratified fates of retreating homeowners. National patterns suggest that white families with government buyout money move to wealthier and whiter areas, while residents of majority-minority neighborhoods are more likely to move to neighborhoods that are majority Black or majority Hispanic and socially vulnerable.

This is an exciting moment to be working as an academic or researcher in climate justice, what I believe is a still nascent field. Much as it was true for environmental justice just a couple of decades ago, the production of knowledge and analytical frameworks requires a multidisciplinary and multidimensional effort. This book thus comes at a pivotal moment. As the editors acknowledge, the very concept of climate justice is marked by a dynamic and complex set of issues that require "critical pluralism." The chapters in the book brilliantly demonstrate that this pluralism is already evident across disciplines and entry points into the topic of climate justice.

In building and expanding this field, we must remember the imperative that researchers should and can work within, alongside, and in support of the communities that are seeking justice from the ground up. These communities speak for themselves but, importantly, are also speaking out on behalf of all of us and Mother Earth. They are doing so, as environmental justice advocates and researchers have done, not only in their own geographic and political spaces but through a multiscale lens, as this book's editors emphasize. They recognize that the issues and challenges they face, as well as policy actions they seek, are simultaneously local, global, and regional.

We can see this multiscalarity in practice in the Climate Justice pavilions each year at international sites of multilateral climate negotiations where nation-states and others come together to form international consensus on climate action. Notably, the first of these pavilions was created in 2022 by three longstanding environmental justice organizations: WE ACT, the Deep South Center for Environmental Justice, and the Bullard Center. While academics debate the precise contours of and differences between environmental justice and climate justice, these fathers and mothers of the movement, who still sit at the helm of the oldest environmental justice organizations, show us that this is largely an academic debate. What these historic leaders teach us and what this book embodies is that the lived experience of frontline communities around the world are not rooted in such distinctions. Rather, they are rooted in the kinds of critical engagement, deep analysis, and production of knowledge and solutions that will deliver justice of all types to those who are most often denied it.

1

When Natural Scientists, Social Scientists, and Humanists Meet to Discuss Climate Justice

REBECCA S. MARWEGE, NIKHAR GAIKWAD, AND JOERG M. SCHAEFER

THE MULTIDISCIPLINARY CHALLENGE OF CLIMATE JUSTICE

In December 2023 a group of more than two hundred scientists from over ninety research institutions in twenty-six countries published the Global Tipping Points report, which warned that five critical natural thresholds in the climate system had already been surpassed and that crossing more tipping points would lead to existential harm to both human societies and the natural environment.[1] Similarly, in January 2024 a group of researchers warned that the Greenland ice sheet was melting at a rate that was 20 percent faster than previously thought, and that the further melting of glaciers could lead to the collapse of one of the planet's key ocean currents, the Atlantic Meridional Overturning Circulation, with devastating consequences for global weather patterns, ecosystems, and global food security.[2] Despite these dire scientific warnings, pledges made at the Copenhagen climate meeting by industrialized countries to provide $100 billion a year by 2020 in climate financing to poor nations remain woefully unfulfilled.[3]

These studies are just two examples in a long series of scholarly warnings of the catastrophic consequences of a heating climate and the pressing need for social, political, and economic action on the climate crisis. Beyond the environmental and physical dimensions, the relationship between climate justice, climate resilience, and just transition raises a cluster of questions that transcend traditional disciplinary boundaries: How can climatic vulnerability and resilience be determined? How are communities around the world already experiencing the effects of climate change? How do affected marginalized communities themselves refer to climate justice issues? What are the historical legacies of colonialism, how do they contribute to climate injustices today, and what can we learn from them for the future? The political, anthropological, and philosophical dimensions of these questions speak closely to urgent work being conducted, among others, in the atmospheric, ecological, and physical sciences on a rapidly changing climate and the continued impact of greenhouse gas (GHG) emissions.

What is striking, however, is that despite the recognition of the dramatic scale of the problem of climate change, no single, agreed-upon definition or concept of climate justice exists. As we shall argue, the dynamic expansion of the climate justice field and the proliferation of the critical usage of the term "climate justice" reflects the pressing need to act on climate change but also poses a challenge to academic scholarship as it runs a serious risk of losing meaning across scholarly conversations.

Figure 1.1 shows an exponential increase in the academic use of the term "climate justice" in recent years. While the term became popular in scholarly writing in 2010, its usage has skyrocketed since 2018. At the same time, figure 1.2 documents that while about 25 percent of the academic references to the term arise in disciplines such as environmental sciences and environmental studies, the remainder of references are evident in a range of disciplines, such as political science, law, economics, anthropology, sociology, business, and management. The pie chart highlights the multidisciplinary usage of the term "climate justice" in scholarly writing.

While the increase in the use of the term underscores its growing importance in both the academic and public discourse, without multidisciplinary engagement, references to climate justice risk becoming a hollow buzzword. A recent commentary in *Nature Climate Change* contends, for example, that attempts by natural scientists to conceptualize and develop frameworks for climate justice have obscured the transdisciplinary origins of environmental justice and neglected colonialism's role in contributing to the climate crisis.[4] As we will argue, it is important not only for climate scientists to engage with the philosophical interpretation and the social drivers of the climate crisis, but also for scholars in the humanities and social sciences to continuously engage with climate science.

From a philosophical perspective, efforts to frame climate justice have explored which justice paradigms are most suitable to conceptualize the unprecedented scale of harm caused by climate change. Distributive justice, for example, emphasizes that those with the highest economic capacity bear greater responsibility for compensating climate-induced costs, while corrective justice frames climate injustice in terms of historical accountability for existing pollution. Approaches that focus on future generations invoke the rights of those most affected by climate change, both now and in the future.[5] Critical race theory contributes to this discourse

Articles and Books That Mention "Climate Justice" by Publication Year

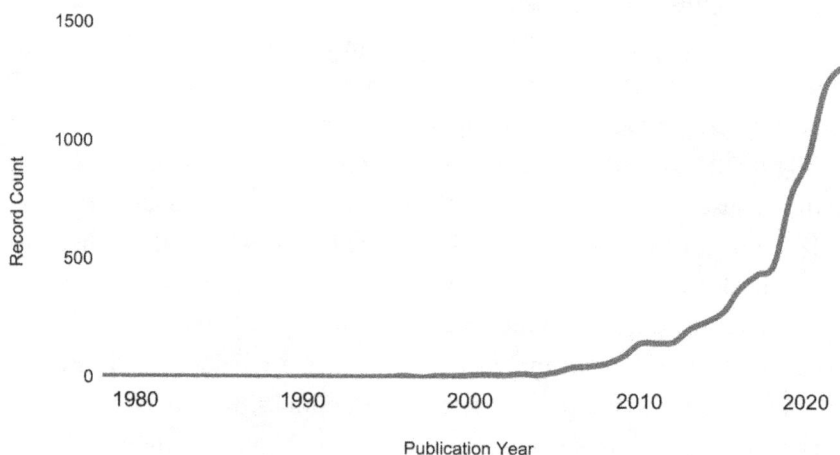

FIG 1.1

Evolution of citations of the term "climate justice," based on a search of book chapters and articles in the Web of Science Core Collection, illustrating an exponential increase over the past decade.

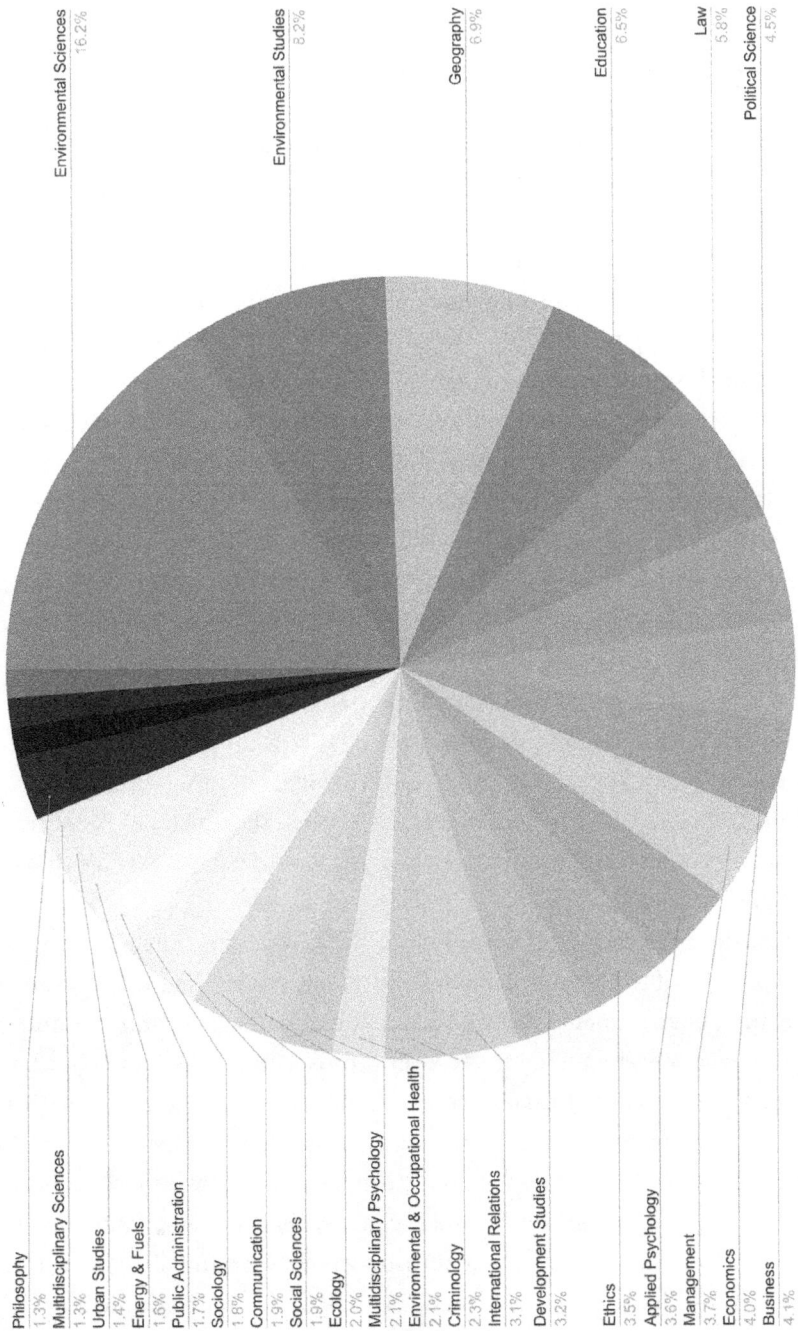

The pie chart labels, reading clockwise from top:

- Environmental Sciences 16.2%
- Environmental Studies 8.2%
- Geography 6.9%
- Education 6.5%
- Law 5.8%
- Political Science 4.5%
- Business 4.1%
- Economics 4.0%
- Management 3.7%
- Applied Psychology 3.6%
- Ethics 3.5%
- Development Studies 3.2%
- International Relations 3.1%
- Criminology 2.3%
- Environmental & Occupational Health 2.1%
- Multidisciplinary Psychology 2.1%
- Ecology 2.0%
- Social Sciences 1.9%
- Communication 1.9%
- Sociology 1.8%
- Public Administration 1.7%
- Energy & Fuels 1.6%
- Urban Studies 1.4%
- Multidisciplinary Sciences 1.3%
- Philosophy 1.3%

FIG. 1.2

Disciplinary breakdown of the usage of the term "climate justice," based on all published book chapters and articles in the Web of Science Core Collection. The pie chart plots only those disciplines that contribute more than 1 percent of the total references to "climate justice" in all cited works. The unlabeled pie slices from left to right are Social Work (1.3%), Engineering, Environmental (1.3%), Anthropology (1.2%), and Area Studies (1.2%).

by emphasizing the intersections of structural discrimination, particularly the unique challenges faced by Black women, and by examining how race and identity intersect with power structures to perpetuate systemic environmental discrimination. Furthermore, Indigenous scholars advocate for greater emphasis on Indigenous philosophies to reconceptualize relationships among humans and with the more-than-human environment.[6] These philosophical debates highlight the structural foundations of the climate crisis and underscore its epistemological dimensions, particularly concerning whose voices are prioritized in knowledge creation.

Meanwhile, the social sciences provide distinct and valuable frameworks for interpreting the economic, social, and political dimensions of the climate crisis through the lens of justice considerations. For instance, perceptions of at least three types of climate injustice engender policy resistance and erode policy legitimacy in the realm of climate politics: first, uneven exposure to the effects of climate change across nations and communities within nations; second, unequal distribution of climate change costs; and third, uneven allocation of the benefits of mitigation and adaptation policies, which often exacerbate rather than resolve preexisting inequalities.[7] Such frameworks can be used to understand a range of societal phenomena.

Consider, for example, how national-level economic effects of climate change can influence international efforts targeting climate cooperation. Warming environments can wreak economic damage on some countries while conversely enhancing the economic prospects of other countries, such as those with colder climates. Research indicates that the cool, relatively wealthy nations of North America and Eurasia have historically contributed most to GHG emissions and in turn have caused the most economic damage to other nations from climate change, but they have simultaneously registered economic gains from anthropogenic warming. By contrast, the world's poorest countries have made virtually zero contributions to global emissions but have been hardest hit economically from the emissions of richer economies.[8]

This pattern can be seen starkly when evaluating the global economic impacts of US emissions. Figure 1.3 illustrates how US emissions between 1990 and 2014 augmented the economic outputs of so-called Global North countries while causing deleterious effects on the economic outputs of so-called Global South countries. Patterns such as these help explain why developing countries have insisted that international climate agreements incorporate principles of historical responsibility and mechanisms of redress, such as loss and damage financing, while richer countries such as the United States have resisted such demands. In this manner, considerations of climate (in)justice lie at the core of societal efforts to combat the climate crisis—whether at the international, national, local, or individual level. By shedding light on how structural injustices intersect with climate policies and actions, social scientific approaches can therefore offer important tools to analyze and address the climate crisis (see fig. 1.3).

Of course, humanistic and social scientific analyses would be incomplete without the contributions of the natural sciences, which provide essential insights into natural and climatic phenomena and the anthropogenic drivers of climate change. Engagement with the natural sciences is therefore

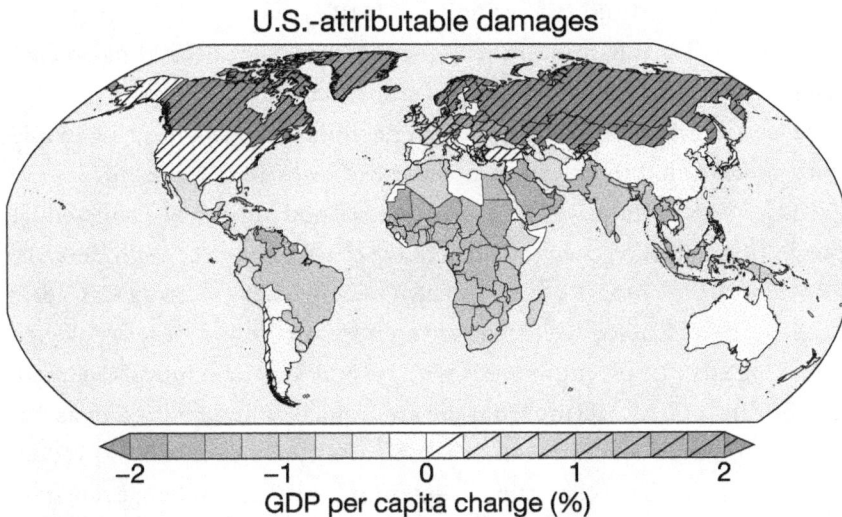

FIG. 1.3

GDP per capita changes in countries around the world that can be attributed to US GHG emissions between 1990 and 2014. Hatched shades reflect positive changes in countries' GDP per capita; unhatched shades reflect negative changes. (Copyright: Callahan and Mankin 2022, Creative Commons Attribution 4.0 International License.)

indispensable for identifying responsibility for the climate crisis and evaluating trade-offs among different policy responses. However, assessing the climate crisis through the lens of the natural sciences also requires an engagement with the humanities and social sciences to highlight both concerns of epistemological justice, such as whether the analysis should be extended to the more-than-human to include multispecies angles, as well as facilitate internal critiques of exclusionary dynamics within scientific practices that could be improved to advance climate justice.

Rather than assessing climate justice through a monodisciplinary lens, we argue for an expansive yet convergent understanding and application of the concept. This approach necessitates fostering deeper exchanges among the humanities, social sciences, and natural sciences without prioritizing one discipline over another. Two prominent topics of contemporary debate that exemplify the amalgamation of ethical, social, political, and scientific perspectives and underscore the importance of a shared discourse in understanding and critiquing these issues are "geoengineering" and "measuring, reporting, and verification" (MRV) policy tools. Both of these have been hailed as harbingers of utopian (or dystopian, depending on one's predilections) salvation in the face of climate change.

Scientists originally discussed the possibilities of geoengineering in military contexts, but the idea was popularized by Lyndon Johnson's Science Advisory Committee in 1965, which discussed and ultimately dismissed the possibility of albedo geoengineering in the climate context.[9] With the exacerbation of the climate crisis in recent years, geoengineering has again been widely discussed. Geoengineering refers to a range of technologies. These include carbon engineering with carbon dioxide removal technologies or direct GHG capture from the atmosphere (often referred to as "air capture"), ocean fertilization, ocean alkalinity enhancement to reduce atmospheric GHG concentration by intensifying the gas exchange rate from the atmosphere to the ocean, as well as solar engineering. The term "solar engineering" usually describes Solar Radiation Management and includes technologies such as albedo enhancement (increasing the reflectiveness of clouds), space reflectors, or the dissemination of stratospheric aerosols.[10] Geoengineering discussions were at one point infamously associated with the fantasy of the so-called scientist with the global thermostat, based on the idea that eventually engineers could regulate global and even local temperatures by such techniques.[11]

Since then, the scientific discussion has evolved, and recent IPCC reports offer a measured and more critical view of geoengineering for a variety of reasons. These include, first, the lack of justice considerations regarding, for example, how geoengineering modifications could potentially disrupt local and regional ways of living.[12] Second, there is a paucity of research thoroughly evaluating geoengineering's side effects, such as the impact of increased sulfur aerosol concentrations in the atmosphere and the danger of destabilizing ecospheres on the earth's surface. To give a specific example, recent modeling studies have shown that the widely discussed interventions of cloud whitening and aerosol injection to the stratosphere have the potential to reduce heat in some local areas in the short term, such as in California, where cloud-whitening experiments are already realized on a smaller scale. However, these interventions could have notable and potentially disastrous consequences on heat, wind, and precipitation patterns in regions that are geographically farther away, likely hitting the poorest areas the hardest.[13]

Therefore, humanistic and social scientific approaches to climate justice are central for evaluating the scientific viability and appropriateness of geoengineering as a tool to combat the climate crisis going forward. But the scope for such intersecting approaches is much broader. As scientists have pointed out, even with substantial reductions in overall emissions, already existing GHG in the atmosphere will continue to contribute to rising temperatures if not removed from the atmosphere.[14] An argument could therefore be made that some types of geoengineering technologies are relevant for rectifying overshoot and legacy emissions until the global economy has been transformed and decarbonized. The complexity of geoengineering demonstrates that to account for the scientific, technical, social, and ethical ramifications of these new technologies, we need to go far beyond a narrow, monodisciplinary approach to climate justice.

Another, often less critically viewed concept is that of MRV, which was introduced at the Bali Action Plan at the Conference of the Parties 13 in 2007 and describes a set of guidelines for both developed and developing countries that aim to improve transparency around climate mitigation efforts. Measurement here describes quantifying both climate mitigation efforts and their consequences, such as the removal of emissions through carbon sinks and overall emission reductions. Reporting describes the requirement to provide an update about these efforts every four years through national communication and biennial update reports every two years. Verification describes that parties to the Bali Action Plan not only measure and report on their climate mitigation efforts but also agree to allow for international consultation and analysis to assess the biennial update reports.[15]

This framework provides an important step in the right direction to promote state compliance, accountability, and transparency around carbon mitigation plans. However, accurately quantifying the required scale of emission reduction efforts is incredibly challenging and requires scholarly conversations across various scientific and social scientific disciplines. A key obstacle is that in practice, existing empirical estimates of the damages from climate change, which would indicate the need for stronger climate *mitigation* efforts, tend to be confounded by *adaptation* efforts already undertaken by communities, as research in economics has shown.[16]

For example, albacore fisheries have adapted to extreme weather events through the use of harvest control rules, which periodically determine catch limits and can be adapted to changing weather patterns. These controls can, however, mask the financial, social, and environmental costs of climate change for fisheries and fish populations around the world, including economies of Pacific small developing states that rely on the sustainable management of these fisheries.[17] Thus, estimates of revealed damages are biased downward because they do not fully account for human efforts to adapt to climate change. These estimates risk downplaying the responsibility of the main contributors to the climate crisis, such as countries in the Global North, in compensating for these damages and therefore demonstrate the limits of MRV policy tools to advance climate justice.

These are only a few examples that underscore the difficulty of accurately capturing the social and environmental costs of anthropogenic climate change: Making the problem measurable is just one among many steps in deciding how to confront it, identifying who should pay for climate-related costs, and how those who are harmed most could get more influence in the decision-making process, at both the local and global levels. Thus, they illustrate the limits of a traditional monodisciplinary approach to research related to climate justice. In turn, this book calls for supporting synergies between different disciplines, incentivizing the creation of transdisciplinary research grant proposals and projects, and overcoming the myopic teaching of climate justice within individual disciplines.

GENESIS AND ARGUMENT OF THIS BOOK

Grappling with the paradox of the growing ubiquity and simultaneous lack of clarity of the term climate justice in public discourse, we started an interdisciplinary research network on climate justice at Columbia University's Climate School in 2020, which became the genesis of this book. At the time, Columbia University was, in a way, simultaneously both ideally suited and ill-positioned as a home for climate justice conversations. Columbia and its flagship Earth Institute were already highly active in the transdisciplinary field of climate studies, and the Climate School was just being founded with the vision to become the leading academic force for actionable scientific and public policy work on the climate crisis. Yet most conversations about research on climate change were still taking place within disciplinary silos and homes. This book attempts to bring together some of the main insights that we have gleaned from the experience of promoting multidisciplinary conversations on climate justice.

We initially considered our climate justice network to be a *naive experiment* to bridge the disconnect between the natural sciences, social sciences, and humanities on the topic of climate justice. As transdisciplinary conversations grew across our network, we realized that the traditional framework of scholarly thinking fails to capture how dynamic and fast the climate crisis spreads through every part of society—at the global, national, and local levels—and how climate change is already dramatically exacerbating socioeconomic inequalities both within and across countries. Exchanging different disciplinary perspectives is therefore crucial to account for the fact that while the climate crisis has environmental and physical dimensions, it also has profound ethical, social, and political ramifications. Concepts such as historical responsibilities, compensation to redress past, present, and future losses, and "just transition" are, for example, central to contextualizing our scientific understanding of climate change and informing how states and societies should respond to climate change.

In our experience, as more disciplines took part, offered perspectives, and became integrated into our discussions, the clearer the challenges and complexities of climate justice became, and the more convinced we became of the benefits of multidisciplinary approaches to understanding climate justice. Reflecting such diversity of perspectives, this book provides an academic snapshot of climate justice debates across different disciplines at this point in time, with the aim to provide a broader conceptual framework that can be built on in future years. We are circumspect in our claims and aware of the limits of our approach. For example, due to the institutional context and academic focus of our research network, we are missing the engagement of nonacademics and activists, as well as communities and representatives of governments of those most affected by climate change.[18]

Our goal is therefore not to tell those most affected by climate change what the "correct" view of climate justice is, but simply to provide insight into the synergies and struggles of climate justice debates across academia. Additionally, we do not give a comprehensive insight into *all* academic disciplines (a nearly impossible task), but we offer a snapshot of the disciplines we have worked with to demonstrate that the more disciplines that are included, the better chance we have to make sense of climate justice. Instead of one fixed term, we propose to conceptualize climate justice as an emerging and continuously developing multidisciplinary concept that requires "critical pluralism," or the need for the acknowledgment of difference, as well as an "open, participatory, and intersubjective process of communication" that allows us to achieve a common understanding across these differences.[19]

Drawing from our conversations as well as the chapters presented in this book, we identified five different dimensions of analysis that resonate throughout the very different themes and topics that the different chapters discuss—the temporal, spatial, agential, structural, and ontological dimensions. While time and space are externally given dimensions, we as humans intervene in them to make them meaningful; an understanding of these interventions is necessary to make sense of the causes and impacts of climate change. To address these causes and impacts, we need to ask whose agency is required to transform and rethink the existing structural backdrop to climate change. At the same time, to answer these questions, we also need to consider the ontological dimension, which allows us to think through different ways of viewing the problem of climate change in the first place, and the relevance and meaning of climate justice for different communities more generally. Focusing on the academic context, this book therefore argues that, counterintuitively, it is methodological and conceptual disagreement between different disciplines that is required to illuminate the different angles of the concept of climate justice and move toward a convergent understanding of the term.

More specifically, we can conceptualize these different dimensions in the following ways:

- *Temporal:* Climate justice as a concept is not static and cannot be defined holistically or in its entirety at any point in time. It has always been and remains a moving target, influenced by (a) a widening and intensification of the climate crisis and changes in the physical and natural scientific understanding of climate change and its impacts on human communities and the natural environment, (b) our developing understanding of the historical legacies of systems of oppression such as settler colonialism and the influence they continue to yield today, and (c) our evolving understanding of the role present and future generations should play in climate justice considerations.
- *Spatial:* With exponentially rising temperatures, more and more regions are severely affected by climate change. Even though climate change affects all of us, sensitivities to climate disturbances differ radically across and within geographical and geopolitical regions. For example, climate justice means something different to citizens in the Global North, who have historically contributed the largest

share of GHG emissions, than it does to those living in the Global South, who are most affected yet have the least resources to deal with the effects of climate change. Even within these broader regions, climate justice raises distinct concerns of urgency and equity for different groups, such as frontline communities and those residing in fossil fuel–producing regions.[20] Additionally, climate adaptation efforts, such as albedo engineering, as we discussed previously, can have unforeseen consequences in regions far from where they are initially applied. This highlights the importance of including an expanded spatial lens in climate justice considerations.

- *Agential:* Climate justice raises questions about the different types of agency that play a role in responding to the climate crisis. For example, one key figure who is often referenced in climate justice discourses is Greta Thunberg and her one-child protest that inspired a larger youth climate movement. While references to Thunberg emphasize the potential for change through individual action, collective agency approaches focus on the role groups, such as movements, Indigenous nations, or states, play in pushing for climate agreements. Beyond this human-centered focus, multispecies approaches underline the need to consider more-than-human agency that is exerted by the environment itself, and how to include this type of agency in human political discourses. For example, as chapter 10 of this book points out, climate justice requires us to think about the impact of climate change on marine organisms in the Arctic and how we reconceptualize our relationship with the environment in response to this crisis.

- *Structural:* The broader structures within which agents operate both provide opportunities and restrict how agents push for climate justice with their actions. Structures can take several forms. First, built environments and material structures, such as buildings, ports, roads, and other infrastructural constructions like seawalls and levees, all fundamentally contribute to GHG emissions but also hold the potential for mitigation and adaptation to climate change. Second, nonmaterial structures, such as international agreements (e.g., the Paris Agreement or the Kyoto Protocol), global trading and financial systems, and domestic legislation (for example, the Inflation Reduction Act in the United States), provide opportunities for political responses to the climate crisis; at the same time, these institutional structures risk succumbing to path dependency and reinforcing the status quo. Both material and nonmaterial structures often reflect historical legacies, such as the influence of settler colonialism on unjust power structures today, and therefore raise the question of how newly decolonized structures can be shaped that center climate justice considerations.

- *Ontological:* Ontological considerations underline the role that different worldviews play in identifying the causes, impacts, and justice considerations of climate change. For example, chapter 13 in this book describes how climate science can be both an important tool for marginalized groups to demand financial compensation and political recognition from their own state agencies but also risks perpetuating authoritarian worldviews that sideline the experiences of those who are already most affected by climate change. An inclusive understanding of climate justice therefore requires not only the inclusion of different worldviews, but also an awareness of whose stories remain untold.

Applying these different principles can help us think through the climate justice issues at stake in very different case studies and examples. For example, the chapter on urban studies (chapter 4) describes how colonizing states in the Global North bear historical responsibility for the majority of the world's transportation emissions due to their past and continued funding of deeply inequitable and car-oriented carbon-intensive infrastructure in the Global South. In doing so, authors Jacqueline Klopp and Festival Boateng point to both the temporal dimension, with the legacy of colonialism still blocking investments in popular transport modes today, as well as the spatial dimension, with similar transport modes imposed across different geographies generating detrimental environmental and climate impacts at both the local and global levels. Considering the structural dimension, the authors emphasize that colonized cities were designed according to European standards, which favored personalized transport (thus constructing material structures that still contribute to climate injustices today), and how these tendencies were reinforced by international institutions such as the World Bank, which further encouraged carbon-intensive infrastructure. To remedy these tendencies, the authors highlight the agency dimension, asserting that those who "run transportation" must be part of global conversations to finance alternative transport systems that would significantly reduce GHG emissions. This also raises questions of the ontological dimension, or different worldviews, as Klopp and Boateng point out that "modern" ideas of transportation often draw on the experience of cities in the Global North that include automobility rather than favor alternative modes of transport embraced by local communities, such as "informal" rickshaw and minibus systems.

A very different example is offered in chapter 11, on oceanography. While the chapter encapsulates a much longer time span and broader geography than the chapter on urban studies, the five different dimensions can help to think through climate justice issues here, too. The chapter describes how, over millions of years, the Late Cretaceous coastline was transformed into fertile grounds due to the activity and subsequent death of phytoplankton that contributed to the creation of clay beds, which provided ideal conditions for farming and gave the region the name "Black Belt." The authors, Kailana Acosta and Gisela Winckler, thereby illustrate how time interacted with space to transform the geological and biological conditions that would affect subsequent anthropogenic interventions. While for hundreds of years the Black Belt region was maintained and sustainably farmed by Indigenous people, the authors highlight how, with the onset of colonization, the region became the center of enslaved agriculture. They describe how this created the structural conditions of discrimination and exclusion that continue to marginalize Black and Indigenous communities today. Focusing on a lack of agency, they also describe how the descendants of those who were exploited in the pursuit of industrialization and the beginnings of anthropogenic climate change continue to be disproportionately affected by climate injustices today. Ontologically, the exploitation by colonial powers of natural patterns that led to fertile land demands a critical introspection into the role the natural sciences can play in both exposing these relationships but also learning from them and engaging with the worldviews of those who have served to protect these grounds over hundreds of years, such as the Indigenous inhabitants of the region.

Together, these different dimensions provide a prism through which we can assess the interactions of the spatial, temporal, agential, structural, and ontological dimensions of climate justice in very different instances where climate justice is at play. Beyond the examples discussed in this book, we hope that they can provide a matrix through which to analyze other climate justice–related examples in comparative, historical, and future cases.

ARGUMENT IN HISTORICAL AND SCHOLARLY CONTEXT

Humans have grappled with the impact of environmental crises on their way of life and survival for centuries, if not millennia, with any major religious text containing references to environmental crises. However, as Sheila Foster describes in her prelude, the emergence of the term "environmental justice" is often attributed to Warren County, North Carolina, in 1982, where residents protested the dumping of 120 million pounds of PCB-contaminated soil in a majority Black county. While there had been several environmental protests before, the protests in Warren County highlighted the intersectionality of continued systemic discrimination against Black residents and underscored the relationship between environmental injustice and environmental racism.[21] Broadening the concept of environmental justice to include the Indigenous land rights movement and social and economic justice concerns, in 1991 the First National People of Color Environmental Leadership Summit in Washington, DC, drew attention to the environmental harms experienced by marginalized communities, including Black, Asian, Latino, and Indigenous Americans.[22]

During this period, more radical and eco-anarchist movements also started to arise, such as Greenpeace in 1971, which started as a direct action group that protested nuclear testing, as well as the emergence of the Animal Liberation Front and Earth First in the 1980s, which introduced militant tactics such as monkeywrenching and the destruction of private property. These movements opposed reformist approaches, which they critiqued for failing to change the underlying conditions of environmental destruction, but were often met with extraordinary levels of state repression.[23] For example, Joseph Dibee, one of the last free members of the Earth Liberation Front, was finally caught by the FBI in Cuba in 2018 and indicted on a charge of participating in a series of arsons in the 1990s, after spending years as a fugitive in Syria, Russia, and Mexico.[24]

Simultaneously, many different environmental movements emerged across the globe, such as the Chipko Andolan movement in India, which opposed commercial logging efforts, as well as the anti-deforestation movement in Brazil, including the Brazilian Rubber Tappers campaign from 1977 to 1988, which sought to prevent the clearing of rainforests for cattle farms.[25] Activists of these movements were often met with violence and state repression, with Chico Mendes, the founder of the Rubber Tapper movement, being killed by cattle ranchers in 1988. This underscores the high costs members of these movements have faced for engaging in environmental activism.[26]

Around the same time, the term "climate justice" slowly started to emerge in public discourse, with one of the first mentions in the *Greenhouse Gangsters vs. Climate Justice* report compiled by the UK-based Corporate Watch Group in 1999, which exposed the political influence of the petroleum industry. The report advocated tackling the underlying causes of global warming and highlighted the destructive climatic effects of fossil fuel consumption. About a decade later, in 2010, the climate justice–focused "World People's Conference on Climate Change and the Rights of Mother Earth" in Cochabamba produced a further set of climate justice–centered principles that focused on Indigenous perspectives and anticolonialism, denounced the use of carbon markets, and suggested a Universal Declaration on the Rights of Mother Earth, as well as the establishment of an International Climate and Environmental Justice Tribunal. Importantly, however, these principles were marginalized at UN-sponsored Climate Conferences (COP), as well as the Paris Agreement of 2015, which highlighted the continued exclusion of Indigenous perspectives at these international forums.[27]

Nonetheless, the Paris Agreement is still widely regarded as an initial breakthrough in global climate governance, as it required both developed and developing countries to set emission reduction targets to limit global temperature rise to below 2°C. Further progress was made in 2022 at COP27 in Sharm el-Sheikh with the establishment of a Loss and Damage Fund to compensate poor and vulnerable countries for climate-related harm. This fund marked the first formal acknowledgment of the obligation of wealthier nations to provide financial assistance to more vulnerable countries in adapting to climate change.[28] Nonetheless, the fund has been criticized for its lack of specificity, leaving loopholes that allow wealthier nations to evade meaningful financial transfers.[29]

Situating these uneven exposures and responsibilities in their historical context, Ashley Dawson et al. have argued that as "the climate emergency deepens, discourses used for centuries to dehumanize others while legitimating the extractive violence of capitalism, colonialism, and patriarchy increasingly target those whose lives are most menaced and who are displaced by fossil capitalism."[30] Responses to the climate crisis must go beyond international agreements and seek to restructure the extractive relations at the bottom of these injustices. In *Climate Futures—Reimagining Global Climate Justice*, Kum-Kum Bhavnani et al. therefore advocate for reimagining climate justice through the lenses of culture, Indigeneity, and race. They emphasize the need to envision a different future by critically engaging with colonial pasts and capitalist presents.[31]

The call for reimagining climate justice raises the question of how such a transformation could be realized in practice. Iokine Rodriguez et al., for instance, argue that the manifestation of socioenvironmental conflict exposes the inherent injustices of the status quo and thus presents a crucial first step toward a "just" transformation. However, they also emphasize that such transformations are contingent on comprehensive strategies that affect institutional frameworks, social networks, as well as more general discourses and ways of seeing the world.[32] In line with this view, scholars engaging with postgrowth and degrowth economies have, in recent years, challenged the ideology of unending growth and critiqued neoliberal environmentalism, which seeks to reconcile economic growth with climate mitigation policies.[33] Instead, these scholars advocate for reimagining the current moment as an "opportunity to redefine the direction of 'development' whereby human rights and the environment take precedence over economic growth."[34]

Similarly, trying to rethink this moment as one not only of crisis but also of missed opportunity, this book scrutinizes the academic context as a traditional place of knowledge creation. It provides an introspection into how different academic disciplines have interpreted the underlying ideas of climate justice. We do not intend for this book to be an end-all discussion of the role of academia in pursuing climate justice, but rather to demonstrate the wide variety of perspectives encompassed in academia and encourage more critical reflection into how barriers to knowledge can be torn down and coproduction of knowledge can be advanced in the future. The book seeks to start a critical dialogue about what different academic disciplines, including the natural sciences, social sciences, and humanities, can contribute to this unprecedented project of transforming not only social but also economic and political relations.

In this context, we are aware of our elite positionality with the book project starting at a US Ivy League university. We are particularly mindful of Christine Winter's challenge in her book *Subjects of Intergenerational Justice*, where she argues that "Western ontologies . . . are incompatible with justice generated from Indigenous ontological positions. . . . Neither can claim to be 'universal.'"[35] Similarly, we recognize not only the ontological differences between Indigenous and Western perspectives but also the radically different ways in which climate change is experienced across the Global South and Global North. Additionally, we acknowledge the significant contributions made by both social movements and scholars in advancing our understanding of the multiple facets of climate justice. Our aim is therefore simply to provide insights into the different debates surrounding climate justice across academic disciplines, explicitly without the goal or ambition to present these as universal concepts. On the contrary, the variety of approaches included in this volume raises the question of how we can achieve a "critical pluralism" in the field of climate justice without risking that climate justice loses all conceptual and definitional meaning.[36]

PLAN FOR THE BOOK

To this end, the book seeks to answer how climate justice is conceptualized across different disciplines, how academic research can contribute to achieving climate justice within and beyond academia, and how, looking forward, the practice of academic research can be adapted to engage with a diverse set of stakeholders who will be affected most by a changing climate.

In the process of editing the book, we met with the chapter contributors for a one-day workshop, during which we developed a set of guiding questions. Authors were free to decide to what extent they wanted to respond to these in their chapters:

1. In conceptualizing climate justice in your discipline/research area, how do you think about answering the following questions: Climate justice is for whom? Climate justice is for what end?
2. What is the difference between conceptualizations of climate justice and the reality of climate justice in your area of specialization?
3. In your discipline/research area, have you encountered climate justice from the perspective of justice issues related to climate change itself, or justice issues related to scientific and societal efforts to combat climate change, or both?
4. Do you think about climate justice with respect to mitigation or adaptation, or both? Are there tensions between a more local or more global research focus? How does your discipline deal with exponentially accelerating change and societal lack of action in response to it?
5. Looking forward, do you think climate justice is a useful concept? What are some key takeaways from your research?

Based on these broad questions, contributors then decided to either provide an overview of their discipline's approach to climate justice, such as the chapters on political theory (chapter 2), political science (chapter 3), sociology (chapter 6), and migration studies (chapter 7), or to illustrate their discipline's contribution to the field of climate justice through a deeper analysis of one specific issue, such as the health of migrant workers by chapter 8 on environmental health sciences, or the discussion of the concept of "energy justice" as part of climate justice by chapter 9 on comparative literature. Additionally, some contributors chose to reflect on their experiences in working with different communities on the topic of climate justice, such as chapter 12 on physical science, which explores the practical and ethical considerations of different career paths for natural science students, as well as chapter 15, which explores the imperative but also the challenges of community involvement in coastal flood risk reduction planning.

To move beyond a narrow disciplinary approach, we decided to structure the book according to some common themes the different chapters address. While each chapter addresses many more important themes than we can highlight here, the different chapters pick up on considerations introduced by the previous and following chapters and therefore illustrate the need for a multidisciplinary conversation on climate justice. We arranged them so that more conceptual discussions precede chapters with a stronger applied focus. Nonetheless, all chapters integrate both conceptual analysis and practical examples, fostering a balance that reflects the complexity of the subject.

The first set of chapters explores the theoretical and historical foundations of environmental and climate justice, emphasizing how structural conditions perpetuate existing injustices. Chapter 2, on political theory, examines the concept of "justice" through various theoretical lenses—such as distributive, participatory, recognition, and relational justice—and discusses their relevance to environmental and climate justice issues. Asking what responses to climate injustices could look like, chapter 3, on political science, focuses on the distributive aspect of climate justice to illustrate how unequal costs for climate change adaptation and mitigation differ across the global, national, and local levels and what political factors stand in the way of achieving more progressive climate action. Chapter 4, on urban studies, focuses on cities in both the Global South and Global North, demonstrating how historical legacies can explain existing environmental injustices and exacerbate climate justice concerns. Chapter 5, on atmospheric science, examines the trade-offs between environmental and climate justice, drawing attention to unintended consequences that can arise in addressing one at the expense of the other.

The second set of chapters examines the disproportionate impact the climate crisis has on marginalized communities. These chapters emphasize the interplay between structure and agency, illustrating how existing structures constrain the agency of vulnerable communities while amplifying their exposure to harm. Chapter 6 on sociology explores how social systems and structural inequalities contribute to climate injustice. Complementing this, chapter 7, on migration studies, demonstrates how climate change magnifies existing *injustices* in global migration and displacement patterns and argues that to rectify these, systems of inequality and oppression will need to be addressed comprehensively. Chapter 8, on environmental health science, extends these discussions by highlighting the intersecting injustices faced by agricultural workers. It underscores how climate change intensifies these injustices, raising justice concerns that go beyond environmental issues and require comprehensive policy responses at both local and global levels.

The third set of chapters critically examines the limitations of a "business-as-usual" approach and places the current climate crisis into the context of colonial history and its legacy on global capitalism, and different forms of climate injustices today. Introducing the concept of "energy justice," chapter 9, on comparative literature, explores how climate justice can be understood across multiple spatial and temporal scales. Chapter 10, on history, builds on this idea by cautioning against "climate reductionism" and emphasizing the importance of multispecies justice. Complementing these discussions, chapter 11, on oceanography, draws attention to the entanglement of natural and social histories and how these affect who continues to be excluded from climate justice debates today. The chapter calls for introspection into how science can center climate justice in research and praxis. Similarly providing introspection into science, chapter 12, on physical science, considers career choices in the private industry as a "hard case" for assessing the influence individual scientists can have on addressing the climate crisis. Acknowledging that many aspects of the climate system are now well understood but that collective action remains insufficient, the authors offer practical considerations for scientists and recent graduates when making career decisions.

The fourth set of chapters critically examines who shapes climate justice narratives and advocates for broadening the discourse to encompass diverse linguistic, ontological, and cosmological perspectives. Chapter 13, on anthropology, highlights alternative narratives of the climate crisis by exploring how pomelo farmers in southeastern China describe climate anomalies. This perspective challenges the notion of climate change as a singular reality and underscores the importance of incorporating alternative forms of knowledge into the conversation. Adding to this discussion, chapter 14, on religion, argues for a "pluriversal" understanding of climate justice that recognizes the influence of religious worldviews and practices in shaping responses to the climate crisis. Chapter 15 presents a case study of New York Flood Protection, one of the largest flood barrier projects in history, demonstrating how the inclusion of communities in decision-making processes is both normatively and practically essential for achieving effective and equitable climate solutions.

In the conclusion, chapter 16, we pick up on some of the common themes across the chapters and elaborate on our call for deeper multidisciplinary collaboration and the need to transform academic structures to center concerns for climate justice.

But our goals are more ambitious. The experiment of this book has shown that if academia wants to play a leading role in responding to the climate crisis, multidisciplinary research structures will be essential to overcome siloed scholarly conversations on climate justice. Only through this can we transform the monodisciplinary status quo and develop the necessary groundbreaking research to prevent further climate harm.

By allowing for more cross-disciplinary engagement, scholars and educators will be better equipped to grasp what climate justice means and how to develop dynamic research and teaching programs to counter this major crisis. These will need to respond to growing demand in the humanities and social sciences to learn about the scientific underpinnings of the climate crisis and vice versa, allowing science students to recognize and grasp the justice dimensions of the climate crisis.

Taking into account that more and more critical global tipping points of the climate system are going to be crossed, we believe that the current moment presents one of the greatest threats to humanity we will ever experience. However, it also offers the opportunity to rethink existing academic and research structures and advance meaningful and timely research on climate justice.

CHAPTER SUMMARIES

In chapter 2, Mary Witlacil discusses the different conceptual dimensions of environmental justice and climate justice. She makes the case for moving beyond a "trivalent" approach to justice, which relies on a distributive, participatory, and recognition-based definition of justice, to an Indigenous and intergenerational account of climate justice that centers relational and epistemological concerns.

In chapter 3, Zara Riaz and Page Fortna add the perspective of political science. They discuss how framing climate justice as an issue of political (re)distribution highlights the dilemmas and disjuncture between the goal of achieving just distributive outcomes and the realities of the structure of political incentives that shape whether and how that goal is achievable. They describe how these disjunctures manifest at both global and domestic levels and apply both to climate inaction and to action.

In chapter 4, Jacqueline Klopp and Festival Boateng discuss the role of sustainable urban development and the relevance of climate justice to understanding responsibilities for climate harm in the urban context in the Global South.

In chapter 5, atmospheric scientist Róisín Commane warns of the consequences not only of climate change, but also adaptation and mitigation. Specifically, Commane argues that any climate mitigation policy needs to be fully assessed for both its climate and its environmental justice impacts before it gets adopted.

Jennifer Givens and Mufti Nadimul Quamar Ahmed highlight in chapter 6 the role sociology can play in identifying and addressing the social drivers and their unequal impacts on the climate crisis. Specifically, the authors focus on the global scale through an analysis of concepts such as climate debt, environmental load displacement, and ecologically unequal exchange theory.

In chapter 7, Alex de Sherbinin, David Wrathall, Susana Adamo, Sara Pan-Algarra, and Elena Giacomelli discuss how superimposing climate change on south-to-north migration demonstrates that existing justice issues remain. Their chapter illustrates how the historical impact of emissions in the Global North on citizens of the Global South—in terms of loss and damage from extreme events, declining crop yields, the impacts of extreme heat, etc.—renders the justice aspects of migration even more stark.

In chapter 8, Lewis Ziska, Jeffrey Shaman, Emily Weaver, and Ami Zota discuss the relevance of environmental health sciences to understanding the implications climate change will have on the health of agricultural workers and the climate injustices these impacts present. They contend that addressing the plight of migrant workers, including those from twelve to eighteen years of age, is necessary not only in a humanitarian sense but also to ensure that farm production and food security will be sustainable in an uncertain climate.

Chapter 9, by Jennifer Wenzel, considers the question of climate justice by juxtaposing it with the closely related imperatives of environmental and energy justice. These terms are often invoked together, sometimes even as synonyms. But to what extent are they contradictory, incompatible, or incommensurable, particularly when considered at multiple scales of space and time? To answer these questions, Wenzel juxtaposes Andreas Malm's influential account of the birth of "business as usual" in *Fossil Capital* with analyses of the carbon economy that emphasize the inextricability of capitalism and colonialism at a global scale.

The contribution by Julia Lajus and Emma Gilheany in chapter 10 approaches climate justice through the lens of a multispecies environmental history and anthropology of the Arctic. Lajus and Gilheany focus on how climate change has created new and different mobilities for people and other species, using the migration of species as a way to examine four main concerns for Arctic Indigenous communities: adaptation and mitigation in relation to climate change, changes in toxicity related to climate, diminishing of sea ice, and regime shifts in the oceans.

Chapter 11, by Kailani Acosta and Gisela Winckler, portrays the succession and interconnectivity of geology, biology, race, agriculture, and climate justice through an analysis of how the Late Cretaceous coastline preconditioned agricultural use and led to the creation of cotton fields, which resulted in the emergence of the Black Belt region as a center of both agriculture and slavery. The authors argue that the patterns still contribute to environmental and climate injustice today. To revise these patterns, they suggest introspection on patterns of exclusion within science, which they argue is fundamental to creating a more inclusive future.

Adam Sobel and Melanie Bieli, in chapter 12, reflect on the potential to have a positive impact on climate change as a scientist working either in academia or in the private sector. They advocate for continuous self-reflection and assessment of one's theory of change against the available evidence, but at the same time recognize that many of the most important mechanisms of change are difficult to measure.

In chapter 13, anthropologist Sheng Long explores the epistemological dimension of climate (in)justice. Long focuses on alternative narratives that articulate environmental challenges in the absence of Western or modern scientific terminology, discussing how Hakka-speaking pomelo farmers in southeastern China understand climate anomalies using their established agricultural knowledge, even without terms such as "climate change" or "global warming."

Chapter 14, by Courtney Bender and Raffaella Taylor-Seymour, complicates the understanding of climate justice as a secular concept. The authors propose two distinct paths toward thinking about and including religion in the pursuit of climate justice. The first approach focuses on the practical activities of so-called religious actors operating and contending with climate change in a variety of circumstances. The second path focuses on religion in a more expansive register, gesturing to the many ways that religious thought has established, worked with, and sought to address core ethical and justice questions of scale, environmental community, and existential futurity through theological debate, ritual, spiritual experience, and engagements with the unseen. They argue that such efforts are establishing more transformative forms of climate justice from the ground up.

Paul Gallay's chapter 15 highlights the need for increased coproduction in flood risk reduction policymaking. Gallay offers options for overcoming systemic barriers to effective coproduction, which include the failure to provide sufficient time and resources to plan collaboratively, histories of mistrust due to past failure to center community knowledge, and governments' tendency to frame resilience planning around narrow transactional goals rather than systemic transformation or intersectional problem-solving.

In the final chapter, we, the editors, synthesize key points of convergence and divergence across the diverse approaches taken in the preceding chapters to conceptualize and operationalize climate justice in the natural sciences, social sciences and humanities. The conclusion underscores the costs associated with the status quo in climate justice scholarship and pedagogy, the urgency to center climate justice across disciplines, and the benefits to be gained from multidisciplinary perspectives on climate justice. In addition, it makes a series of actionable recommendations on how more interdisciplinary research and education can be facilitated.

Notes

1. In September 2023 a study in the journal *Science Advances* also found that six of the nine planetary boundaries had been transgressed with catastrophic consequences of the Earth system's biosphere. See Richardson et al., "Earth Beyond Six of Nine Planetary Boundaries," 1. See also Lenton et al., "Quantifying the Human Cost of Global Warming."
2. Greene et al. "Ubiquitous Acceleration in Greenland Ice Sheet Calving," 527.
3. Gaikwad et al., "Climate Action from Abroad."
4. Coolsaet et al., "Acknowledging the Historic Presence of Justice."
5. Boran, "On Inquiry Into Climate Justice," 30–33.
6. Thomas, *The Intersectional Environmentalist*, 32; Winter, *Subjects of Intergenerational Justice.*
7. Dolsak and Prakash, "Three Faces of Climate Justice," 283.
8. Callahan and Mankin, "National Attribution of Historical Climate Damages," 15.
9. President's Science Advisory Committee, *Restoring the Quality of Our Environment.*
10. Biello, "What Is Geoengineering."
11. Hulme, *Can Science Fix Climate Change?*, 35.
12. Intergovernmental Panel on Climate Change, "Special Report on Global Warming of 1.5 _C (Sr15)"; Core Writing Team, *Climate Change 2023 Synthesis Report Summary for Policymakers.*
13. Bednarz et al., "Injection Strategy—a Driver of Atmospheric Circulation"; Wan et al., "Diminished Efficacy of Regional Marine Cloud Brightening.
14. Khan, "MRV as a Tool for Achieving Just Outcomes."
15. United Nations Framework Convention on Climate Change, *Handbook on Measurement, Reporting and Verification*, 11.

16. Shrader, "Improving Climate Damage Estimates."
17. Merino et al., "Adaptation of North Atlantic Albacore Fishery"; Bell et al., "Pathways to Sustaining Tuna-Dependent Pacific Island Economies."
18. Exploring the concept of climate justice from an interdisciplinary angle, Mandy Meikle, Jake Wilson, and Tahseen Jafry, in their survey article "Between Mammon and Mother Earth," have discussed contradictions between different conceptions of climate justice in both academic and activist discourses, with a particular focus on the divisions between developing and developed countries, as well as industrialized nations and the demands of indigenous peoples. See Meikle et al., "Climate Justice: Between Mammon and Mother Earth," 488.
19. Schlosberg, *Environmental Justice and the New Pluralism*, 9, 17.
20. Gaikwad et al., "Creating Climate Coalitions."
21. Schlosberg, *Defining Environmental Justice*, 47.
22. Schlosberg and Collins, "From Environmental to Climate Justice," 359.
23. Best and Nocella II, eds., *Igniting a Revolution*, 16–21.
24. Wolfe, "The Rise and Fall of America's Environmentalist Underground."
25. Bhatt, "The Chipko Andolan," 7; Barbosa de Almeida, "The Politics of Amazonian Conservation," 171.
26. Alexander Dunlap and Andrea Brock therefore argue that the state and private companies are central to enforcing environmental destruction and political repression. Dunlap and Brock, eds., *Enforcing Ecocide*, 3–11.
27. Tokar, "Evolution and Continuing Development of the Climate Justice Movement," 19–20.
28. United Nations Environmental Programme (UNEP), "What You Need to Know About the Cop27."
29. Gaikwad et al., "Climate Action from Abroad."
30. Dawson et al., "Urban Climate Insurgency—an Introduction."
31. Bhavnani et al., eds., *Climate Futures—Reimagining Global Climate Justice*.
32. Rodriguez et al., *Grassroots Struggles for Alternative Futures*, 10, 14.
33. Hickel, *Less Is More*.
34. Tahseen, Mikulewicz, and Helwig, "Introduction: Justice in the Era of Climate Change," 4.
35. Winter, *Subjects of Intergenerational Justice*.
36. Schlosberg, *Defining Environmental Justice*, 47.

BIBLIOGRAPHY

Armiero, Marco, Ethemcan Turhan, and Salvatore Paolo De Rosa, eds. *Urban Movements and Climate Change—Loss, Damage and Radical Adaptation*. Amsterdam: Amsterdam University Press, 2023.

Barbosa de Almeida, Mauro. "The Politics of Amazonian Conservation: The Struggles of Rubber Tappers." *Journal of Latin American Anthropology* 7, no. 1 (2022): 170–219.

Bednarz, Ewa M., Amy H. Butler, Daniele Visioni, Yan Zhang, Ben Kravitz, and Douglas G. MacMartin. "Injection Strategy—a Driver of Atmospheric Circulation and Ozone

Response to Stratospheric Aerosol Geoengineering." *Atmospheric Chemistry and Physics* 23, no. 21 (2023): 13665–84.

Bell, Johann D., Inna Senina, Timothy Adams et al. "Pathways to Sustaining Tuna-Dependent Pacific Island Economies During Climate Change." *Nature Sustainability* 4 (2021).

Best, Steven, and Anthony J. Nocella II, eds. *Igniting a Revolution—Voices in Defense of the Earth*. Stirling: AK Press, 2006.

Bhatt, Prasad Chandi. "The Chipko Andolan: Forest Conservation Based on People's Power." *Environment and Urbanization* 2, no. 1 (1990): 7–18.

Bhavnani, Kum-Kum, John Foran, Priya A. Kurian, and Debashish Munshi, eds. *Climate Futures—Reimagining Global Climate Justice*. London: Zed Books, 2019.

Biello, David. "What Is Geoengineering and Why Is It Considered a Climate Change Solution?" *Scientific American*, 2010. https://www.scientificamerican.com/article/geo engineering-and-climate-change/.

Boran, Idil. "On Inquiry Into Climate Justice." In *Routledge Handbook of Climate Justice*, ed. Jafry Tahseen. London: Routledge, 2019.

Callahan, Christopher W., and Justin S. Mankin. "National Attribution of Historical Climate Damages." *Climatic Change* 172, no. 40 (2022): 1–19.

Core Writing Team. *Climate Change 2023 Synthesis Report Summary for Policymakers*. IPCC (Geneva, 2023).

Corporate Watch Group. *Greenhouse Gangsters vs. Climate Justice* (1999).

Coolsaet, Brendan, Julian Agyeman, Prakash Kashwan et al. "Acknowledging the Historic Presence of Justice in Climate Research." *Nature Climate Change* (2024). https://doi.org /10.1038/s41558-024-02218-5.

Dawson, Ashley, Marco Armiero, Ethmecan Turhan, and Roberta Biasillo. "Urban Climate Insurgency—an Introduction." *Social Text 150* 40, no. 1 (2022): 1–20.

Dolsak, Nives, and Aseem Prakash. "Three Faces of Climate Justice." *Annual Review of Political Science* 25 (2022): 283–301.

Dunlap, Alexander, and Andrea Brock, eds. *Enforcing Ecocide*. Cham: Palgrave Macmillan, 2022.

Gaikwad, Nikhar, Federica Genovese, and Dustin Tingley. "Climate Action from Abroad: Assessing Mass Support for Cross-Border Climate Transfers." *International Organization* 79, no. 1 (2025): 146–72.

Gaikwad, Nikhar, Federica Genovese, and Dustin Tingley. "Creating Climate Coalitions: Mass Preferences for Compensating Vulnerability in the World's Two Largest Democracies." *American Political Science Review* 116, no. 4 (2022): 1165–83.

Greene, Chad A., Alex S. Gardner, Michael Wood, and Joshua K. Cuzzone. "Ubiquitous Acceleration in Greenland Ice Sheet Calving from 1985 to 2022." *Nature*, no. 625 (2024): 523–28.

Hickel, Jason. *Less Is More: How Degrowth Will Save the World*. London: Penguin Random House, 2020.

Hulme, Michael. *Can Science Fix Climate Change? A Case Against Climate Engineering*. Cambridge: Polity Press, 2014.

Intergovernmental Panel on Climate Change. *Special Report on Global Warming of 1.5 _C (Sr15)*. 2018. https://www.ipcc.ch/sr15/.

Khan, Anu, "MRV as a Tool for Achieving Just Outcomes in the Carbon Removal Sector." Society of Fellows and Heyman Center for the Humanities, 2023. https://sofheyman.org/media/videos/mrv-as-a-tool-for-achieving-just-outcomes-in-the-carbon-removal-sector.

Lenton, Timothy M., Chi Xu, Jesse F. Abrams et al. "Quantifying the Human Cost of Global Warming." *Nature Sustainability* 6 (2023): 1237–47.

Meikle, Mandy, Jake Wilson, and Tahseen Jafry. "Climate Justice: Between Mammon and Mother Earth." *International Journal of Climate Change Strategies and Management* 8, no. 4 (2016): 488–504.

Merino, Gorka, Haritz Arrizabalaga, Igor Arregui et al. "Adaptation of North Atlantic Albacore Fishery to Climate Change: Yet Another Potential Benefit of Harvest Control Rules." *Frontiers in Marine Science* 6 (2019): 1–14.

President's Science Advisory Committee. *Restoring the Quality of Our Environment: Report of the Environmental Pollution Panel*. Washington, DC: The White House, 1965.

Richardson, Katherine, WIll Steffen, Wolfgang Lucht et al. "Earth Beyond Six of Nine Planetary Boundaries." *ScienceAdvances* 9, no. 37 (2023): 1–16.

Rodriguez, Iokine, Mariana Walter, and Leah Temper, eds. *Grassroots Struggles for Alternative Futures*. London: Pluto Books, 2024.

Schlosberg, David. "Defining Environmental Justice: Theories, Movements and Nature." Oxford: Oxford University Press, 2007.

Schlosberg, David, and Lisette B. Collins. "From Environmental to Climate Justice: Climate Change and the Discourse of Environmental Justice." *Wire's Climate Change* 5 (2014): 359–74.

Shrader, Jeffrey. "Improving Climate Damage Estimates by Accounting for Adaptation." SSRN, 2023.

Tahseen, Jafry, Michael Mikulewicz, and Karin Helwig. "Introduction: Justice in the Era of Climate Change." In *Routledge Handbook of Climate Justice*, ed. Jafry Tahseen. London: Routledge, 2019.

Thomas, Leah. *The Intersectional Environmentalist: How to Dismantle Systems of Oppression to Protect People + Planet*. New York: Voracious, 2022.

Tokar, Brian. "On the Evolution and Continuing Development of the Climate Justice Movement." In *Routledge Handbook of Climate Justice*, ed. Jafry Tahseen. London: Routledge, 2019.

United Nations Environmental Programme. "What You Need to Know About the Cop27 Loss and Damages Fund." 2022. https://www.unep.org/news-and-stories/story/what-you-need-know-about-cop27-loss-and-damage-fund.

United Nations Framework Convention on Climate Change. *Handbook on Measurement, Reporting and Verification for Developing Country Parties*. 2014.

Wan, Jessica S., Chih-Chieh Jack Chen, Simone Tilmes, Matthew T. Luongo, Jadwiga H. Richter, and Katharine Ricker. "Diminished Efficacy of Regional Marine Cloud

Brightening in a Warmer World." *Nature Climate Change* (2024). https://doi.org/10 .1038/s41558-024-02046-7.

Winter, Christine J. *Subjects of Intergenerational Justice—Indigenous Philosophy, the Environment and Relationships.* London: Routledge, 2022.

Wolfe, Matthew. "The Rise and Fall of America's Environmentalist Underground." *New York Times Magazine*, May 26, 2022. https://www.nytimes.com/2022/05/26/magazine /earth-liberation-front-joseph-mahmoud-dibee.html.

The Intersection of Climate, Environment, and Justice

Theory, Politics, Science, and Urban Perspectives

2

Introduction to Theories of Environmental and Climate Justice

MARY E. WITLACIL

P olitical theory is a humanistic form of inquiry that seeks to make sense of the political world by engaging in critical and historically informed interpretation. Political theorists ask questions that grapple with the "meaning and significance" of the political—be it events, practices, crises, or ideas—to develop concepts and context that illuminate otherwise opaque aspects of our political world.[1] Many of these concepts are "essentially contested," or open to persistent debate and revision.[2] Political theory and philosophy have much in common, often drawing on the same resources and asking similar questions. When the emphasis is explicitly political, theorists and philosophers focus on the role of power, forms of government, institutions, and economics, while other forms of philosophical inquiry foreground questions pertaining to ethics, knowledge, being, and reason. Given the provisional and reiterative nature of political theory, political theorists are, as Sheldon Wolin once said, "intent on posting warnings" or suggesting tactics, rather than offering "predictions" or solutions.[3] This chapter explores several conceptualizations of justice drawn from political theory and philosophy, how they have been used to theorize environmental and climate justice, and tactics for addressing injustice wrought by the climate crisis.

Environmental and climate justice build on the myriad approaches to dealing with questions of justice in political thought and philosophy. Before activists of color and Indigenous communities illuminated how environmental inequity could reinforce injustice, theories of justice did not explicitly attend to the relationship between race, class, and environmental damage. Environmental justice theorizing has responded to and learned from environmental justice movements to expand the sphere of concern within the main threads of justice theorizing.[4] Many scholars now incorporate "'the trivalent view' of justice" in their theorization, which includes distributive, participatory, and recognition justice (discussed shortly).[5] Historically, environmental justice emerged to challenge unjust distributions of environmental harms and benefits, alternatively conceived as eco-racism, environmental classism, and environmental privilege.[6] More recently, scholars have called for expanding environmental justice to include the capabilities approach, Indigenous conceptions of justice, and epistemic injustice.[7] Environmental justice is considered along anthropocentric and ecocentric axes, where the former emphasizes human-centered justice, and the latter foregrounds ecological justice.[8] Climate justice has become an increasingly vital intervention in justice studies, but while the climate is an aspect of the environment, climate justice and environmental justice are distinct frameworks for grappling with their respective forms of injustice.

An organizing point of emphasis in climate justice is that those who have contributed the least to greenhouse gas emissions are likely to suffer the worst consequences of climate change. Theories of climate justice consider the political and ethical implications for how particular subjects or entities become recipients of the burdens or benefits associated with climate change. As such, theorists of climate justice attend to the implications of climate policy (e.g., mitigation and adaptation), who should bear responsibility, the subjects of justice, the role of individual versus collective action, and the form of justice best attuned to the climate crisis. Climate justice scholarship incorporates the aforementioned theories of justice, along with an emphasis on intergenerational justice, which reckons with the generational delay between climate destructive behavior (e.g., burning fossil fuels) and consequences of the climate crisis (e.g., rising sea levels, harsher and less predictable weather, weakening of the Atlantic Meridional Overturning Circulation or AMOC).[9]

Subjects of climate justice include those who are most vulnerable to climate change—e.g., Indigenous peoples, those on the climate frontlines, workers in the carbon economy, future humans—as well as the more-than-human world, ecological systems, and the planet. Theories of justice that adopt a relational ethic drawn from Indigenous worldviews are better able to conceive of justice for the more-than-human world.[10]

Throughout this chapter, I untangle the strands of justice woven throughout political theory. Each section pivots on a different approach to justice. For each approach, I discuss how it was originally theorized, prior to exploring how theorists of environmental justice and climate justice incorporate said approach into their analysis. This chapter advocates for moving beyond the trivalent approach to justice—or distributive, participatory, and recognition—toward an Indigenous intergenerational climate justice that is relational, is epistemologically sensitive, and attends to the more-than-human world.[11]

THEORIES OF JUSTICE CONFRONT THE ENVIRONMENT AND THE CLIMATE

Distributive Justice

Distributive justice is concerned with equality and the just distribution of societal goods and harms.[12] To develop a theory of distributive justice, John Rawls developed a thought experiment to ask how one might conceive of justice if one emerged into the world as a thinking person without any prior knowledge of social positionality. This is essentially a social tabula rasa, or what Rawls refers to as "original position": a person existing without awareness of their class, culture, geography, race, gender, or sexual identity. Rawls argued that when justice is considered from this perspective, people would agree that justice corresponds with equality, and the equal distribution of social goods. He did not endeavor to theorize a truly egalitarian society, but a society in which there was "fair equality of opportunity." As a result, he defended a certain degree of inequality, in which people should benefit from their accomplishments, and in cases when the unequal distribution of societal goods has the capacity to benefit all members of society.[13]

Owing to considerations of equal distribution of resources, distributive justice has long been a salient framework for theorizing environmental justice.[14] To wit, the US Environmental Protection Agency defines environmental justice as "[protection] from disproportionate and adverse human health and environmental effects," while ensuring "equitable access to a healthy, sustainable, and resilient environment."[15] Distributive elements embedded in this definition include "disproportionate . . . effects" and "equitable access." Further, activists, environmental justice advocates, and scholars often incorporate distributive claims in their emphasis "on how the distribution of environmental risks mirrors the inequity in socioeconomic and cultural status."[16] Distributive environmental justice foregrounds the relationship between marginalized communities and the siting of polluting industries, land use patterns or zoning laws, and infrastructure disparities (e.g., aging lead water pipes in Flint, Michigan).[17] Environmental injustice refers not only to the disproportionate siting of hazardous waste sites and polluting industries in low-income neighborhoods and communities of color, but also to inequitable access to environmental benefits like parks or trail systems, and the unequal rate of environmental regulatory enforcement. To achieve distributive environmental justice would require "environmental equity among communities" within the United States and worldwide.[18]

Distributive climate justice emphasizes equity in the distribution of climate change related burdens and benefits, as well as who should bear responsibility for adaptation and mitigation (for more discussion of the distributive framework, see chapter 3). Climate justice scholars often assess the extent to which policies of mitigation, adaptation, and decarbonization will unevenly distribute benefits and burdens.[19] Mitigation refers to altering policies, economic relations, and institutions with the goal of avoiding climate change, while adaptation refers to using policy and institutional change to minimize the disastrous effects of climate change, when it might "undermine people's entitlements."[20] (See also chapter 5 for how climate policies may reinforce environmental injustice.) Distributive climate justice foregrounds equity in the allocation of responsibility for climate policy, the distribution of climate consequences, and the burdens associated with meeting climate targets. Some policy approaches to adaptation and mitigation will exact greater financial and environmental costs on populations vulnerable to climate change; for example, carbon taxes on fossil fuel companies often result in rate hikes for consumers.[21] Poor and working-class people pay more for energy as a proportion of their income, which positions them to bear a disproportionate burden for mitigation.[22] To address the disproportionate burden placed on underprivileged and minoritized communities, Simon Caney develops the concepts of the "just burden" and the "just target," where the former refers to a fair distribution of burdens associated with climate policy, and the latter refers to a fair distribution of the impacts of climate change and the impacts of climate policy.[23]

Distributive climate justice advocates for avoiding inequitable and/or unfair allocation of burdens, benefits, and responsibility for adaptation and mitigation. While this approach necessitates consideration of one's social position, culture, class, and geography, the distributive approach does not traditionally foreground cultural recognition or psychological consequences for being a member of a minoritized community. It relies on an ideal (rather than material, structural, or relational) conceptualization of justice that defers to an individualistic experience of the goods of liberal democracies. As Iris Marion Young argues, distributive accounts of justice overemphasize merit and wealth, without attending to how social processes and institutions reinforce the misallocation of material and nonmaterial goods (e.g., "power, opportunity, or self-respect"). Young articulates a more expansive view of justice based on the politics of difference, which emphasizes how institutions and processes can reinforce domination or enable liberation.[24] In *Black Rights/White Wrongs*, Charles Mills critiques Rawls by arguing that the ideal original position is assumed to be a white male, and because this presumes whiteness and maleness as the ideal position, it is a problematic vantage from which to consider injustice.[25] Finally, distributive justice prioritizes property relations, which establishes unequal social relations as a core principle of society, and views the environment as primarily a resource to plunder. For this reason, theories of recognition emerged to address this concern.

Justice as Recognition

Recognition moves beyond theories of distributive inequity to understand the relationship between identity, social status, and injustice. Recognition theorists emphasize how social identity is constructed dialogically, in relation to others, which requires reciprocity in recognition.[26] By contrast, misrecognition denigrates individual life by devaluing one's identity; constrains an individual's capacity to flourish; and establishes the conditions for distributive injustice. For Charles Taylor and Axel Honneth, recognition has an explicitly psychological component and is essential for developing self-esteem and self-worth, which is why Taylor believes that recognition "is a vital human need."[27] Honneth argues that one requires recognition to participate in the political realm, whereas misrecognition is a delegitimizing force that results in feelings of "disrespect or humiliation."[28] Recognition emphasizes how justice requires the "right to be different," and to be respected as different.[29] Both of these are protected rights acknowledged by the United Nations Declaration on the Rights of Indigenous People in 2007.[30] If a politics of difference rests at the core of recognition, Nancy Fraser advocates for "a *critical* theory of recognition" sensitive to issues of political economy and redistribution.[31]

To avoid misrecognition, scholars of environmental justice have advocated for expanding distributive environmental justice to include respect for difference and due recognition of social status, class, culture, and geography, to avoid replicating inequities related to one's social position. Misrecognition often enables the inequitable distribution of environmental burdens, where misrecognition refers to a lack of respect for one's bodily autonomy, one's cultural relationship to land, and/or inability to validate one's identity.[32] In environmental justice, misrecognition happens when one's culture or community continuity is threatened by the disproportionate siting of environmental burdens, due to hazardous waste, industry, mining, or pollution; as well as when one's knowledge, culture, or experience with an environmental burden is ignored or misrepresented. In 2016 a youth-led Indigenous movement (#NoDAPL) sought to protect clean drinking water from the development of the Dakota Access Pipeline, a natural gas pipeline. Even if conditions of distributive justice had been achieved during the planning phases for the pipeline, it would have resulted in misrecognition for the Standing Rock Sioux Nation.[33] This is because the Dakota Access Pipeline threatened not only Indigenous territorial sovereignty but also Indigenous epistemologies, which emphasize the sacredness of land and water and the interconnectedness of all living beings.[34] Any pipeline leaks, whether or not on Indigenous land, would have contaminated the land or water and disrespected Indigenous kinship relations with the Earth. To overcome misrecognition, decision-makers, industries, and politicians need to validate community concerns about cultural erasure, health risks, and environmental consequences due to pollution and industrial sites. To avoid reinforcing the subordinated status of vulnerable populations, decision-makers must consider social status, culture, and geography in siting decisions for projects that may cause environmental harm and human health risks. Recognition, however, may not sufficiently capture disrespect and contestation over knowledge claims. Epistemic injustice is a "necessary complement" to theories of recognition and participatory justice because it emphasizes the wrongs done to individuals "as knowers." Gwen Ottinger calls instead for careful knowing, based on a feminist ethic of care, as an antidote to epistemic injustice. Careful knowing entails recognizing and incorporating the expertise of "marginalized knowers" in environmental justice assessments and policy.[35] Finally, given that recognition validates the importance of difference, scholars have drawn on recognition to consider the more-than-human world as subjects and recipients of justice.[36]

While recognition has received its due in theories of environmental justice, it remains an undertheorized approach to climate justice. One framework for considering misrecognition in climate justice distinguishes between formal and discursive misrecognition in climate injustice. Misrecognition, as discussed by Tor Benjaminsen et al., contributes to formal injustice through a lack of inclusion in decision-making, structural disinclusion in the political process, and misrecognition by formal institutions. Discursive injustice happens when discourses used to frame policy proposals, recommendations, and assessments misrepresent a group or individual subject to climate policy. For instance, when a climate mitigation project is framed as a win-win without consideration of social costs, discursive misrecognition illuminates how local communities may be harmed more than they benefit.[37] Similarly, in the case of geoengineering, there is a tendency to deploy "expertise imperialism," in which experts—who are presumed to know best—have the formal power to research or implement potentially risky climate solutions, like solar radiation management (SRM), without sufficiently explaining the methods or consequences, or giving affected communities a seat at the table.[38] Most forms of SRM are hotly contested and hypothetical (for now), with intense scrutiny about implementing *or even researching* strategies like stratospheric aerosol injection—which would mimic the cooling effect of a volcanic eruption by injecting sulfur into the stratosphere. A further form of discursive misrecognition happens when experts and those in power justify geoengineering research based on how it might help a reified group of people, like "the poor."[39] This framing misrecognizes differences among who is included in the discursive category of "the poor," rather than viewing affected parties as subjects who deserve informed consent and formal recognition.

Recognition justice emphasizes how respect for cultural and epistemological difference is essential for psychological well-being. However, when scholars overemphasize the psychological aspects of recognition, Fraser warns that they miss the material effects of injustice, or how misrecognition can cause institutionalized modes of subordination based on one's social and economic status. To overcome misrecognition requires viewing all individuals "as full partners in social interaction" regardless of their social status, identity, or socio-economic position.[40] For this reason, Fraser advocates for a form of justice that includes representation, redistribution, *and* recognition.[41] Given that recognition prioritizes subjectivity and identity when considering justice, a vital extension of recognition is the topic of participatory justice, which I discuss next.

Participatory Justice

If recognition demands that one must consider difference—in terms of culture, geography, class, and social status—then participatory justice seeks to overcome issues of misrecognition through informed consent about risks and benefits, and meaningful involvement in decision-making processes. Participation, however, is not enough. For this reason, participation must include considerations of social status, culture, class, geography, gender, and Indigeneity to achieve "participatory parity."[42] Participatory parity occurs when all members of society—regardless of social and economic status—are allowed to participate as peers, without reifying or tokenizing their identity/ies or disregarding their social position or status as knowers.[43]

Participatory justice emphasizes how structural inequity plays a role in who participates in decision-making and institutional practices, and whether their participation affects political and institutional outcomes.[44] To achieve participatory parity in environmental justice, individuals in underrepresented, vulnerable, and marginalized communities must be included in environmental decision-making. Frequently, those included in decision-making already reap environmental benefits, whereas those in fenceline communities are too politically disenfranchised to resist environmental burdens. In this way, forms of injustice have the insidious tendency to "maintain and reinforce each other." While misrecognition and distributive injustice exacerbate participatory injustice, some scholars argue that participatory justice rests on its own merits, and that we should be concerned about unjust institutional practices even when they produce just outcomes.[45] For example, the US National Environmental Protection Act (NEPA) mandates public involvement during the assessment phase of proposed actions (e.g., for projects that could entail environmental risks). During the public commenting phase for addressing uranium mining and transportation permits in northern Arizona, there were only three public meetings (in Fredonia and Flagstaff, and Tuba City, a town in the Navajo Nation), two of which were held during work hours in locations inconvenient to the vast and disparately populated Indigenous nations (the Hopi, Havasupai, and Diné) affected by the mines.[46] While this policy aspires to address participatory parity, it can result in unjust outcomes when public commenting periods are just a box to check and do not consider location accessibility and meeting times, thereby excluding those in rural locations without reliable transportation and those unable to get away from work. This practice disproportionately affects people most likely to face environmental risks, including those in racially minoritized, low income, and

Indigenous communities. Embedded within calls for participatory parity are interrelated concerns about equality, representation, and who deserves to be included in an expanded vision of representation.

Theories of climate justice often focus on the vulnerability to climate impacts faced by marginalized communities. It is a triple injustice that those who are more vulnerable to climate change are often those who have contributed the least to greenhouse gas emissions and are the least likely to be included in decision-making about climate policy. Participatory justice is one way to address these inequities. Calls for participatory climate justice include developing decision-making processes for "informed consent through inclusive public participation" as well as procedures "to correct the harms that [climate adaptation and mitigation] policies might impose on citizens."[47] Participatory justice implies an "obligation" to invite those affected by climate policy into the decision-making process.[48] However, inclusive participation and procedural parity are often not meaningfully incorporated into climate decision-making processes. For instance, vulnerable populations may have the formal opportunity to voice concerns about adaptation policies, but politicians have the ultimate power to decide whether to acknowledge or sideline their concerns in policy implementation. Rather than top-down climate policy implementation, Breena Holland envisions a more horizontal approach to "procedurally just adaptation," which gives affected communities formal inclusion in the decision-making process and the ability to control the implementation of decisions in their communities.[49] Participatory climate justice entails meaningful involvement of affected parties and vulnerable populations to influence decision-making about adaptation (including geoengineering), mitigation, and energy transitions.

Participatory justice may be an essential element of both environmental and climate justice, but—without addressing concerns about distribution and recognition, or subsequent concerns about capabilities and intergenerational justice—it does not ensure that the threshold for justice has been met. For this reason, participatory justice is an essential extension to other theories of justice. This is because without including recognition and distributive concerns, it is difficult to assess if equitable representation has been achieved. For instance, Christine Winter argues that meaningfully including Indigenous people in decision-making—when their philosophies and worldviews are incompatible with those in dominant culture and settler colonial states—would require cultural humility and a willingness of settler states to accommodate a plurality of worldviews.[50] Environmental personhood in Aotearoa might be a step in the right direction because it provides recognition of Māori relational principles to the natural world, as well as reparations for the violence and injustice caused by the settler state of New Zealand. Providing representation for the nonhuman world is one way to accommodate Indigenous philosophy and move toward participatory parity. Having addressed the basis for the trivalent-approach to justice, I turn toward a discussion of the capabilities approach: a liberal understanding of justice based on the necessary preconditions for leading a life of dignity.

The Capabilities Approach

Beyond theories of recognition, distribution, and participation, scholars have grappled with how justice bears on an individual's capacity to flourish and achieve their goals in life. The capability approach is a liberal theory of justice pioneered by Amartya Sen and Martha Nussbaum. Nussbaum explains, "The capabilities approach . . . is an outcome-oriented approach" that focuses on the necessary conditions for an individual to flourish and lead a life of human dignity.[51] Sen argues that one's capacity to flourish is constrained or enabled by the social, economic, and political conditions in which one lives, all of which are compounded by international development, access to quality education, health care, and safety. For Sen, justice directly correlates to freedom, which refers to the ability "to choose a life one has reason to value."[52] Returning to Nussbaum's theory of human dignity, Holland points out that capabilities provide a framework for viewing "well-being as *multi-dimensional*."[53] To this end, Nussbaum identifies several capabilities that are essential and irreducible for having the capacity to live a dignified life. Among these "central human capabilities" are bodily health, bodily integrity, affiliation (or social connections), play, the ability to live alongside other species, and control over one's environment.[54] At issue is whether one has the opportunity to lead a life of human dignity and the capacity to thrive.[55] As Winter argues, "Dignity confers subjectivity," which determines both rights and obligations.[56]

If the capabilities approach necessitates considering opportunities needed to thrive, then a livable and healthy environment is essential for human flourishing. Accordingly, Holland argues that the environment is instrumentally valuable for human life and should be viewed as a "meta-capability."[57] Without a healthy environment—free from arsenic in the soil, lead in the water, and exhaust in the air—it would not be possible to address the other capabilities. (For more discussion of environmental health, see chapter 8). While cultural preservation is an element of recognition for Indigenous communities, Schlosberg notes that the capability to care for the land and nonhuman kin is essential for Indigenous cultural survival. In contrast to Holland's emphasis on the instrumental value of the environment, he views the capabilities approach as an avenue for theorizing ecological justice and the *intrinsic* value of nature.[58] To this end, the capabilities approach must overcome the individualistic foundation for the theory and move toward a community-level understanding of who is worthy of consideration.[59] By expanding capabilities to the community or system level, ecosystems could be agents that have particular needs and capabilities in order to function and flourish.[60] This could bear on the health and functioning of the environment and humans as well. If the capabilities approach can include community-level capabilities, it could adopt a more relational approach to environmental justice compatible with Indigenous worldviews.[61] As previously mentioned, dignity determines subjectivity, which entails rights and obligations. By drawing on Indigenous philosophy, a relational capabilities approach could emphasize the obligations and duties humans have to all living beings and the more-than-human world, just as our ecological kin should be considered subjects of justice.[62] By caring for the environment as kin, humans give living and nonliving entities the capacity to flourish, just as the environment returns the necessary conditions for humans to flourish.

A stable climate is a vital precondition for the flourishing of all life on Earth. Scholars of climate justice and the capabilities approach argue that Nussbaum's capabilities list should be expanded to include a stable and hospitable climate without dangerous natural disasters due to climate change.[63]

The capabilities approach could assess how adaptation policies affect the needs, capabilities, and vulnerabilities of communities threatened by climate change.[64] Climate change exacerbates and disturbs the threshold levels of capabilities for climate-vulnerable populations, which is why Holland argues that the approach should include environmental and climactic sustainability as a "meta-capability."[65] Other scholars turn to the capabilities approach as the basis for intergenerational justice, or the idea that current generations bear responsibility for ensuring that future generations inherit a livable planet and just world. Given the lag time between carbon dioxide accumulation in the atmosphere and the felt consequences of climate change, Edward Page suggests that intergenerational justice is compatible with the capability approach, because present generations ought to avoid undermining the capacity for future generations to thrive.[66] As such, intergenerational climate justice represents a tension between maintaining access to the capabilities necessary to thrive in the present and ensuring that the climate crisis does not disrupt these capabilities in the future.[67]

Not everyone agrees that the capabilities approach adequately addresses social justice and climate change, or the resulting needs for both intergenerational and Indigenous justice. When intergenerational justice relies on the capabilities approach, which implies a linear temporality that foregrounds *human* flourishing, it precludes a cyclical temporality and relational approach that allows for the flourishing of human and more-than-human kin into the distant future. For this reason, the capabilities approach is incompatible with Indigenous philosophy and cosmologies.[68] Additionally, while a stable climate and sustainable environment are essential preconditions for leading a life of human dignity, the environment is a tertiary and instrumental concern in Nussbaum's framing of the capabilities, rather than the basis for all other capabilities.[69] Finally, while one goal of the capabilities approach is to address the possibility for freedom, it does so while relying on universalizing Western/liberal ideals and providing an individualistic (rather than collectivistic) assessment of the preconditions for human flourishing. As such, the theory risks minimizing the effect of structural and economic inequities on injustice and the importance of respecting difference for overcoming injustice. In the closing section, I turn to a discussion of intergenerational justice and how Indigenous philosophy addresses problems associated with other approaches to justice.

RELATIONALITY AND INTERGENERATIONAL CLIMATE JUSTICE

This chapter has covered capabilities and the trivalent approach to justice, including distribution, participation, and recognition. At the beginning of the chapter I argued for an Indigenous intergenerational climate justice that is relational, is epistemologically sensitive, and attends to the more-than-human world. By placing Western approaches to justice in conversation with Māori and Aboriginal philosophy, Winter develops a theory of intergenerational Indigenous environmental justice, well-poised to meet the demands of my argument.[70] Rather than relying on classical liberalism and capitalism as sources of justice, she offers an ethos based on Indigenous worldviews to overcome the limitations associated with liberal intergenerational environmental justice (i.e., the capabilities approach). She argues that liberal intergenerational justice is individualistic, has a limited capacity to think beyond the present, is anthropocentric, and maintains an instrumental worldview. Rather than emphasizing property relations, Winter suggests a place-based custodial relationship to the land, which recognizes the integrity of environmental systems and the obligations humans have to the natural world. In place of individualism, she advocates a collective approach to consider justice and obligations. Instead of progressive temporalities, she conceives of time in a cyclical manner, where the past and future are always already present in the current moment. Finally, Winter foregrounds a relational ethic that incorporates insights from Māori philosophy, where entities in the natural world are ancestors and kin with humans, which obliges humans to care for the more-than-human world. Environmental personhood in Aotearoa, which offers rights to environmental entities (Te Urewera, Mount Taranaki, and the Whanganui River), is an example of how Indigenous intergenerational justice might be put into practice, because it honors how Māori-iwi (people) view the forest, mountains, and rivers as ancestors.[71]

The insights from Indigenous intergenerational environmental justice are productive for theorizing a truly just climate justice. Climate change is a problem of excess and hubris caused by the Western world, settler colonialism, and capitalist economic systems. Indigenous intergenerational justice foregrounds an ethic of relationality that extends the temporal scale considered for the obligations and conditions of justice. This approach emphasizes the relationships that each of us has to one another and the natural world. It beckons us to consider how—even though individual efforts are unlikely to dramatically alter CO_2 emissions—we can have a net positive impact on the climate by acting in solidarity to reduce fossil fuel use, stop overconsumption, minimize the use of harmful technologies, and avoid environmentally harmful behavior. For truly meaningful change, the more radical insights of Winter's approach would need implementation, rather than paying lip-service to Indigenous philosophy and cosmologies. Changing our relationship to land from one of property ownership to a custodial obligation would upend the Western preference for extractive exploitation of the environment. Incorporating Indigenous temporalities into climate justice, would demand renegotiating climate agreements to avoid warming above 2° Celsius (with sanctions for countries in the West that did not drastically reduce fossil fuel consumption), an international preference for mitigation over adaptation, and a globally just transition. Foregrounding a relational ethic would call for environmental entities to have a seat at the table or at least including their representatives in climate policy negotiations. While leading approaches to climate justice are far from the radical ethos theorized by Winter, the ethic of intergenerational Indigenous environmental justice expands what is possible in this nonideal world.

• • •

This chapter has examined how theories of environmental and climate justice enlarge the scope and focus of the prevailing approaches to justice. Both environmental and climate justice are vital extensions to theories of justice. A stable climate and healthy environment are essential baselines for meeting the requirements of justice in any form. Whether a theory of justice emphasizes anthropocentric or ecocentric values, recognition or distribution, human dignity or relationality, or individual or collective subjectivity, without a livable climate there is no world in which to theorize justice. Consider this a warning.[72]

Notes

1. Grant, "Political Theory, Political Science, and Politics," 581.
2. Connolly, *The Terms of Political Discourse*, 10.
3. Wolin, *Politics and Vision: Continuity and Innovation*.
4. Mah, "Toxic Legacies and Environmental Justice."
5. Holland, "Procedural Justice in Local Climate Adaptation," 395; Hourdequin, "Climate Change, Climate Engineering," 274; Kortetmäki, "Reframing Climate Justice," 320–34; Pellow, *What Is Critical Environmental Justice?*; Schlosberg, *Defining Environmental Justice*.
6. Cone, "Whose Earth Is It Anyway?," 36–46; Nixon, *Slow Violence and the Environmentalism of the Poor*; Park and Pellow, *The Slums of Aspen*.
7. Ottinger, "Careful Knowing as an Aspect of Environmental Justice"; Winter, *Subjects of Intergenerational Justice*.
8. Okereke and Charlesworth, "Environmental and Ecological Justice"; Schlosberg, *Defining Environmental Justice*; Stevis, "Whose Ecological Justice?"
9. Winter, "Does Time Colonise Intergenerational Environmental Justice Theory?"; Winter, *Subjects of Intergenerational Justice*; Ferrari, "What Would Happen If the Atlantic Meridional Overturning Circulation Collapses?"
10. See also chapter 14 in this volume.
11. Thank you to Kaleigh Karageorge for invoking David Schlosberg's comments at WPSA 2024, when he suggested it was time to move beyond the trivalent approach.
12. Kaswan, "Distributive Environmental Justice."
13. Rawls, *A Theory of Justice*, 17, 65–70, quote on 68; Schlosberg, *Defining Environmental Justice*, 13; Kaswan, "Distributive Environmental Justice."
14. Schlosberg, *Defining Environmental Justice*.
15. Environmental Protection Agency, "Environmental Justice."
16. Schlosberg, *Defining Environmental Justice*, 55.
17. Kaswan, "Distributive Environmental Justice."
18. Harrison, *From the Inside Out*, 13.
19. Sardo, "Responsibility for Climate Justice."
20. Caney, "Climate Change," 666.
21. Dolšak and Prakash, "Three Faces of Climate Justice."
22. Dolšak and Prakash.
23. Caney, "Climate Change," 667.
24. Young, *Justice and the Politics of Difference*.
25. Mills, *Black Rights/White Wrongs*.
26. Coolsaet and Néron, "Recognition and Environmental Justice," 52–63.
27. Taylor, "The Politics of Recognition," 26; Honneth, "Recognition and Justice," 351–64.
28. Honneth, "Recognition and Justice," 352.
29. Coolsaet and Néron, "Recognition and Environmental Justice," 52.

30. Coolsaet and Néron; United Nations, *United Nations Declaration on the Rights of Indigenous Peoples*.

31. Fraser, "Recognition or Redistribution?," 167.

32. Schlosberg, *Defining Environmental Justice*, 60.

33. Coolsaet and Néron, "Recognition and Environmental Justice."

34. Coolsaet and Néron; Estes and Dhillon, eds., *Standing with Standing Rock*; Gilio-Whitaker, *As Long as Grass Grows*.

35. Ottinger, "Careful Knowing," 2. See also chapter 13 in this volume.

36. Ottinger; Whyte, "The Recognition Paradigm of Environmental Injustice"; Schlosberg, *Defining Environmental Justice*; Coolsaet and Néron, "Recognition and Environmental Justice."

37. Benjaminsen et al., "Recognising Recognition in Climate Justice."

38. Hourdequin, "Climate Change, Climate Engineering," 37, referencing Allan Buchanan, "Social Moral Epistemology," *Social Philosophy & Policy* 19 (2002): 126–52.

39. Hourdequin.

40. Fraser, "Rethinking Recognition."

41. Fraser, "Reframing Justice in a Globalizing World."

42. Fraser, "Rethinking Recognition," 116.

43. Fraser; Ottinger, "Careful Knowing."

44. Holland, "Procedural Justice in Local Climate Adaptation."

45. Bell and Carrick, "Procedural Environmental Justice," quote on 102.

46. A note on language: Diné, which means "people" in Navajo, is the preferred name for the people who come from Dinétah, the traditional homeland of the Navajo. The Navajo Nation is the official name for the tribal nation. Energy Fuels Resources (USA) INC, Class II Air Quality Permit Numbers 62877, 62878, 63895 for the Canyon, EZ, and Arizona 1 Mines, Respectively, August 16, 2016, https://static.azdeq.gov/permits/energyfuelsres_rs.pdf.

47. Dolšak and Prakash, "Three Faces of Climate Justice."

48. Hourdequin, "Climate Change, Climate Engineering," 273.

49. Hourdequin, 198.

50. Winter, *Subjects of Intergenerational Justice*.

51. Nussbaum, "Beyond the Social Contract," 12.

52. Sen, *Development as Freedom*, 74.

53. Holland, "Capabilities, Well-Being, and Environmental Justice." Emphasis in original.

54. Nussbaum, *Frontiers of Justice*, 76–78.

55. Nussbaum, "Capabilities and Social Justice."

56. Winter, *Subjects of Intergenerational Justice*, 190.

57. Holland, "Justice and the Environment," 319; Holland, "Environment as Meta-Capability."

58. Schlosberg, *Defining Environmental Justice*.

59. Schlosberg; Winter, *Subjects of Intergenerational Justice*.

60. Schlosberg, *Defining Environmental Justice*; David Schlosberg, "Climate Justice and Capabilities."
61. Winter, *Subjects of Intergenerational Justice.*
62. Winter.
63. Page, "Intergenerational Justice of What: Welfare," 464.
64. Schlosberg, "Climate Justice and Capabilities."
65. Holland, "Justice and the Environment."
66. Page, "Intergenerational Justice of What."
67. Winter, *Subjects of Intergenerational Justice.*
68. Winter.
69. Holland, "Justice and the Environment."
70. Winter, *Subjects of Intergenerational Justice.*
71. To learn more about environmental personhood and the rights of nature, see Stone, *Should Trees Have Standing?*; Winter, "A Seat at the Table"; and Winter, *Subjects of Intergenerational Justice.*
72. I owe a debt of gratitude to the editors of this volume—Rebecca Marwege, Nikhar Gaikwad, and Joerg Schaeffer—for welcoming my contribution, hosting a generative book workshop, and providing attentive feedback on my chapter. My sincere thanks to the anonymous reviewer whose careful reading made my chapter unimaginably better. Finally, thank you to Jeff Feng, Jennifer Givens, David Herrera, Kaleigh Karageorge, Sean Parson, Scott Ritner, and Cynthia Witter, who all offered insightful feedback on earlier drafts.

BIBLIOGRAPHY

Bell, Derek, and Jayne Carrick. "Procedural Environmental Justice." In *The Routledge Handbook of Environmental Justice*, ed. Ryan Holifield, Jayajit Chakraborty, and Gordon Walker, 101–12. London: Routledge, 2017.

Benjaminsen, Tor A., Hanne Svarstad, and Iselin Shaw of Tordarroch. "Recognising Recognition in Climate Justice." *Institute of Development Studies (IDS)*, October 28, 2021. https://opendocs.ids.ac.uk/opendocs/handle/20.500.12413/16912.

Caney, Simon. "Climate Change." In *The Oxford Handbook on Distributive Justice*, ed. Serena Olsaretti, 664–88. Oxford: Oxford University Press, 2018. https://doi.org/10.1093/oxfordhb/9780199645121.013.23.

Caney, Simon. "The Struggle for Climate Justice in a Non-Ideal World." *Midwest Studies In Philosophy* 40, no. 1 (September 2016): 9–26. https://doi.org/10.1111/misp.12044.

Caney, Simon. "Two Kinds of Climate Justice: Avoiding Harm and Sharing Burdens." *Journal of Political Philosophy* 22, no. 2 (June 2014): 125–49. https://doi.org/10.1111/jopp.12030.

Cone, James H. "Whose Earth Is It Anyway?" *CrossCurrents* 50, no. 1/2 (2000): 36–46. http://www.jstor.org/stable/24461228.

Connolly, William E. *The Terms of Political Discourse.* 3rd ed. Princeton: Princeton University Press, 1993.

Coolsaet, Brendan, and Pierre-Yves Néron. "Recognition and Environmental Justice." In *Environmental Justice: Key Issues*, ed. Brendan Coolsaet, 52–63. London: Routledge, 2020.

Dolšak, Nives, and Aseem Prakash. "Three Faces of Climate Justice." *Annual Review of Political Science* 25, no. 1 (May 12, 2022): 283–301. https://doi.org/10.1146/annurev-polisci-051120-125514.

Environmental Protection Agency. "Environmental Justice." February 6, 2024. https://www.epa.gov/environmentaljustice.

Estes, Nick, and Jaskiran Dhillon, eds. *Standing with Standing Rock: Voices from the #NoDAPL Movement*. Minneapolis: University of Minnesota Press, 2019.

Ferrari, Raffaele. "What Would Happen If the Atlantic Meridional Overturning Circulation (AMOC) Collapses? How Likely Is It?" *Ask MIT Climate*, November 7, 2024. https://climate.mit.edu/ask-mit/what-would-happen-if-atlantic-meridional-overturning-circulation-amoc-collapses-how-likely.

Fraser, Nancy. "Recognition or Redistribution? A Critical Reading of Iris Young's *Justice and the Politics of Difference*." *Journal of Political Philosophy* 3, no. 2 (June 1995): 166–80.

Fraser, Nancy. "Reframing Justice in a Globalizing World." *New Left Review* (2005): 69–88.

Gani, Jasmine K., and Rabea M. Khan. "Positionality Statements as a Function of Coloniality: Interrogating Reflexive Methodologies." *International Studies Quarterly* 68, no. 2 (March 14, 2024). https://doi.org/10.1093/isq/sqae038.

Gilio-Whitaker, Dina. *As Long as Grass Grows: The Indigenous Fight for Environmental Justice, from Colonization to Standing Rock*. New York: Beacon Press, 2019.

Grant, Ruth W. "Political Theory, Political Science, and Politics." *Political Theory* 30, no. 4 (2002): 577–95. https://doi.org/10.1177/0090591702030004007.

Harrison, Jill Lindsey. *From the Inside Out: The Fight for Environmental Justice Within Government Agencies*. Cambridge: MIT Press, 2019.

Holland, Breena. "Environment as Meta-Capability: Why a Dignified Human Life Requires a Stable Climate System." In *Ethical Adaptation to Climate Change: Human Virtues of the Future*, ed. Allen Thompson and Jeremy Bendik-Keymer, 145–64. Cambridge: Cambridge University Press, 2012. https://doi.org/10.7551/mitpress/9780262017534.003.0008.

Holland, Breena. "Capabilities, Well-Being, and Environmental Justice." In *Environmental Justice: Key Issues*, ed. Brendan Coolsaet, 64–77. New York: Routledge, 2021.

Holland, Breena. "Justice and the Environment in Nussbaum's 'Capabilities Approach': Why Sustainable Ecological Capacity Is a Meta-Capability." *Political Research Quarterly* 61, no. 2 (June 2008): 319–32. https://doi.org/10.1177/1065912907306471.

Holland, Breena. "Procedural Justice in Local Climate Adaptation: Political Capabilities and Transformational Change." *Environmental Politics* 26, no. 3 (May 4, 2017): 391–412. https://doi.org/10.1080/09644016.2017.1287625.

Honneth, Axel. "Recognition and Justice: Outline of a Plural Theory of Justice." *Acta Sociologica* 47, no. 4 (December 2004): 351–64. https://doi.org/10.1177/0001699304048668.

Hourdequin, Marion. "Climate Change, Climate Engineering, and the 'Global Poor': What Does Justice Require?" Ethics, Policy & Environment 21, no. 3 (2018): 270–88. https://doi.org/10.1080/21550085.2018.1562525.

Kaswan, Alice. "Distributive Environmental Justice." In *Environmental Justice: Key Issues*, ed. Brendan Coolsaet, 21–36. London: Routledge, 2020.

Kortetmäki, Teea. "Reframing Climate Justice: A Three-Dimensional View on Just Climate Negotiations." *Ethics, Policy & Environment* 19, no. 3 (September 2016): 320–34. https://doi.org/10.1080/21550085.2016.1226238.

Mah, Alice. "Toxic Legacies and Environmental Justice." In *Environmental Justice; Key Issues*, ed. Brendan Coolsaet, 121–31. New York: Routledge, 2021.

Mills, Charles. *Black Rights/White Wrongs: The Critique of Racial Liberalism.* Oxford: Oxford University Press, 2017.

Newell, Peter, Shilpi Srivastava, Lars Otto Naess, Gerardo A. Torres Contreras, and Roz Price. "Toward Transformative Climate Justice: An Emerging Research Agenda." *WIREs Climate Change* 12, no. 6 (November 2021): e733. https://doi.org/10.1002/wcc.733.

Nixon, Rob. *Slow Violence and the Environmentalism of the Poor.* Cambridge: Harvard University Press, 2011.

Nussbaum, Martha C. "Beyond the Social Contract: Capabilities and Global Justice. An Olaf Palme Lecture, Delivered in Oxford on 19 June 2003." *Oxford Development Studies* 32, no. 1 (2004): 3–18. https://doi.org/10.1080/1360081042000184093.

Nussbaum, Martha. "Capabilities and Social Justice." *International Studies Review* 4, no. 2 (2002): 123–35. https://doi.org/10.1111/1521-9488.00258.

Nussbaum, Martha. *Frontiers of Justice: Disability, Nationality, Species Membership.* Cambridge: Harvard University Press, 2006.

Okereke, Chukwumerije, and Mark Charlesworth. "Environmental and Ecological Justice." In *Advances in International Environmental Politics*, ed. Michele M. Betsill, Kathryn Hochstetler, and Dimitris Stevis, 328–55. London: Palgrave Macmillan, 2014. https://doi.org/10.1057/9781137338976_13.

Ottinger, Gwen. "Careful Knowing as an Aspect of Environmental Justice." *Environmental Politics*, March 16, 2023, 1–20. https://doi.org/10.1080/09644016.2023.2185971.

Page, Edward A. "Intergenerational Justice of What: Welfare, Resources or Capabilities?" *Environmental Politics* 16, no. 3 (June 2007): 453–69. https://doi.org/10.1080/09644 010701251698.

Park, Lisa Sun-Hee, and David N. Pellow. *The Slums of Aspen: Immigrants vs. the Environment in America's Eden.* New York: New York University Press, 2011.

Pellow, David Naguib. *What Is Critical Environmental Justice?* Cambridge: Polity Press, 2017.

Rawls, John. *A Theory of Justice*, rev. ed. Cambridge: Harvard University Press, [1971] 1999.

Sardo, Michael Christopher. "Responsibility for Climate Justice: Political Not Moral." *European Journal of Political Theory* 22, no. 1 (January 2023): 26–50. https://doi.org/10.1177/1474885120955148.

Schlosberg, David. "Climate Justice and Capabilities: A Framework for Adaptation Policy." *Ethics & International Affairs* 26, no. 4 (2012): 445–61. https://doi.org/10.1017/S0892679412000615.

Schlosberg, David. *Defining Environmental Justice.* Oxford: Oxford University Press, 2007.

Schlosberg, David, and David Carruthers. "Indigenous Struggles, Environmental Justice, and Community Capabilities." *Global Environmental Politics* 10, no. 4 (November 2010): 12–35. https://doi.org/10.1162/GLEP_a_00029.

Schlosberg, David, and Lisette B. Collins. "From Environmental to Climate Justice: Climate Change and the Discourse of Environmental Justice." *WIREs Climate Change* 5, no. 3 (May 2014): 359–74. https://doi.org/10.1002/wcc.275.

Sen, Amartya. *Development as Freedom*. New York: Anchor Books, 1999.

Stevis, Dimitris. "Whose Ecological Justice?" *Strategies: Journal of Theory, Culture & Politics* 13, no. 1 (2000): 63–76. https://doi.org/10.1080/10402130050007520.

Stone, Christopher D. *Should Trees Have Standing? Law, Morality, and the Environment*. 3rd ed. Oxford: Oxford University Press, 2010.

Taylor, Charles. "The Politics of Recognition." In *Multiculturalism: Examining the Politics of Recognition*, ed. Amy Gutmann, 25–73. Princeton: Princeton University Press, 1994.

United Nations. *United Nations Declaration on the Rights of Indigenous Peoples*. Resolution adopted by the General Assembly on September 13, 2007.

Whyte, Kyle. "The Recognition Paradigm of Environmental Injustice." In *The Routledge Handbook of Environmental Justice*, ed. Ryan Holifield, Jayajit Chakraborty, and Gordon Walker, 113–23. London: Routledge, 2017.

Winter, Christine J. "A Seat at the Table: Te Awa Tupua, Te Urewera, Taranaki Maunga and Political Representation." *Borderlands Journal* 20, no. 1 (January 1, 2021): 116–39. https://www.jstor.org/stable/48767815.

Winter, Christine J. *Subjects of Intergenerational Justice: Indigenous Philosophy, the Environment and Relationships*. London: Routledge, 2022.

Wolin, Sheldon S. *Politics and Vision: Continuity and Innovation in Western Political Thought*, exp. ed. Princeton: Princeton University Press, 2004.

3

The Politics of Climate Justice

ZARA RIAZ AND PAGE FORTNA

A future of unmitigated climate catastrophe will have tremendous implications for justice. Broadly speaking, the most disadvantaged countries in the world, and the most disadvantaged communities within countries, will feel the effects "first and worst." As Mary Witlacil notes in chapter 2, it is a "triple injustice" that the adverse effects of climate change will be greatest for those who contributed the least to the crisis and those who are least likely to be included in climate policy decision-making.[1] But a rapid transition to decarbonize to prevent climate catastrophe will require major economic and social disruptions that will also have justice implications. How communities and polities navigate among the interests of those who stand to gain and those who stand to lose from this transition has implications for the likelihood that humanity will stave off the worst effects of anthropogenic climate change. In this chapter, we discuss how the contributions of political science can help us understand how to manage these competing interests.

Politics is classically defined as "who gets what, when, and how."[2] Political science is then the scientific study of how power to shape these outcomes is allocated and the conditions under which it can be redistributed to enhance or mitigate existing inequalities.[3] In this chapter, we pay particular attention to how the distribution of power across individuals and groups shapes state-based climate policy, at either national or subnational levels.

As scholars within the comparative politics and international relations subfields, we focus on positivist approaches to studying climate justice and politics but point readers to work in political theory for normative approaches to the topic, including crucial issues of intergenerational justice, adjudication among different principles for assessing responsibility ("polluter pays" vs. "beneficiary pays" vs. "ability to pay"), and individual vs. political action to solve the climate crisis, among others.[4]

Positivist studies of climate justice may relate to its participatory, representative, or distributive dimensions (corresponding to the "trivalent view" of justice discussed in the previous chapter).[5] However, the discipline's contributions to debates on participatory and representative dimensions of climate justice remain limited, despite the importance of participation and representation in government in the broader discipline.

We argue that, notwithstanding the limitations of distributive approaches discussed in the previous chapter, a distributive justice lens can help explain the nature of the climate impasse. Namely, it can help us understand why progress on climate change has been so slow, and what can be done about it. Given that addressing climate change requires shifting long-standing power imbalances and creates new groups of winners and losers, a distributive framework helps explain where resistance comes from, why it has succeeded, and the conditions under which it can be weakened. Distributive justice refers to the equitable distribution of the benefits and costs of climate action and inaction. Work on distributive politics is thus implicitly about climate justice, even if the term is not always explicitly invoked.[6]

We present interdisciplinary evidence to emphasize the injustices associated with current and projected climate inaction. These inequities stem from differences in the physical impacts of climate change, adaptation abilities, the costs associated with mitigation, and historical responsibility for contributions to the climate crisis. They are seen across levels of analysis: globally between the Global North and South, within states between richer and poorer communities, and at the individual level, with some better able than others to afford adaptation technology (such as air conditioning) or to relocate.

We next present two frameworks—distributive politics and collective action dilemmas—that political scientists frequently employ to understand obstacles to climate reform and the conditions under which they can be overcome. Then, we focus specifically on the distributive framework, reviewing literature that explains the nature of opposition to climate reforms—primarily in industrialized democracies, emphasizing how special interest groups and constituents who stand to lose from decarbonization reforms have slowed climate progress. We conclude with a brief discussion of the limitations of distributional approaches and a call for political scientists to broaden their engagement across disciplinary boundaries to overcome these limitations.

CLIMATE CHANGE AND THE DISTRIBUTIVE IMPLICATIONS OF INACTION

As noted elsewhere in this volume, failing to act on climate change will exacerbate the dire and highly unequal costs that poorer countries, which have contributed significantly less to global emissions than industrialized nations, are already facing. The disparity in costs that developed and developing countries will face due to rising temperatures, sea level, and more frequent and severe storms is stark. These differential costs have implications for countries' growth levels, health and education systems, demographic patterns, and security. At the subnational level, within both the developed and developing world, unmitigated climate change will affect poorer communities and individuals more than richer ones. In other words, the costs of climate change are not distributed equally.

Cross-National Implications (Global South Versus Global North)

Rising temperatures will inflict much greater damage on economic growth in warmer climates, while the domestic economies of colder countries, such as those in northern Europe, Russia, and Canada, may even benefit from global warming up to a certain point.[7] While there is debate and uncertainty about the national-level impacts, there is consensus that as the tropical zone of the Earth expands and temperate zones shift north, countries in the Global South will be hurt much more than those in the Global North.[8] The terms "Global North" and "Global South" do not correspond to the Northern and Southern Hemispheres but are used as shorthand to distinguish developing states from industrialized states. The "Brandt line" running between Mexico and the United States through the Mediterranean Sea, looping over the Middle East, India, and China, and squiggling down to exclude Japan, Australia, and New Zealand, is often taken as the dividing line.[9] Maps showing relative vulnerability to climate change correspond eerily well with the Brandt line.[10]

Beyond direct effects on agriculture, rising temperatures will also have indirect implications for growth and development through their impacts on health and education systems. For example, in South Sudan, the government recently closed schools due to extreme heat, affecting 2.2 million students. The same heat wave overheated hospital machinery. In Malawi, the government has shortened the school day in certain areas due to rising temperatures.[11] Over time, effects are likely to worsen education attainment rates and health outcomes, undermining the growth of countries already experiencing poverty and underdevelopment. Mortality rates from rising global temperatures are projected to increase in many poorer countries, which are disproportionately in warmer climates, while those in industrialized, and colder, countries in the Global North may even decrease. Estimates of temperature related death rates from anthropogenic climate change suggest a 17 percent increase in Accra, Ghana, for example, but a 15 percent decrease in Berlin, Germany, as the number of very cold days decreases.[12] These discrepancies will be exacerbated by less developed countries' inability to adequately respond to disasters when they occur.[13] However, as Jennifer Givens and Mufti Ahmed argue in chapter 6, some countries in the Global South, such as Bangladesh, may be better endowed than those in the North with the "social cohesion—people helping each other" necessary to cope with climate catastrophe.

The number of people exposed to climate events such as rising sea levels and flooding in poorer countries is also expected to increase greatly. Recent estimates indicate 630 million people currently live on land that will be under high tide lines by the end of this century if emissions are not curbed (compared to 190 million under a low carbon emissions scenario). In countries such as Bangladesh, Vietnam, and many small island developing states (SIDS), over 10 percent of the population is projected to live on land affected by chronic coastal flooding or permanent inundation by 2100.[14] In coastal cities across these continents, poor urban development and weak adaptation capacity likely make flooding particularly threatening. Rising sea levels and increasingly severe storms threaten the very existence of SIDS, a category that includes some of the poorest countries in the world, such as Haiti, Timor-Leste, Tuvalu, and São Tomé and Príncipe.

An extensive literature also studies how climate change may affect violent conflict through its impact on conflict stressors such as those mentioned earlier.[15] Climate-driven economic effects, such as loss of agricultural incomes due to adverse climatic conditions, may exacerbate actual or perceived economic inequality and increase the likelihood of conflict.[16] Empirically, there is increasing evidence that climatic conditions are associated with conflict; however, our understanding of the specific mechanisms underpinning the relationship remains limited.[17] If the unequal effects of climate change lead to unequal risks of violent conflict, the "losers" from climate inaction will not only be those who fare worse economically or in terms of health outcomes, but those killed, maimed, or displaced by warfare, exacerbating climate injustice. The relationship goes in the other direction as well. Societies torn apart by conflict and war will be less able to adapt to climate change. Violence and climate change may thus compound the ill effects of each other, with devastating implications for those caught in such spirals.[18]

While the distinction between Global North and Global South countries is commonly made in academic and policy discourse, it also glosses over considerable variation within these countries. For example, rapidly industrializing Global South countries, such as China, India, and Brazil are large greenhouse gas emitters, and also possess much different adaptation capacities than Small Island Developing States. Thus, it is important to note that while the North/South distinction provides a useful lens to examine inequities in climate impacts adaptation and mitigation capabilities, climate change impacts should not be viewed solely through this dichotomy.

Subnational Implications

Unequal distributions of the costs associated with climate inaction also exist at subnational levels globally. Rising sea levels and flooding in coastal cities across the Global South documented above are likely to inundate coastal slums, which are often built on low-lying marshy mangrove swamps that are unsuitable for formal development.[19] Repeated and severe flooding in urban slums with weak infrastructure can not only destroy such infrastructure but also increase the prevalence of diseases, further exacerbating poverty.

In the United States, marginalized communities are disproportionately more vulnerable to climate change impacts. For example, future flood risks in the United States are projected to affect Black communities disproportionately, while current flooding is mostly affecting poor communities with a predominantly white population.[20] Heat island effects in American cities, which make heat waves much more dangerous, are greater for people of color and those living below the poverty line than for nonwhite Hispanics and richer Americans, whose neighborhoods include more trees and parks.[21]

In both the Global North and South, Indigenous communities are also especially vulnerable to climate change, given that their livelihoods often are intricately linked to the natural environment. Climate change is thus likely to have a wide range of adverse consequences on both material and nonmaterial outcomes for these communities.[22] Similarly, those performing jobs outside, including construction and agricultural workers, are also particularly negatively affected by the rising temperatures and increased extreme weather events.

Inequalities exist across genders as well, particularly in South Asia and sub-Saharan Africa, where agriculture is the most important employment sector for women, and events such as drought or erratic rainfall can exacerbate existing economic and social precarity.[23] When disaster strikes, women are less likely to move.[24]

THE DISTRIBUTIVE POLITICS OF CLIMATE CHANGE

In distributive accounts of climate politics, scholars emphasize how conflicts over the material and/or ideological costs and benefits of proposed climate policies—or lack of policies—shape outcomes.[25] The nature of the political impasse of climate action quickly becomes clear: Countries, communities and individuals will face significant and deleterious costs if urgent action is not taken. And these costs will be distributed in a highly uneven manner. However, action to mitigate climate change will require imposing significant costs on certain countries and sectors, creating a new set of climate "losers." Climate politics thus remains volatile, as support for or opposition toward climate action is deeply intertwined with perceptions of injustice.[26]

Despite some cross-national variation in the implementation of policies, governments have largely failed to enact climate reforms needed to avert the disastrous effects of climate change.[27] We argue that a distributive politics framework helps explain both the climate impasse as well as how policies can better respond to the concerns of those harmed by climate action.

We first describe the distributive politics framework in greater detail, including how it differs from other prominent explanations of climate inaction in political science. We then review literature that explains how obstacles to climate policy proposals can be overcome by first addressing adversely affected voters' concerns and then examining the role of special interest groups. The literature focuses primarily on democracies. While the theories may apply in nondemocracies too, how much distributive concerns influence policymaking likely differs across regime types. Existing research is also situated disproportionately in the Global North, an epistemic bias that raises its own issues of climate justice.[28]

A Distributive Politics Framework

Political scientists typically employ one of two frameworks to understand action and inaction surrounding climate policy. In a collective action framework, actors are better off cooperating with each other to achieve a common group goal, such as cutting global emissions, which benefits everyone, but often fail to do so because cooperative behavior is costly for individual actors (whether states, subnational polities, or individual people). Because everyone in the group has incentives to free-ride, or enjoy the benefits of cooperation without taking costly action themselves, the collective good remains underprovided absent selective incentives, such as rewards for cooperating or sanctions for defecting.[29] While the collective action framework has implications for our understanding of climate justice, particularly as free-riding incentives can lead to inaction, we focus our review primarily on the second framework, distributive politics, which places greater emphasis on how competing interests and power imbalances across groups affect climate policy.

Two main sets of questions guide the study of distributive politics.[30] First are questions about how, when, and to which identifiable localities or groups politicians allocate government goods and services. A common assumption in the discipline is that allocation decisions are driven by electoral incentives (in democracies), and more generally by politicians' attempts at political survival. Second, distributive politics frameworks ask whether politicians' allocations of goods and services reflect the needs of constituents, i.e., whether policies distribute welfare benefits according to the preferences of the "median voter," or whether special interest groups capture disproportionate shares.[31] Together, these dimensions— accountability and responsiveness—lie at the heart of understanding "who gets what, when, and how" and thus reflect the "study of politics writ large."[32]

We review the growing distributive politics literature as it pertains to climate change. Our goal is not to debate the merits of collective action versus distributive frameworks or make claims about how they are intertwined, but instead to argue that the growing emphasis on the redistributive nature of climate policy has brought concerns of climate justice—either implicitly or explicitly—to the forefront of the political science literature examining climate policy roadblocks.

Preferences Over Climate Action

In both Global North and South countries, action to mitigate climate change will distribute costs and benefits unevenly at the domestic level, with already marginalized groups likely to face the brunt of these costs. Decarbonization creates significant costs for individuals, communities, and corporations engaged in fossil fuel industries, who will lose their jobs, lifeblood, and profits. In the United States, coal-dependent communities are often already socially and economically disadvantaged, and decarbonization policies—without proper redress—could significantly exacerbate these existing inequalities. The geographic clustering of fossil-fuel industries can pose acute political challenges; politicians are especially likely to attend to these concerns given that targeting voters concentrated in single districts (e.g., in "coal country") is an easier strategy for winning legislative seats than targeting voters whose discontent is widely diffused geographically.[33]

In democratic systems, the political feasibility of any climate policy, or policy more generally, depends on how the public perceives and ultimately reacts to it.[34] Assuming that politicians seek to stay in office and that voters can sanction them electorally, politicians will only support policy proposals unlikely to generate backlash among their constituents. A growing literature in political science attempts to understand the determinants of support for climate policies, and the compensation packages that accompany decarbonization to generate buy-in from injured communities. We identify three main insights emerging from this literature on the determinants of policy support.

First, political science literature highlights that policy vulnerability matters for shaping preferences over climate policies.[35] However, support for decarbonization policies can be increased with redress mechanisms that affected communities perceive as fair. For example, Daniel Yuichi Kono documents how generous unemployment benefits can increase support for the

American Clean Energy and Security Act among legislators from districts with high carbon-intensive employment.[36] In the United States and India, individuals exposed to job loss via green policies but not to climate change impacts directly are more likely to prefer direct material transfers and less adaptation investment relative to those who face direct climate vulnerability.[37] Understanding these preferences is important from a distributive justice perspective, as it gives insight into how those affected by transition policies perceive the fairness of various real and potential policies.

Political scientists have also begun to grapple with whether compensation for job losses associated with decarbonization should flow to individuals or communities. Qualitative research in sociology, anthropology, and history establishes that coal communities in both the United States and India have a strong sense of a shared identity. Because employment in coal is often generational and associated with distinct values, such as traditional lifestyles and deep community bonds in the United States and with working and postcolonial identities in India, decarbonization threatens not only the material but also the psychological interests of these groups in both countries. In localities where group identification is highly salient, individuals may be skeptical of policies that do not address the threats to shared nonmaterial interests likely to emerge from implementing green energy policies.[38]

The second main insight emerging from the political science literature on voters' preferences is that *who* implements climate policies and *how* they are implemented will matter greatly. In many developing countries, the state does not have the capacity to launch transition programs such as the European Green Deal or the American Rescue Plan that provide targeted funding to affected communities. "Loss and damage" funding, through which industrialized countries responsible for historical emissions pledge to assist developing countries, is meant both to compensate them for the effects of climate catastrophe and to help finance their transition to greener economies, helping to overcome capacity constraints. For example, at the UN Climate Change Conference in 2023 (COP 28), countries pledged almost $700 million to the Fund for Responding to Loss and Damage (FRLD), though it falls dramatically short of the estimated $580 billion in climate-related damages that vulnerable nations are projected to face by 2030.[39]

Alongside inadequate follow-through on financing pledges, citizens in countries in the Global South may also be wary of foreign agencies' involvement given the numerous inefficiencies associated with decades of foreign assistance. For example, a study of foreign aid delivering "just transition" policies to affected coal miners in South Africa finds that respondents reject both the government and national and international aid organizations as the parties to deliver the aid. Instead, they prefer independent actors such as the judiciary and local NGOs, which are more likely to be accountable to local communities.[40]

In developed countries, capacity constraints are less salient, yet policy implementation may still face roadblocks. Integrating perspectives of workers, communities, and companies engaged in fossil-fuel industries across the United States, Alex Gazmararian and Dustin Tingley argue that policy credibility is a major barrier to climate progress.[41] Affected actors' lack of support for climate policies—even when these policies might include compensation packages that voters perceive as just—stems from concerns that the government may not deliver on its promises of certain types of compensation or that policies will be reversible. Attention to the specific features of policies—such as their reversibility, level of community input, coalition base, and transparency—will be important for overcoming constraints facing climate policies.[42]

Third, an emerging literature documenting the real-world electoral consequences of green policies shows that they can lead to electoral backlash among those adversely affected—such as car users in Milan or those living near wind energy facilities in Ontario—but policies that include explicit measures to usher in a "just transition" can minimize backlash, as well as increase electoral support. For example, Diane Bolet et al. examine the electoral consequences of the "Just Transition Agreement" negotiated by the incumbent Spanish Socialist Party (PSOE) and find that the party's vote share in the 2019 election was significantly higher than in localities that were not subject to the agreement. Their qualitative evidence suggests that not only were the redistributive measures, such as early retirement schemes, social assistance, and additional public goods funding, important for generating public support, but the social dialogue that contributed to building trust between the industry participants and government was also crucial.[43]

Institutions, Special Interests, and Climate Policymaking

Policies to avert the deleterious consequences of climate change threaten companies that have amassed large profits from fossil-fuel production, and whose assets will be "stranded" by a green energy transition. As Leah Stokes states, "We shouldn't expect [these] groups facing an existential threat from climate policy—fossil fuel companies and electric utilities—to go quietly into the night."[44] Political scientists demonstrate how special interest groups successfully challenged policy reforms to reduce carbon pollution levels across various contexts, highlighting that entrenched power (im)balances between proponent and opponent interest groups play a significant role in explaining why opponents have been so successful.[45] Given that the status quo over policy is shaped by carbon-intensive industries that benefit from inaction and that have for centuries cultivated significant wealth and power, it will require significant time and effort for clean-energy coalitions to contest this power.

Corporations resist reforms using several political tools. Their incumbent position allows them to work directly with politicians and regulators, and to influence party positions through contributions and by manipulating public opinion to advance perspectives that deny scientific consensus on climate change. Further, the "double-representation" of antireform proponents on the left and the right in advanced industrialized economies— including labor and capital constituencies—has favored the status quo by ensuring that carbon polluters have a voice regardless of who is in power.[46]

While these accounts emphasize the inherently distributive and "deeply material" nature of the climate policy impasse, they also help explain the conditions under which reformers are more likely to succeed. For example, Matto Mildenberger argues that cross-national variation in policymaking institutions can help explain variation in reforms.[47] In countries such as Norway, with corporatist political systems (i.e., those in which interest group representation is institutionalized through the state), carbon polluters have direct access to policymaking settings. Consequently, policies are designed in close consultation with these actors, and reforms often minimize costs to carbon polluters. In countries with pluralist structures, such as the United States (in which interest groups are not institutionally linked to the state), policy proposals may reflect polluters' interests when these groups have allies in power, but they may have minimal influence when climate policy advocates are in power. This variability in the latter means that climate policy reform often becomes highly politicized, as corporations have incentives to mobilize conflict into the public domain to control the political agenda and reduce electoral incentives to pass climate reforms.

Other political institutions that can moderate the probability of successful climate reform include veto points, and electoral institutions that give Green parties greater representation.[48] Together, these interest-based and institutional accounts provide valuable insights into why acting to mitigate the distributive injustices associated with climate inaction has proven so challenging, and when those challenges can be overcome.

Implications of Climate Action for Marginalized Groups

Climate action poses adverse effects for many other entities, including small states in the Global South and other communities or individuals who rarely have the political power to contest or participate in the policymaking process, unlike special interest groups who have amassed large profits, or voters residing in electorally valuable constituencies. In this section, we discuss implications of climate action for such groups.

Countries in the Global South will face high costs of transitioning away from fossil fuels, as many depend heavily on fossil fuels to increase economic growth and development and provide affordable energy.[49] For the Global North, which relied on carbon-intensive industries to reach current development levels, to ask poorer countries to transition away from fossil fuels rapidly to reach zero emissions global targets is seen as deeply unfair. Further, developing countries have lower economic, technical, and institutional capacities to transition away from fossil fuels quickly and are less able than developed states to compensate fossil fuel workers who lose their jobs.[50] As Nigeria's environmental minister Ishag Salako stated, "Asking Nigeria, or indeed, asking Africa, to phase out fossil fuels is like asking us to stop breathing without life support."[51]

In advanced democracies, energy transitions are also likely to distribute costs unevenly within subpopulations, even among those not involved in the fossil fuel sector. For example, carbon taxes unjustly target the poor who spend a higher percentage of their incomes on energy, and new renewable energy projects could pose pollution, health, and cultural costs in Native American tribal lands.[52] As these examples highlight, policies to address the injustices of climate catastrophe can create new injustices, though these pale in comparison to the injustices associated with inaction.

Unfortunately, unequal costs of climate policies are also likely to manifest at lower levels. There can be stark differences in the gains and losses associated with climate policies within communities and within households. For example, in Saint Louis, Senegal, a city trying to cope with rising sea levels and coastal erosion, building a seawall requires displacing those living closest to the shore. Those who must relocate are mostly being redirected toward an inland displacement camp, which has created challenges for a population that depends primarily on fishing for its livelihood.[53] Typically, mitigation policies pose starker distributive implications, but as this case highlights, adaptation policies too can create or multiply the sets of "losers" who face distributive costs within communities already facing climate threats.

• • •

The politics of climate change are complex. Climate change is already hitting the poorest and most vulnerable countries and communities "first and worst." These injustices will be exacerbated if states do not act urgently to mitigate its worst effects. But while they will be far exceeded by the injustices wrought by climate catastrophe, there are nonetheless potential inequities in policies to combat climate change. The politics of who benefits and who loses out from climate change and from policies to mitigate it—that is, the distribution of costs and benefits—helps to explain the considerable variation across national and subnational governments in the ambition and success of climate reforms. As this review has highlighted, distributive politics provides one framework for understanding why opposition to climate reforms has been so strong, despite the urgency of the problem. This framework underlines the importance of considering not only aggregate costs and benefits as determinants of policy success, but also the allocation of these costs and benefits across individuals, groups, and industries. In doing so, it offers insights into both the conditions under which climate reforms are likely to meet strong opposition and where they are likely to succeed and multiply.

While a distributive politics framework can help illuminate these conditions, it also faces limitations. Specifically, implicit in a distributive politics framework is the assumption that individuals and groups can participate in the policymaking process to shape policy outcomes in their interests. However, many who are currently facing or who will face adverse consequences of climate inaction or action lack representation in the decision-making processes that affect them (see chapter 2). Though procedural justice and political representation have long represented critical research agendas across various political science subfields, much of this work remains unlinked to the discipline's literature on climate change. Interrogating existing theories to illuminate the procedural and representative dimensions of climate justice will be a fruitful avenue for future work in the political science of climate change.

Alongside further study of the procedural dimensions of climate justice, we also call on political scientists to deepen interdisciplinary collaborations. Given that political scientists are increasingly asking which policies are likely to generate buy-in from affected individuals and groups, collaborations with other disciplines are needed to better understand how perceptions of policies may interact with local diversity or inequalities. Anthropologists can help political scientists understand how the discourse and lived experience of weather and climate anomalies, whether or not referred to in the Western "scientific" language of "climate change," might affect individuals' and communities' attitudes toward climate policy and whether it is considered just (see chapter 13). Taking both Western and local and Indigenous knowledge of the impacts of climate change into account is crucial for understanding the political viability of just climate policy, and its efficacy (see chapter 10).[54] Political scientists, geographers, and demographers can help each other understand the justice and political implications of climate mobility, for both migration origins and destinations, and the obstacles those attempting to move will face in the form of restrictive migration policy, particularly in the Global North. Work on climate in political science often defaults to economic and material well-being as determinative of interests, overlooking religion and spirituality as value and identity systems that powerfully motivate politics, and conceptions of fairness and responsibility.[55] Political science also tends to focus on the interests of either individuals (humans) or the state, whereas Raffaella Taylor-Seymour and Courtney Bender argue in chapter 14 of this volume that religion has the potential to inspire new political possibilities based on "understandings of community, responsibility, and relationality" that are more adept at crossing the borders of nation-states and incorporating the "interests" (our term, not theirs) of nonhumans.

At the same time, political science can contribute to other disciplines as they grapple with the obstacles to reaching just climate outcomes. For example, sociology shines light on social and structural inequalities, often emphasizing (to a greater degree than political science tends to) the intersection of race, class, gender, sexuality, indigeneity, and other identities that underlie climate injustice, while political science can help fill the gap in understanding how vested political interests and structures create barriers to a future with low carbon emissions and well-being for all (see chapter 6). Such interdisciplinary engagement remains crucial for broadening the perspectives through which we understand the obstacles to climate action and their justice implications.

Notes

1. Dolšak and Prakash, "Three Faces of Climate Justice"; Barrett, "Local Level Climate Justice?"
2. Lasswell, "Politics: Who Gets What, When and How."
3. By "scientific study," we are referring to the use of the scientific method, or the use of formulating and evaluating falsifiable theories to learn about the world.
4. Normative approaches (including that in chapter 2) focus on prescriptive thought (what "ought" to be), while positivist approaches rely on description and analysis. Although we do not specifically review normative approaches to climate justice in

political science, examining positivist approaches can still inform normative work by highlighting potential constraints to achieving ideals of justice. For a review, see Lane, "Political Theory on Climate Change."

5. See chapter 2 for a detailed overview of the distinctions between these forms of justice and how the term "climate justice" emerged.

6. Dolšak and Prakash, "Three Faces of Climate Justice."

7. Bastien-Olvera et al., "Persistent Effect of Temperature on GDP"; Farajzadeh et al., "The Impact of Climate Change on Economic Growth."

8. Burke et al., "Climate and Conflict"; Bilal and Känzig, "The Macroeconomic Impact of Climate Change"; Nath et al., "How Much Will Global Warming Cool Global Growth?"

9. Lees, "The Brandt Line After Forty Years."

10. Burke et al., "Climate Change and Economic Production by Country," accessed January 13, 2025, https://web.stanford.edu/~mburke/climate/map.php.

11. Dahir, "Extreme Heat Wave Pushes."

12. Carleton et al., "Valuing the Global Mortality Consequences."

13. Millner and Dietz, "Adaptation to Climate Change and Economic Growth in Developing Countries."

14. Kulp and Strauss, "New Elevation Data Triple Estimates."

15. Homer-Dixon, "On the Threshold"; Burke et al., "Climate and Conflict"; Vesco et al., "Natural Resources and Conflict."

16. Koubi, "Climate Change and Conflict."

17. In addition to ambiguity over causal mechanisms, the literature is characterized by debates over whether short-term changes in weather patterns or climate variability adequately operationalize climate change, which is a long-term and large-scale shift in the world's weather patterns. See Koubi.

18. Raising many of the same points as those who focus on climate justice, scholars have critiqued "securitization" of the climate crisis for prioritizing the interests of the Global North, ignoring colonial legacies and development inequalities, and for state-centrism, prioritizing the security of the nation-state over human security concerns. Jayaram, "Shifting Discourses of Climate Security"; Daoudy, "Rethinking the Climate–Conflict Nexus."

19. Goyal and Silva, "Lessons from Coastal Slums."

20. Wing et al., "Inequitable Patterns of US Flood Risk."

21. Hsu et al., "Disproportionate Exposure to Urban Heat Island."

22. Levy and Patz, "Climate Change, Human Rights."

23. Raney et al., "The Role of Women in Agriculture."

24. Bhatta et al., "Climate-Induced Migration in South Asia."

25. Aklin and Mildenberger, "Prisoners of the Wrong Dilemma."

26. Dolšak and Prakash, "Three Faces of Climate Justice," 295.

27. Bergquist et al., "Combining Climate, Economic, and Social Policy." For variation in policy implementation, see Finnegan, "Institutions, Climate Change"; Compston

and Bailey, "Political Strategy and Climate Policy"; Harrison, "The Road Not Taken"; Harrison and Sundstrom, eds., *Global Commons, Domestic Decisions.*

28. Mongkolnchaiarunya, "Epistemic Exclusion in Climate Science."
29. Olson Jr., *The Logic of Collective Action*; Ostrom, *Governing the Commons.*
30. Golden and Min, "Distributive Politics Around the World."
31. The median voter can be considered the voter with the "middle of the spectrum" viewpoint on a particular issue. In certain conditions, targeting this voter increases a politician's probability of winning.
32. Lasswell, "Politics"; Kramon and Posner, "Who Benefits from Distributive Politics?," 461.
33. Rickard, "Electoral Systems, Voters' Interests"; Stokes, "Electoral Backlash Against Climate Policy."
34. Drews and van den Bergh, "What Explains Public Support for Climate Policies? A Review of Empirical and Experimental Studies."
35. Bechtel et al., "Interests, Norms and Support"; Bayer and Genovese, "Beliefs About Consequences from Climate Action"; Kono, "Compensating for the Climate"; Bergquist et al., "Combining Climate, Economic, and Social Policy"; Meckling et al., "Winning Coalitions for Climate Policy."
36. Kono, "Compensating for the Climate."
37. Gaikwad, Genovese, and Tingley, "Creating Climate Coalitions."
38. Gaikwad, Genovese, and Tingley.
39. Bhandari et al., "What Is 'Loss and Damage' from Climate Change?"
40. Mohlakoana et al., "Varieties of Just Transition."
41. Gazmararian and Tingley, *Uncertain Futures.*
42. Colantone et al., "The Political Consequences of Green Policies"; Stokes, "Electoral Backlash Against Climate Policy."
43. Bolet et al., "How to Get Coal Country to Vote."
44. Stokes, *Short Circuiting Policy*, 4.
45. Stokes; Matto Mildenberger, *Carbon Captured*; Hughes and Urpelainen, "Interests, Institutions, and Climate Policy."
46. Stokes, *Short Circuiting Policy*
47. Mildenberger, *Carbon Captured.*
48. Karapin, *Political Opportunities for Climate Policy*; Hughes and Urpelainen, "Interests, Institutions, and Climate Policy."
49. Jakob and Steckel, "The Just Energy Transition."
50. Weko and Goldthau, "Bridging the Low-Carbon Technology Gap?"
51. Plumer and Bearak, "Nations at Climate Summit Agree."
52. Dolšak and Prakash, "Three Faces of Climate Justice," 286.
53. *Christian Science Monitor,* "'The Ocean Is What We Know.' Can Senegal Woo Climate Refugees Inland?," December 20, 2022, https://www.csmonitor.com/World/Africa/2022/1201/The-ocean-is-what-we-know.-Can-Senegal-woo-climate-refugees-inland.

54. See also Mongkolnchaiarunya, "Epistemic Exclusion in Climate Science."
55. Notwithstanding a subfield, and an organized section of the American Political Science Association, on "Religion and Politics." For a recent interdisciplinary look at religion and climate politics, see Berry, ed., *Climate Politics and the Power of Religion.*

BIBLIOGRAPHY

Aklin, Michaël, and Matto Mildenberger. "Prisoners of the Wrong Dilemma: Why Distributive Conflict, Not Collective Action, Characterizes the Politics of Climate Change." *Global Environmental Politics* 20, no. 4 (2020): 4–27.

Barrett, Sam. "Local Level Climate Justice? Adaptation Finance and Vulnerability Reduction." *Global Environmental Change* 23, no. 6 (2013): 1819–29.

Bastien-Olvera, Bernado A., Francesco Granella, and Frances C. Moore. "Persistent Effect of Temperature on GDP Identified from Lower Frequency Temperature Variability." *Environmental Research Letters* 17, no. 8 (2022): 084038.

Bayer, Patrick, and Federica Genovese. "Beliefs About Consequences from Climate Action Under Weak Climate Institutions: Sectors, Home Bias, and International Embeddedness." *Global Environmental Politics* 20, no. 4 (2020): 28–50.

Bechtel, Michael M., Federica Genovese, and Kenneth F. Scheve. "Interests, Norms and Support for the Provision of Global Public Goods: The Case of Climate Co-Operation." *British Journal of Political Science* 49, no. 4 (2019): 1333–55.

Bergquist, Parrish, Matto Mildenberger, and Leah C. Stokes. "Combining Climate, Economic, and Social Policy Builds Public Support for Climate Action in the US." *Environmental Research Letters* 15, no. 5 (2020): 054019.

Berry, Evan, ed. *Climate Politics and the Power of Religion.* Bloomington: Indiana University Press, 2022.

Bhandari, Preety, Nate Warszawski, Deirdre Cogan, and Rhys Gerholdt. "What Is 'Loss and Damage' from Climate Change? 8 Key Questions, Answered." World Resources Institute, November 4, 2024. https://www.wri.org/insights/loss-damage-climate-change.

Bhatta, Gopal Datt, Pramod Kumar Aggarwal, Santosh Poudel, and Debbie Anne Belgrave. "Climate-Induced Migration in South Asia: Migration Decisions and the Gender Dimensions of Adverse Climatic Events." *Journal of Rural and Community Development* 10, no. 4 (2015).

Bilal, Adrien, and Diego R. Känzig. "The Macroeconomic Impact of Climate Change: Global vs. Local Temperature." National Bureau of Economic Research, 2024. https://www.nber.org/papers/w32450.

Bolet, Diane, Fergus Green, and Mikel González-Eguino. "How to Get Coal Country to Vote for Climate Policy: The Effect of a 'Just Transition Agreement' on Spanish Election Results." *American Political Science Review* (2023): 1–16.

Burke, Marshall, Solomon M. Hsiang, and Edward Miguel. "Climate and Conflict." *Annual Review of Economics* 7, no. 1 (August 1, 2015): 577–617. https://doi.org/10.1146/annurev-economics-080614-115430.

Carleton, Tamma, et al. "Valuing the Global Mortality Consequences of Climate Change Accounting for Adaptation Costs and Benefits." *Quarterly Journal of Economics* 137, no. 4 (November 2022): 2037–2105. https://doi.org/10.1093/qje/qjac020.

Colantone, Italo, Livio Di Lonardo, Yotam Margalit, and Marco Percoco. "The Political Consequences of Green Policies: Evidence from Italy." *American Political Science Review* 118, no. 1 (2024): 108–26.

Compston, Hugh, and Ian Bailey. "Political Strategy and Climate Policy." In *Turning Down the Heat*, ed. Hugh Compston and Ian Bailey, 263–88. London: Palgrave Macmillan UK, 2008. https://doi.org/10.1057/9780230594678_15.

Dahir, Abdi Latif. "Extreme Heat Wave Pushes South Sudan to Close Schools." *New York Times*, March 20, 2024. https://www.nytimes.com/2024/03/20/world/africa/extreme-heat-south-sudan-schools-climate.html.

Daoudy, Marwa. "Rethinking the Climate–Conflict Nexus: A Human–Environmental–Climate Security Approach." *Global Environmental Politics* 21, no. 3 (2021): 4–25.

Dolšak, Nives, and Aseem Prakash. "Three Faces of Climate Justice." *Annual Review of Political Science* 25, no. 1 (May 12, 2022): 283–301. https://doi.org/10.1146/annurev-polisci-051120-125514.

Drews, Stefan, and Jeroen C.J.M. van den Bergh. 2015. "What Explains Public Support for Climate Policies? A Review of Empirical and Experimental Studies." *Climate Policy* 16 (7): 855–76. doi:10.1080/14693062.2015.1058240.

Enarson, Elaine, Alice Fothergill, and Lori Peek. "Gender and Disaster: Foundations and Directions." In *Handbook of Disaster Research*, by Havidán Rodríguez, Enrico L. Quarantelli, and Russell R. Dynes, 130–46. New York: Springer New York, 2007. https://doi.org/10.1007/978-0-387-32353-4_8.

Farajzadeh, Zakariya, Effat Ghorbanian, and Mohammad Hassan Tarazkar. "The Impact of Climate Change on Economic Growth: Evidence from a Panel of Asian Countries." *Environmental Development* 47 (2023): 100898.

Finnegan, Jared J. "Institutions, Climate Change, and the Foundations of Long-Term Policymaking." *Comparative Political Studies* 55, no. 7 (June 2022): 1198–1235. https://doi.org/10.1177/00104140211047416.

Gaikwad, Nikhar, Federica Genovese, and Dustin Tingley. "Creating Climate Coalitions: Mass Preferences for Compensating Vulnerability in the World's Two Largest Democracies." *American Political Science Review* 116, no. 4 (2022): 1165–83.

Gazmararian, Alexander F., and Dustin Tingley. *Uncertain Futures: How to Unlock the Climate Impasse*. Cambriege: Cambridge University Press, 2023.

Golden, Miriam, and Brian Min. "Distributive Politics Around the World." *Annual Review of Political Science* 16, no. 1 (May 11, 2013): 73–99. https://doi.org/10.1146/annurev-polisci-052209-121553.

Goyal, Anubhav, and Maria Matos Silva. "Lessons from Coastal Slums of Global South Toward Flood Resilience." In *Greening Our Cities: Sustainable Urbanism for a Greener Future*, ed. Alessandra Battisti, Cristina Piselli, Eric J Strauss, Etleva Dobjani, and Saimir Kristo, 347–59. Advances in Science, Technology & Innovation. Cham: Springer Nature Switzerland, 2024. https://doi.org/10.1007/978-3-031-49495-6_25.

Harrison, Kathryn. "The Road Not Taken: Climate Change Policy in Canada and the United States." *Global Environmental Politics* 7, no. 4 (2007): 92–117.

Harrison, Kathryn, and Lisa McIntosh Sundstrom. *Global Commons, Domestic Decisions: The Comparative Politics of Climate Change*. Cambridge: MIT Press, 2010.

Homer-Dixon, Thomas F. "On the Threshold: Environmental Changes as Causes of Acute Conflict." *International Security* 16, no. 2 (1991): 76–116.

Hsu, Angel, Glenn Sheriff, Tirthankar Chakraborty, and Diego Manya. "Disproportionate Exposure to Urban Heat Island Intensity Across Major US Cities." *Nature Communications* 12, no. 1 (2021): 2721.

Hughes, Llewelyn, and Johannes Urpelainen. "Interests, Institutions, and Climate Policy: Explaining the Choice of Policy Instruments for the Energy Sector." *Environmental Science & Policy* 54 (2015): 52–63.

Jakob, Michael, and Jan Christoph Steckel. "The Just Energy Transition." Background Paper for the WWF, 2016, 622539162–569.

Jayaram, Dhanasree. "Shifting Discourses of Climate Security in India: Domestic and International Dimensions." *Third World Quarterly*, February 19, 2024, 1–19. https://doi.org/10.1080/01436597.2024.2314003.

Karapin, Roger. *Political Opportunities for Climate Policy: California, New York, and the Federal Government*. Cambridge: Cambridge University Press, 2016.

Kono, Daniel Yuichi. "Compensating for the Climate: Unemployment Insurance and Climate Change Votes." *Political Studies* 68, no. 1 (February 2020): 167–86. https://doi.org/10.1177/0032321719836066.

Koubi, Vally. "Climate Change and Conflict." *Annual Review of Political Science* 22, no. 1 (May 11, 2019): 343–60. https://doi.org/10.1146/annurev-polisci-050317-070830.

Kramon, Eric, and Daniel N. Posner. "Who Benefits from Distributive Politics? How the Outcome One Studies Affects the Answer One Gets." *Perspectives on Politics* 11, no. 2 (2013): 461–74.

Kulp, Scott A., and Benjamin H. Strauss. "New Elevation Data Triple Estimates of Global Vulnerability to Sea-Level Rise and Coastal Flooding." *Nature Communications* 10, no. 1 (2019): 1–12.

Lane, Melissa. "Political Theory on Climate Change." *Annual Review of Political Science* 19, no. 1 (May 11, 2016): 107–23. https://doi.org/10.1146/annurev-polisci-042114-015427.

Lasswell, Harold. "Politics; Who Gets What, When and How, New York." New York: Whittesey House, 1936.

Lees, Nicholas. "The Brandt Line Ar Forty Years: The More North–South Relations Change, the More They Stay the Same?" *Review of International Studies* 47, no. 1 (2021): 85–106.

Levy, Barry S., and Jonathan A. Patz. "Climate Change, Human Rights, and Social Justice." *Annals of Global Health* 81, no. 3 (2015): 310–22.

Meckling, Jonas, Nina Kelsey, Eric Biber, and John Zysman. "Winning Coalitions for Climate Policy." *Science* 349, no. 6253 (September 11, 2015): 1170–71. https://doi.org/10.1126/science.aab1336.

Mildenberger, Matto. *Carbon Captured: How Business and Labor Control Climate Politics*. MIT Press, 2020.

Millner, Antony, and Simon Dietz. "Adaptation to Climate Change and Economic Growth in Developing Countries." *Environment and Development Economics* 20, no. 3 (2015): 380–406. https://www.jstor.org/stable/26391934.

Mohlakoana, Nthabiseng, Muhammed Lokhat, Nives Dolšak, and Aseem Prakash. "Varieties of Just Transition: Public Support in South Africa's Mpumalanga Coal Community for Different Policy Options." *PLOS Climate* 2, no. 5 (2023): e0000205.

Mongkolnchaiarunya, Jittip. "Epistemic Exclusion in Climate Science: Why We Grow the Wrong Trees in the Wrong Places." Harvard University, 2024.

Nath, Ishan B., Valerie A. Ramey, and Peter J. Klenow. "How Much Will Global Warming Cool Global Growth?" *NBER Working Paper Series*, 2023. https://econweb.ucsd.edu/~vramey/research/NRK_GlobalWarming_GlobalGrowth.pdf.

Olson Jr., Mancur. *The Logic of Collective Action: Public Goods and the Theory of Groups.* Vol. 124. Cambridge: Harvard University Press, 1971.

Ostrom, Elinor. *Governing the Commons: The Evolution of Institutions for Collective Action.* Cambridge: Cambridge University Press, 1990.

Plumer, Brad, and Max Bearak. "In a First, Nations at Climate Summit Agree to Move Away from Fossil Fuels." *New York Times*, December 13, 2023.

Raftery, Adrian E., Alec Zimmer, Dargan MW Frierson, Richard Startz, and Peiran Liu. "Less than 2 C Warming by 2100 Unlikely." *Nature Climate Change* 7, no. 9 (2017): 637–41.

Raney, Terri, Gustavo Anríquez, André Croppenstedt et al. "The Role of Women in Agriculture." ESA Working Paper no. 11–02 (March 2011). https://ageconsearch.umn.edu/record/289018/?v=pdf.

Rickard, Stephanie J. "Electoral Systems, Voters' Interests and Geographic Dispersion." *British Journal of Political Science* 42, no. 4 (2012): 855–77.

Stokes, Leah C. "Electoral Backlash Against Climate Policy: A Natural Experiment on Retrospective Voting and Local Resistance to Public Policy." *American Journal of Political Science* 60, no. 4 (October 2016): 958–74. https://doi.org/10.1111/ajps.12220.

Stokes, Leah C. *Short Circuiting Policy: Interest Groups and the Battle Over Clean Energy and Climate Policy in the American States.* New York: Oxford University Press, 2020.

Vesco, Paola, Shouro Dasgupta, Enrica De Cian, and Carlo Carraro. "Natural Resources and Conflict: A Meta-Analysis of the Empirical Literature." *Ecological Economics* 172 (2020): 106633.

Weko, Silvia, and Andreas Goldthau. "Bridging the Low-Carbon Technology Gap? Assessing Energy Initiatives for the Global South." *Energy Policy* 169 (2022): 113192.

Wing, Oliver EJ, William Lehman, Paul D. Bates et al. "Inequitable Patterns of US Flood Risk in the Anthropocene." *Nature Climate Change* 12, no. 2 (2022): 156–62.

4

Transport, Justice, and Climate Crisis in Cities

A Paradigm Shift

JACQUELINE M. KLOPP AND FESTIVAL G. BOATENG

C ities are among the most complex of human creations. Increasingly dotting our planet, cities carve out expanding patterns of human habitat and consumption, transforming landscapes, ecologies, and life. Involving dense, intricate human interactions, activities, cultures, infrastructures, and ecologies, as well as deep inequalities and contestations over resources, power, place, and space, the study of cities has helped create, shape, and draw in numerous disciplines. These include architecture, urban planning, preservation, and design but also ecology, public health, history, political science, sociology, and economics—many, if not all, of the disciplines covered in this volume. Given the importance of understanding cities and their impacts holistically, urban studies has emerged as an interdisciplinary field dedicated to the study of cities in all their layered richness.

Unsurprisingly, urban studies is heavily engaged in conversations around climate and climate justice. Cities, especially in the Global North, are associated with voracious appetites for energy and resources. Urbanization is associated with transformation and extraction from local landscapes and Indigenous peoples accompanied by often violent histories of displacement, ecological destruction, and colonization.[1] As an intrinsic part of global capitalist dynamics, some cities—dubbed global cities—draw in vast flows of capital, while others send resources outward, languish, and sometimes shrink. Mobility of people is also part of this dynamic. Cities increasingly are home to migrants—from near and far—in search of security and opportunity, often as their own homelands face extraction and environmental and, increasingly, climate-related devastation.[2] These migrants send financial flows back, creating networks of support and resiliency.

As centers of dense activities, cities are associated with contributing to the climate crisis: Covering a tiny fraction of the Earth's land surface, urban areas are responsible for approximately 70 percent of global CO_2 emissions and 60–80 percent of global energy consumption.[3] They are also associated with devastating and inequitable climate impacts affecting large numbers of people, from flash/surface flooding, heat waves, severe rainstorms, extreme heat days, and droughts to air pollution exposure from wildfires. As Sheila Foster notes in the prelude to this book, these climate impacts often feed into and exacerbate already existing, longstanding, and deep inequalities linked to histories of racism, colonialism, and racial capitalism with the related exclusionary and discriminatory land-use and investment practices. Thus, emerging questions of climate, justice, and the city are able to draw on scholarship around the just city for help in answering them.[4]

With this in mind, this chapter contributes to grappling with climate justice within urban studies by looking at urban transport and mobility. Several reasons justify using the transport sector as a case study to explore climate and justice in cities. First, the sector is one of the biggest contributors of urban carbon emissions, having planetary-scale impacts on livelihoods, health, climate and ecological crises, and justice. Second, historically, how cities develop, and hence generate emissions, has been deeply entwined with available means of movement and the infrastructures built around them, infrastructures that are linked to deep path dependencies and carbon "lock in" that need to be undone to address the climate crisis.[5] Alex Marshall captures this with much clarity in *How Cities Work*, noting that "every city built has grown from the spine of its transportation system, like flesh around bones."[6] The climate crisis, and the layered injustices around it, thus cannot be fully understood without a critical examination of the transport developments in cities as well as the struggles around their often problematic and inequitable impacts.

TRANSPORT AND THE URBAN CLIMATE CRISIS

One key preoccupation around transport or mobility in urban studies is the reality that most contemporary cities have been constructed largely around one major technology: the private car.[7] This has helped lead to cities marked by car-centered road infrastructure and land-use systems and highly stratified, low density, sprawling spatial forms of suburbanization across the globe.[8] The complex factors that drive these dynamics are sometimes called the automobility system, conceptualized as a self-organizing, nonlinear system worldwide involving cars, drivers, roads, petroleum supplies, and many novel objects, technologies, and signs.[9] Automobility systems help engender separation, more and longer travel, and high levels of dependence on individual and motorized modes. Ultimately, this produces many forms of inefficient energy use and damage to both the environment and human health.

In an age of escalating concerns about damage to our climate system, continuously climbing emissions from transportation are a key focus and elicit varied interventions.[10] Currently, an estimated 23 percent of total energy-related CO_2 emissions worldwide come from transport, with the bulk of these GHG emissions coming from heavy- and light-duty vehicles located in cities of the Global North. These vehicles rely predominantly on internal combustion engines that mostly burn fossil fuels like diesel and gasoline.[11]

Trips in these vehicles contribute heavily to alarming amounts of air pollution from incomplete combustion. This pollution comes from both tail pipe exhaust and tires and brakes. Tail pipe exhaust consists of tiny highly toxic particulate matter, producing nitrous and carbon oxides that in turn can create dangerous ground-level ozone.[12] Less known, tires and brakes also produce tiny toxic particles that contribute to both air and microplastics pollution.[13] Within the toxic mix of air pollution from vehicles are carbon pollutants, including tiny bits of sooty material called black carbon that absorb heat and affect our climate. These transport pollutants are concentrated in cities, intense centers of consumption, vehicle movement, and emissions.[14]

Despite current policies and actions, these transport emissions stubbornly continue to climb, a fact that is increasingly acknowledged as a severe public health as well as a climate problem.[15] Less recognized within policy circles is that, embedded within this global transportation emissions trajectory, are numerous, often hidden layered stories of social, environmental, and climate injustices. This chapter provides an overview of some of these injustices and argues that they are not peripheral to the world's transport emissions problem but instead, lay at the very core of the compounding problems we now face in our cities. Hence, if we wish to adequately address these problems, we must do so from a frame that centers justice.

Whether in the United States or globally, and as detailed in chapter 3, injustices emerge from the fact that those most affected by current autocentric urban mobility systems, their emissions, and carbon impacts are the least responsible and also the least consulted, recognized, and resourced. The most vulnerable, ignored, and affected—most often low-income, historically underserved communities and populations, including older people, children, and people with disabilities within those communities—have the fewest resources to address their problems, including resources to adapt to climate change. This chapter unbundles these "three forms of injustice."[16] It renders them visible to support a more holistic justice-based understanding of why we have growing transport emissions with all the serious problems they create for people, equity, and our public and planetary health.

This justice frame, while growing in transport studies through concepts such as access, equity, and transport and mobility justice, is still not the mainstream in the transport planning disciplines most strongly associated with urban studies, planning, and engineering departments.[17] One reason for the neglect of climate and justice issues more generally within these fields stems from the preoccupation with technologies and efficiencies. The recent focus on "smart cities" and optimization tends to exacerbate this problem of looking at issues with "technology goggles" instead of a human-centered approach that focuses on justice, broadly understood, to involve addressing inequalities in capabilities, opportunities, and outcomes, as well as in processes of recognition and deliberation.[18] A "technology goggles" approach tends to focus optimistically on often exciting new technologies (for example, AI, autonomous vehicles, electrification) as silver bullet solutions to complex societal problems. Instead, they might be critically explored as potential aids—if judiciously used—within a well thought out, holistic process that centers people and problems. Especially given the history of the automobile, it should be clear that new technologies need to be seen as also potentially problematic in their own right, raising new challenges and potential inequities. This persistence of "technology goggles" approaches is part of a techno-politics associated with the transport sector that tends to narrow complex urban mobility and access challenges to technical problems of how to get more rapid vehicle flows from point A to point B.

Fortunately, rich anthropological, historical, and political science approaches within urban studies focused on mobility and access have led to the questioning of some of the basic assumptions of transport planning, and this is leading to more interdisciplinary and sophisticated approaches to untangling our transport emissions, justice, and related problems in cities. The climate crisis has also intensified the growing focus on understanding the interrelations between mobility and land-use systems, urban planning, and environmental and climate justice issues, both within countries and between what we will call the "Global North" (colonizing) and "Global South" (colonized).

TRANSPORT, ENVIRONMENTAL INJUSTICES, LOCAL COMMUNITIES, AND CLIMATE JUSTICE IN THE UNITED STATES

With the advent of mass manufacturing and marketing of cars at the start of the twentieth century, more space in cities was given over to roads and highways, both in the Global North and the colonized Global South. These roads and highways were also designed largely to facilitate rapid private vehicle movement. Historically, the emphasis on car-oriented infrastructure encouraged more and more driving in what has since been called "induced demand," and more car dependency to satisfy basic needs. America, home of mass-produced, inexpensive cars, is often seen as the epitome of this car-centric and dependent urban development, but this general trend has been globalized.[19]

In the United States, this rapid increase in private car ownership has been accompanied by a decline in investment in walkability and public transit, which are intrinsically more efficient and lower emissions per capita. More and more carbon-intensive resources have been going into roads and highways. This has left the very poorest—who are often from the same overburdened communities and do not have access to private vehicles—with inadequate public transit services and options. This is an often unrecognized justice issue, and in 2016 the National Association for the Advancement of Colored People (NAACP) made a resolution on "Establishing That Access to Public Transportation System Is a Basic Civil Right." In it they note that "public resources continue to subsidize more and wider roads, free parking and the location and relocation of jobs and services farther away from existing, underfunded transit services and communities, many of them impoverished and/or heavily populated with minorities or disabled persons."[20]

The building of the highway system across the country and through cities in the 1950s and 1960s also entailed massive destruction of neighborhoods and connective urban fabric with long-lasting impacts. One example of many includes the New Orleans Claiborne Expressway, which was cut through the vibrant community of Tremé, widely considered the oldest African American neighborhood in America, decimating a thriving community. At the time, the American sociologist and historian Lewis Mumford penned trenchant, still apt critiques of the singular and destructive focus on building for cars in American cities; in 1966 he famously wrote in the *New York Review of Books*:

> What's transportation for? This is a question that highway engineers apparently never ask themselves: probably because they take for granted the belief that transportation exists for the purpose of providing suitable

outlets for the motorcar industry. To increase the number of cars, to enable motorists to go longer distances, to more places at higher speeds has become an end in itself. Does this over-employment of the motorcar not consume ever larger quantities of gas, oil, concrete, rubber, and steel.[21]

Woven into these developments were profound injustices that historical scholarship has also documented especially well: Highways were cut through the places where people were the least empowered to resist, low-income communities of color, often as a deliberate and racist strategy of marginalization.[22] This created mass displacement and trauma, aptly called "root shock" by public health expert Mindy Fullilove.[23]

In addition to this shock, trauma, and disruption of low-income communities of color, car-centered transport planning led to other deep and enduring environmental injustices. To this day, Black Americans are exposed to higher levels of air pollution from vehicles traversing their inner-city neighborhoods compared to white counterparts, and this leads to higher rates of respiratory and cardiovascular problems, among myriad other health issues.[24] Part of the car-oriented planning model also involved the federal government subsidizing, and hence encouraging, conversion of rural lands to low-density suburban development.[25] This, in turn, led to both higher infrastructure and service costs, with funding often going to sustain wealthier communities, and more emissions as commuting by car to jobs became part of the common daily rhythms in American cities.

This development was also intimately connected to deepening racial segregation and related environmental injustices that have long-lasting legacies in US cities. As white Americans retreated to racially segregated, greener suburbs with less air pollution exposure, low-income communities ended up living in places riven apart by highways and hence inundated with vehicles and emissions (see the prelude to this book). Some of this polluting traffic involves commuters from suburbs and also freight and other heavy-duty trucks, as these same neighborhoods tend to host critical city infrastructure like food distribution points, power plants, industry and increasingly warehouses for the goods ordered online by predominantly wealthier households.[26] This causes disproportionate public health problems in these underserved and overburdened communities, from the effects of intense air and noise pollution to death from vehicle crashes. Another unrecognized justice issue linked to these dynamics is that along with the elderly, low-income people of color continue to be disproportionately at risk of being killed by drivers in vehicles in the United States.[27]

The key point here is that climate change is affecting cities that are already riven by living layers of historical injustices. Without interventions, climate-change impacts will deepen pre-existing injustices as well as create new ones. Recent research suggests that people in low income, underserved neighborhoods of color who suffer from environmental injustices associated with past transport planning decisions are the same people who are more vulnerable to climate impacts from extreme heat to flooding.[28] The concrete and asphalt of road systems and the lack of green space in some inner-city neighborhoods, along with general underinvestment associated with suburbanization and "white flight," tend to lead to higher temperatures during summer months.[29] The asphalt and concrete of highways absorb and reradiate heat, intensifying the effects of heat in what is called the urban heat island effect. This effect can lead to inner cities facing temperatures in summer many degrees higher than leafy suburbs nearby. Coexposure to air pollution and extreme heat associated with climate change also appears to have larger effects beyond the sum of the individual effects, and some evidence suggests that air pollution can lead to lower visibility and higher crashes as well.[30] During flood events, impervious surfaces like concrete and lack of water absorbing green space often also mean water accumulates and can make flooding worse. While this remains an active area of research, climate-related injustices appear to be emerging in a way that connects to the well-documented, layered historical dynamics of transport-related injustices in the United States. It is thus an urgent problem to explore further within urban studies, and to do so means justice must stay centered in our frame.

TRANSPORT AND ENVIRONMENTAL INJUSTICES GLOBALLY AND THE LINK TO CLIMATE JUSTICE

Mass consumption of cars and the rise of emissions intensive car-centric cities with related justice issues are largely associated with the United States, but systems of "automobility" are global.[31] Part of the initial vector for globalization of these automobility systems was colonialism. Along with the well-documented shaping of transport infrastructure for colonial and global interests across the Global South, cars were imported for colonial officials, settlers, and foreign corporations.[32]

Writing about Nairobi, Kenya, Anyamba notes that Europeans designed their city around personalized transport: first horses, bikes, and rickshaws, then the motorcar ruled.[33] By 1928 the young colonial town that emerged rapidly out of a railway stop had five thousand cars, making it the town with possibly the highest per capita levels of private automobile ownership in the world at the time. In Nairobi, as occurred in other colonies in Africa, the majority of people resorted to nonmotorized transport—either foot or bikes—while the colonial elite moved around the city in cars and lived in distant socially and racially segregated suburbs, leading, much like in America, to a focus on expensive road building to connect these far-flung, racially and later socially segregated places with the city core.[34] A persisting gulf, with some variations, was set that would emerge across the formerly colonized Global South, and, as in the United States and Europe, these automobility systems led to chronic and growing problems of congestion, crashes, and air pollution.

The personal car–oriented, highly consuming, and extractive Global North, with its car manufacturing industries, bears the greatest responsibility for pushing automobility systems and their related emissions across the globe. While the culpability of the Global North in historical transport emissions from vehicles located in or used there to feed consumption-based economies is known, the history of Global North–dominated financial institutions like the World Bank in encouraging and funding deeply inequitable, dangerous car-oriented, carbon-intensive infrastructures in the Global South, often at the request of planners and politicians in those countries steeped in ideologies of "automobility," is less recognized.[35] This has unsurprisingly exacerbated in the Global South the same public health and justice problems experienced in places like the United States.

One feature of the injustice dynamics associated with this global funding and investment pattern lay in the fact that the vast majority of people in low- and middle-income countries that receive transport infrastructure loans, especially the poorer classes and vulnerable groups in cities across

Asia, Africa, and Latin America and the Caribbean, rely on walking and various forms of shared mobility as public transport, and these shared modes remain underinvested in or ignored. Most people rely on privately provided, shared mobility services sometimes called "informal," paratransit, or popular to access opportunities and services.[36] Most of these forms of highly used, shared transport modes emerged from the failure of colonial governments to invest in public transport. To fill the vacuum, self-help and Indigenous entrepreneurship produced diverse and often colorful and relatively affordable mobility service businesses that employ large numbers of people.[37] Yet investments have often gone into highways built with contracts with foreign companies and constructed through densely populated cities to primarily serve those with private vehicles.[38] This is done without consideration for either existing shared mobility as public transport or pedestrians, leading to alarmingly high numbers of road deaths, especially along these "upgraded" corridors.[39]

The result of this deeply unjust mismatch between infrastructure and the way the majority of people travel in most Global South cities is a massive set of problems, including poor service and labor conditions in the sector and severe public health problems, from air pollution to crashes at the global level. This is compounded by continued colonial patterns of urban and transport planning practices that often fail to recognize and plan for existing popular modes.[40] Demolitions and destruction from road construction largely to serve personal vehicles used by a relatively small elite have, much like in the United States in the 1950s and 1960s, raised serious human rights concerns.[41] While unexplored from a public health point of view, no doubt this too has led to undocumented "root shock." Thus, we see mirrored on the global scale some of what has been the experience in local, underserved communities in the United States as well.

It is worth emphasizing that one of the ways that this investment landscape has been so skewed is through the dismissal of Indigenous, popular modes, such as motorcycle taxis, rickshaws, and minibus systems, as "informal" and hence out of the planning and investment realm, even though they are ubiquitous and dominant in so many rapidly urbanizing cities across the globe.[42] They are often called paratransit or "informal" transport because of the diverse informalities in their labor relations, service provision, and business operations; less recognized are what causes these diverse informalities, as well as the benefits of these systems that provide employment and aim to fill gaps created by grossly inadequate investment in public transport. Widely used because of availability or costs or both, these businesses often pay for formal licenses and fees, and many informalities emerge in relation to these systems out of unmet needs or corruption. For example, without planning, there may be a need to create an ad hoc bus stop or route or, to continue to provide service, the demand to pay a bribe to police. These are not intrinsic characteristics of these systems but produced by state and planning failures in the transport sector that deeply affect lower-income people, women, children, and older people. Interestingly, this dynamic is also visible in a city such as New York, where poor and expensive bus service in largely immigrant communities has led to the development of "dollar" or commuter vans in response to some of the same features of their Global South counterparts.[43]

Despite providing access and opportunities to people, popular shared transport modes are linked to high numbers of crashes and are often highly polluting. Their regular use means large numbers of passengers and workers are exposed to high levels of pollution with serious health impacts, including cardiovascular and respiratory illnesses.[44] For example, two-stroke motorcycles and auto rickshaws, which are popular because of their accessibility and ability to circumvent traffic, can emit more pollution per mile than passenger cars. Minibus systems, widely used in most cities, often rely on modified fuels and old, poorly maintained vehicles, part of a used-vehicle recycling economy that some argue shifts serious emissions from North America and Europe to Asia or Africa.[45] Given that the poorer segments of urban populations tend to use these modes, improving these systems for safety—which includes reducing disease-causing emissions—is an important justice issue, one that has parallels with the environmental justice issues facing underserved communities in the United States and elsewhere.

While likely to be worse polluters than many private vehicles, on a per capita basis, many of these popular modes, especially the minibuses, may still generate less emissions per capita than private vehicles. However, we have not adequately researched emissions from popular modes, and a great deal of data is missing, meaning current transport contributions to emissions inventories for many metropolitan areas are probably inaccurate. While some existing efforts to quantify popular transport emissions exist, these emissions are not always disaggregated carefully to allow us to isolate the total contribution from popular or "informal" transport even though this is critical for policy purposes.[46] To address the complex set of issues around popular transport emissions, it is critical to properly measure pollutants from these modes and understand their specific contributing share to national and local GHG inventories and the health and equity impacts of this pollution. This is also important for accessing climate finance for decarbonization, for example, by replacing motorcycles with cleaner e-bikes.

Another key area of future research is to understand the impacts of climate change on these popular transport systems. Increasing heat and extreme heat days make passengers using these modes, as well as workers, suffer health and income impacts since they typically do not have means to cool air. Increased likelihood and severity of flooding and flash floods are leading to dangerous conditions and losses as well as disruption of services. One innovative and rare study looked at transport and access disruptions of flooding in Kinshasa and estimated $1.2 million worth of losses a day.[47] Once again, these are climate injustices, and the impacts as well as the connection to growing "loss and damages" conversations are aspects we are just beginning to explore.[48] Much more research is necessary if we are to be able to demonstrate and quantify climate injustice in terms of impacts.

Finally, because of the failure of recognition of "popular transport modes," the key actors who run transportation in most Global South cities are largely absent from global conversations on opportunities to access climate finance to reduce GHG emissions as well as to adapt. This is despite the fact that one model by the International Transport Forum of the OECD suggests that by applying improvements to informal transport similar to those of formal bus systems, we could see carbon dioxide emissions decline by an additional 12 percent by 2050.[49] Goals to improve popular transport modes are currently rarely found in Nationally Determined Commitments, although as part of climate justice advocacy, organizations like the Global

Network for Popular Transport are advocating for this. In addition, climate action plans at all levels appear to ignore the need to both enhance and leverage popular transport resiliency.

Exclusion of popular transport operators, workers, and passengers and their communities in global transport and climate discussions and policy leads to lost opportunities for improvements both for these underserved people and communities and for carbon reduction and resilience. The people who work in, own, and use popular transport suffer from a deep lack of resources as well as a lack of recognition and inclusion in climate change dialogues and decision-making at all levels. This fits the classic definition of climate injustice but is woefully understudied, in part because of the assumption, now slowly being questioned, that these modes will inevitably be replaced by a "modern" system. This imagined "modern" system often draws on the ideal of automobility, despite the fact that purchasing a private vehicle is expensive and out of the reach of most in the Global South.

CONCLUSION: TOWARDS INTER- AND TRANS-DISCIPLINARITY AND SYSTEMS THINKING AND MOBILITY WITHIN THE JUSTICE FRAME

As we face climate and other environmental crises, efforts are being made to move out of automobility systems and the injustices that underpin them, toward more livable, healthy, and equitable cities, reducing transport emissions in the process. To address all the problems with current automobility systems and planning that is shaped by them, a new, more human- versus car-centered urban transport paradigm, based on solid more transdisciplinary scholarship, is spreading globally through collaboration and exchange.[50] This shift, perhaps most visible in cities in Europe, North America, and East Asia, is occurring through movements for change that draw on growing (but still underrecognized) awareness of the public health costs and historical transport and environmental injustices, as well as the shrinking of transport choices that car dependency entails.[51]

This shift involves leveraging ideal characteristics of cities—proximity, efficiency, access, diversity, and density—to move toward high-quality, multimodal transport systems with renewed investment in mass transit and nonmotorized transport, mixed use, and more compact and transit-oriented development.[52] This includes many related approaches: from "15-minute cities" where all basic necessities should be available within a 15-minute walk or ride on a bike or transit to "low-emission or car-free zones" and "complete streets," road diets, and traffic calming, where infrastructure caters to all users. The "avoid, shift, and improve" framework is also gaining traction; this involves working to first avoid trips whenever possible through better land-use and organization, shifting trips from cars to lower-emissions shared, public, and nonmotorized transit, and finally making improvements such as safer, smaller, cars or bikes that are shared, have better fuels and efficiencies, or are electric. Interestingly, these types of interventions have not been fully incorporated into IPCC modeling and our climate-reduction strategies and models, and they may offer important, not fully calculated, benefits for climate protection.[53]

However, what should be clear from historical experience is that justice must stay at the center of this paradigm shift. It cannot be assumed that sustainable or green transportation interventions are automatically just, even if they can lead to overall air pollution, health, and climate benefits; interventions for repairing past injustices and ensuring just impacts in particular need constant monitoring and involve political struggle.[54] For example, more recently electrification has been a major focus as a climate intervention to address transport emissions. Concern is growing, however, around the extent to which a singular focus on electrification represents an attempt to simply adapt automobility systems—with its many other ills—to the climate age, perpetuating injustices and engendering new ones, for example, through new extraction.[55] Such extraction—and the environmental and social harms connected with it, including human rights violations, emissions and water contamination, and depletion—often takes place in vulnerable communities. Here, too, lower-income communities in the Global North and Global South, without specific care, face exclusion from EV policy and benefits while facing the brunt of the negative impacts. Popular transport modes also face potential exclusion from finance and support for electrification and its benefits.[56]

Interdisciplinary work in urban studies and related disciplines with an explicit concern for justice is revealing how current transportation and land use in cities across the globe tend to inscribe and express layered and deep injustices. Further, this work reveals that the processes and policy choices that generate deep injustices are also causing broader harm to our societies and planet. New, more holistic, humanistic and transdisciplinary approaches to urban transport are leading the way to a paradigm shift that is increasingly being seen in the streets and decision-making in our growing cities across the planet.

The climate crisis is a catalyst and driver for more of this work. As transport emissions continue to climb, more people are demanding better explanations for why this is the case and how to surmount this problem. As we have shown, the justice frame has led to new insights and pathways forward out of our entwined and compounding automobility, justice, and environmental challenges in cities. Urban studies and the disciplines focused on transport and planning and theorists of justice and just cities are just beginning to meet and explore the idea of a just and a climate-just city, which, in turn, requires a different kind of view of transport and mobility. Many areas touching on urban transport thus need urgent research and policy attention.[57] One thing is certain: The concept of justice—including climate justice—broadly conceived, with all its distributional, recognition, and procedural dimensions, must remain our compass, guide, and central frame if we are to make both conceptual and practical progress in our fight for truly livable and just cities.

Notes

1. Colten, *An Unnatural Metropolis*; Cronin, *Nature's Metropolis*. See also chapter 5 in this volume.
2. Sassen, *The Global City*. See also chapters 7 and 3 in this volume.
3. Seto et al., "Human Settlements, Infrastructure, and Spatial Planning."
4. Fainstein, *The Just City*; Steele et al., "The Climate Just City."
5. Mattioli et al., "The Political Economy of Car Dependence."
6. Marshall, *How Cities Work*.
7. Wells, *Car Country: An Environmental History*.
8. Angel et al., "The Dimensions of Global Urban Expansion."
9. Urry, "The 'System' of Automobility."
10. Creutzig et al., "Transport: A Roadblock."
11. IPCC, *Climate Change 2022*.

12. See chapter 5.
13. Wang et al., "Evidence of Non-tailpipe Emission Contributions."
14. Liotta et al., "Environmental and Welfare Gains."
15. World Health Organization, "'Transport' in Environment."
16. Dolšak and Prakash, "Three Faces of Climate Justice."
17. Martens, *Transport Justice*; Sheller, *Mobility Justice*.
18. Powell, *Undoing Optimization*; Green, *The Smart Enough City*; Kortetmäki, "Reframing Climate Justice."
19. Peter D. Norton, "Street Rivals: Jaywalking and the Invention of the Motor Age Street," *Technology and Culture* 48, no. 2 (2007): 331–59.
20. National Association for the Advancement of Colored People, "Resolution Establishing That Access to Public Transportation System Is a Basic Civil Right"
21. Mumford, "The American Way of Death."
22. Reft et al., eds., *Justice and the Interstates*.
23. Fullilove, *Root Shock*.
24. Tessum et al., "$PM_{2.5}$ Polluters."
25. Wells, *Car Country.*
26. Fried et al., "Seeking Equity and Justice in Urban Freight."
27. Schmitt, *Right of Way.*
28. IPCC, *Climate Change 2022*; Environmental Protection Agency, "Climate Change and Social Vulnerability."
29. Hoffman, "The Effects of Historical Housing Policies."
30. Rahman et al., "The Effects of Coexposure to Extremes"; Sager, "Estimating the Effect of Air Pollution."
31. Mattioli et al., "The Political Economy of Car Dependence"; Urry, "The 'System' of Automobility."
32. Klopp and Makajuma, "Transportation Infrastructure Integration."
33. Anyamba, *"Diverse Informalities" Spatial Transformations.*
34. Hart, "Of Pirate Drivers and Honking Horns"; Hirst, *The Struggle for Nairobi*; Aligula et al., "Urban Public Transport Patterns in Kenya"; Klopp, "Towards a Political Economy of Transportation Policy."
35. Klopp, "Towards a Political Economy of Transportation Policy"; Porter, "Transport Planning in Sub-Saharan Africa."
36. Behrens et al., "Informal Paratransit in the Global South."
37. Kenda, *Matatu: A History of Popular Transportation*; Hart, *Ghana on the Go*; Global Labor Institute, "Nairobi Bus Rapid Transit."
38. Klopp, "Towards a Political Economy of Transportation Policy."
39. Wales, *The Political Economy of Road Safety.*
40. Njoh, "Planning Rules in Post-colonial States"; Klopp and Cavoli, "Mapping Mass Mini-Bus Transit"; Klopp, "From 'Para-Transit' to Transit?"
41. Amnesty International, "Driven Out for Development."
42. Klopp, "Mapping Mass Mini-Bus Transit."
43. Goldwyn, "Anatomy of a New Dollar Van Route."

44. Ngo et al., "Occupational Exposure to Roadway Emissions"; Guzman et al., "Inequality in Personal Exposure."
45. Boateng and Klopp, "Beyond Bans"; United Nations Environment Program, *Used Vehicles and the Environment.*
46. Kustar et al., "Connecting Informal Transport to the Climate Agenda."
47. He et al., "Flood Impacts on Urban Transit and Accessibility."
48. Cintron-Rodriguez et al., "Loss and Damage Fund."
49. Trouvé et al., "Improving Public Transport and Shared Mobility."
50. Walker, *Human Transit*; Sadik-Khan and Solomonow, *Streetfight.*
51. Hosking et al. "Towards a Global Framework for Transport."
52. Liotta et al., "Environmental and Welfare Gains."
53. Creutzig et al., "Transport"; Liotta et al., "Environmental and Welfare Gains."
54. Wågsæther et al., "The Justice Pitfalls"; Vitrano and Lindkvist, "Justice in Regional Transport Planning."
55. Hosseini and Stefaniec, "A Wolf in Sheep's Clothing"; Boateng and Klopp, "The Electric Vehicle Transition."
56. Edna Odhiambo et al., "The Potential for Minibus Electrification."
57. Hosking et al., "Towards a Global Framework for Transport."

BIBLIOGRAPHY

Aligula, E. M., Z. Abiero-Gairy, J. Mutua, F. Owegi, C. Osengo, and R. Olela. "Urban Public Transport Patterns in Kenya: A Case Study of Nairobi City." Special Report 7, Kenya Institute for Public Policy Research and Analysis (KIPPRA), 2005.

Angel, S., J. Parent, D. L. Civco, A. M. Blei, and D. Potere. "The Dimensions of Global Urban Expansion: Estimates and Projections for All Countries, 2000–2050," *Progress in Planning* 75, no. 2 (2011): 53–107.

Anyamba, T. *"Diverse Informalities" Spatial Transformations in Nairobi: A Study of Nairobi's Urban Process.* Saarbrüken: VDM Verlag, 2008.

Behrens, Roger, Saksith Chalermpong, and Daniel Oviedo. "Informal Paratransit in the Global South." In *The Routledge Handbook of Public Transport*, ed. Corrine Mulley et al. London: Routledge, 2021.

Boateng, Festival Godwin and Jacqueline M. Klopp. "Beyond Bans: A Political Economy of Used Vehicle Dependency in Africa." *Journal of Transport and Land-Use* 15, no. 1 (2022): 651–70.

Boateng, Festival Godwin, and Jacqueline M. Klopp. "The Electric Vehicle Transition: A Blessing or a Curse for Improving Extractive Industries and Mineral Supply Chains?" *Energy Research and Social Science* 113 (2024): 103541.

Cairns, S., J. Greig, & M. Wachs. *Environmental Justice & Transportation: A Citizen's Handbook.* Berkeley: University of California, Institute of Transportation Studies, 2003. https://escholarship.org/uc/item/66t4n94b.

Colton, C. *An Unnatural Metropolis: Wresting New Orleans from Nature.* Baton Rouge: Louisiana State University Press, 2006.

Creutzig, F. et al. "Transport: A Roadblock to Climate Change Mitigation?" *Science* 350, no. 6263 (2015): 911–12.

Cronin, W. *Nature's Metropolis: Chicago and the Great West*. New York: Norton, 1991.

Deka, D. "Environmental Justice, Transport Justice, and Mobility Justice." *International Encyclopedia of Transportation* (2021), 305–10.

Dolšak N., and A. Prakash A. "Three Faces of Climate Justice." *Annual Review of Political Science* 25, no. 1 (2022): 283–301.

Environmental Protection Agency. "Climate Change and Social Vulnerability in the United States: A Focus on Six Impacts" (September 2021). https://www.epa.gov/system/files /documents/2021-09/climate-vulnerability_september-2021_508.pdf.

Fainstein, Susan. *The Just City*. Ithaca, NY: Cornell University Press, 2010.

Fried T., Anne Goodchild, Michael Browne, and Ivan Sanchez-Diaz "Seeking Equity and Justice in Urban Freight: Where to Look?" *Transport Reviews* 44, no. 1 (2024): 191–212.

Fullilove, Mindy Thompson. *Root Shock How Tearing Up City Neighborhoods Hurts America, And What We Can Do About It*. Oakand: New Village Press, 2016.

Global Labor Institute. "Nairobi Bus Rapid Transit: A Labor Impact Assessment." 2019.

Goldwyn, E. "Anatomy of a New Dollar Van Route: Informal Transport and Planning in New York City." *Journal of Transport Geography*, vol. 88 (2020).

Green, B. *The Smart Enough City*. Boston: MIT Press, 2018.

Guzman, L., C. Beltran, R. Morales, and O. Sarmiento."Inequality in Personal Exposure to Air Pollution in Transport Microenvironments for Commuters in Bogotá." *Case Studies on Transport Policy* 11 (2023): 2213–624X.

Hart, J. "Of Pirate Drivers and Honking Horns: Mobility, Authority, and Urban Planning in Late-Colonial Accra." *Technology and Culture* 61, no. 2 (2020): S49–S76.

He, Y. Stephan Thies, Paolo Avner, and Jun Rentschler. "Flood Impacts on Urban Transit and Accessibility—a Case Study of Kinshasa." *Transportation Research Part D: Transport and Environment*, vol. 96 (2021).

Hirst, T. *The Struggle for Nairobi*. Nairobi: Mazingira Institute, 1994.

Hoffman, J. S., V. Shandas, and N. Pendleton. "The Effects of Historical Housing Policies on Resident Exposure to Intra-Urban Heat: A Study of 108 US Urban Areas." *Climate* 8, no. 1 (2020): 12.

Hosking, J., Matthias Braubach, Daniel Buss, Meleckidzedeck Khayesi, Victor Pavarino Filho, and Thiago Hérick de Sá."Towards a Global Framework for Transport, Health and Health Equity." *Environment International*, vol. 169 (2022).

Hosseini, Keyvan, and Agnieszka Stefaniec. "A Wolf in Sheep's Clothing: Exposing the Structural Violence of Private Electric Automobility." *Energy Research & Social Science*, vol. 99 (2023),

IPCC. *Climate Change 2022: Mitigation of Climate Change: Contribution of Working Group III to the Sixth Assessment Report of the Intergovernmental Panel on Climate Change*, ed. P. R. Shukla et al. Cambridge: Cambridge University Press, 2022. https://doi.org/10 .1017/9781009157926.

ITF-OECD. "How Improving Public Transport and Shared Mobility Can Reduce Urban Passenger Carbon Emissions." Paris: OECD, N.d.

Klopp. J. M. "From 'Para-Transit' to Transit? Power, Politics and Popular Transport," ed. Rafael H. M. Pereira and Geneviève Boisjoly. *Advances in Transport Policy and Planning*, vol. 8 (2021): 191–209.

Klopp, J. M. "Towards a Political Economy of Transportation Policy and Practice in Nairobi." *Urban Forum* 23 (2012): 1–21.

Klopp, J. M., and C. Cavoli. "Mapping Mass Mini-Bus Transit in Maputo and Nairobi: Engaging 'Paratransit' in Transportation Planning for African Cities." *Transport Reviews* 39, no. 5 (2019): 657–76.

Klopp, J. M., and G. Makajuma. "Transportation Infrastructure Integration in East Africa in Historical Context." In *Integration of Infrastructures*, ed. Martin Schiefelbusch and Hans-Ludger Dienel, 115–36. Surrey: Ashgate, 2014.

Klopp, J. M., and D. Petretta. "The Urban Sustainable Development Goal: Indicators, Complexity and the Politics of Measuring Cities." *Cities* 63 (2017): 92–97.

Kortetmäki T. "Reframing Climate Justice: A Three-Dimensional View on Just Climate Negotiations." *Ethics, Policy & Environment* 19, no. 3 (2016): 320–34.

Kustar, A., I. Abubakar, Tun Hein, and B. Welle. *Connecting Informal Transport to the Climate Agenda: Key Opportunities for Action*. WRI and VREF (2023).

Lay, M. G. "The History of Transport Planning," in *Handbook of Transport Strategy, Policy and Institutions*, vol. 6, ed. K. J. Button and D. A. Hensher, 156–74. Leeds: Emerald Group Publishing, 2005.

Liotta, C., V. Viguié, and F. Creutzig. "Environmental and Welfare Gains Via Urban Transport Policy Portfolios Across 120 Cities." *Nature Sustainability* 6 (2023): 1067–76. https://doi.org/10.1038/s41893-023-01138-0.

Lucas, K. "Transport and Social Exclusion: Where Are We Now?" *Transport Policy* 20 (2012): 105–13.

Maartens, K. *Transport Justice: Designing Fair Transportation Systems*. New York: Routledge, 2017.

Marshall, Alex. *How Cities Work: Suburbs, Sprawl, and the Roads Not Taken*. Austin: University of Texas Press, 2000.

Mattioli, G., C. Roberts, J. K. Steinberger, and A. Brown. "The Political Economy of Car Dependence: A Systems of Provision Approach. *Energy Research & Social Science* 66 (2020): 101486.

Mumford, L. "The American Way of Death." *New York Review of Books* (April 1966).

Mutongi Kenda. *Matatu: A History of Popular Transportation in Nairobi*. Chicago: University of Chicago Press. 2017.

National Association for the Advancement of Colored People (NAACP). Resolution Establishing That Access to Public Transportation System Is a Basic Civil Right. Accessed August 2025. https://naacp.org/resources/establishing-access-public-transportation-system-basic-civil-right#.

Newman, P., and J. Kenworthy. *The End of Automobile Dependence: How Cities Are Moving Beyond Car-Based Planning*. Washington, DC: Island Press/Center for Resource Economics, 2015.

Ngo, N. S., M. Gatari, B. Yan, S. N. Chillrud, K. Bouhamam, and P. L. Kinney. "Occupational Exposure to Roadway Emissions and Inside Informal Settlements in Sub-Saharan Africa: A Pilot Study in Nairobi, Kenya." *Atmospheric Environment* 111 (June 2015): 179–84.

Njoh, A. *Planning Rules in Post-Colonial States: The Political Economy of Urban and Regional Planning in Cameroon*. New York: Nova Science Publishing, 2001.

Norton, P. D. "Street Rivals: Jaywalking and the Invention of the Motor Age Street. *Technology and Culture* 48, no. 2 (2007): 331–59.

Odhiambo, E., D. Kipkoech, M. Hegazy, A. Hegazy, M. Manuel, H. Schalekamp, and J. M. Klopp. "The Potential for Minibus Electrification in Three African Cities: Cairo, Nairobi and Cape Town." *Volvo Research and Educational Foundations* (VREF), August 2021.

Porter, G. "Transport Planning in Sub-Saharan Africa." *Progress in Development Studies* 7, no. 3 (2007): 251–57.

Powell, A. *Undoing Optimization: Civic Action in Smart Cities*. New Haven: Yale University Press, 2021.

Rahman, M. M., R. McConnell, H. Schlaerth et al. "The Effects of Coexposure to Extremes of Heat and Particulate Air Pollution on Mortality in California: Implications for Climate Change." *American Journal of Respiratory and Critical Care Medicine* 206, no. 9 (November 2022): 1117–27. https://doi.org/10.1164/rccm.202204-0657OC.

Reft, Ryann, Amanda Phillips de Lucas, and Rebecca Retzlaff, eds. *Justice and the Interstates: The Racist Truth About Urban Highways*. Washington, DC: Island Press, 2023.

Sadik-Khan, Janette, and Seth Solomonow. *StreetFight: Handbook for an Urban Revolution*. London: Penguin Books, 2017.

Sager, L. "Estimating the Effect of Air Pollution on Road Safety Using Atmospheric Temperature Inversions. *Journal of Environmental Economics and Management* 98 (2019): 102250.

Sassen, S. *The Global City*. Princeton: Princeton University Press, 1991.

Schafer, K. H., and E. Sclar. *Access for All: Transportation and Urban Growth*. London: Penguin Books, 1975.

Schmitt A. *Right of Way: Race, Class, and the Silent Epidemic of Pedestrian Deaths in America*. Washington, DC: Island Press, 2020.

Schwanen, T. "Towards Decolonised Knowledge About Transport." *Palgrave Commun* 4, no. 79 (2018).

Schwanen, T., D. Banister, and J. Anable. "Scientific Research About Climate Change Mitigation in Transport: A Critical Review." *Transport Research Part A*: Policy Practice 45 (2011): 993–1006.

Seto, K. C., S. Dhakal, H. Blanco et al. "Human Settlements, Infrastructure, and Spatial Planning." In *Climate Change 2014: Mitigation of Climate Change; Contribution of Working Group III to the Fifth Assessment Report of the Intergovernmental Panel on Climate Change*, ed. O. Edenhofer, R. Pichs-Madruga, Y. Sokona et al. New York: Cambridge University Press, 2014.

Sheller, M. *Mobility Justice: The Politics of Movement in an Age of Extremes*. London: Verso Press, 2018.

Steele, W. Jean Hillier, Donna Houston, Jason Byrne, and Diana MacCallum. 2018. "The Climate Just City." *Routledge Handbook of Climate Justice*. New York: Routledge, 2018.

Tessum, C. W., D. A. Paolella, S. E. Chambliss, J. S. Apte, J. D. Hill, and J. D. Marshall. "PM$_{2.5}$ Polluters Disproportionately and Systemically Affect People of Color in the United States." *Science Advances* 7, no. 18 (April 28, 2021): eabf4491. https://doi.org/10.1126/sciadv.abf4491.

United Nations. "Africa and Asia to Lead Urban Population Growth in Next 40 Years—UN Report." 2014. https://news.un.org/en/story/2012/04/408132#.

Urry, J. "The 'System' of Automobility." *Theory, Culture & Society* 21, no. 4–5 (2004): 25–39.

Vitrano, C., and C. Lindkvist. "Justice in Regional Transport Planning Through the Lens of Iris Marion Young." *Planning Practice & Research* (2021).

Wågsæther, K., D. Remme, H. Haarstad, and S. Sareen. "The Justice Pitfalls of a Sustainable Transport Transition." *Environment and Planning F* 1, no. 2–4 (2022): 187–206.

Wales, J. *The Political Economy of Road Safety*. Overseas Development Institute, 2017.

Walker, J. *Human Transit: How Clearer Thinking About Public Transit Can Enrich Our Communities*. Washington, DC: Island Press, 2011.

Wang, Xiaoliang, Steven Gronstal, Brenda Lopez et al. "Evidence of Non-tailpipe Emission Contributions to PM2.5 and PM10 Near Southern California Highways." *Environmental Pollution* 317 (January 15, 2023): 120691. https://doi.org/10.1016/j.envpol.2022.120691.

Wells, C. *Car Country: An Environmental History*. Seattle: University of Washington Press, 2012.

World Health Organization. "'Transport' in Environment, Climate Change and Health." 2024. https://www.who.int/teams/environment-climate-change-and-health/healthy-urban-environments/transport.

5 | The Potential Conflict Between Climate Justice and Environmental Justice

Reducing Greenhouse Gas Emissions
Does Not Improve Air Quality

RÓISÍN COMMANE

There is a widespread assumption that addressing climate change will automatically be good for the environment, but that is not always the case. There are some longstanding, but little discussed, tensions between the steps needed to address climate change in a way that prioritizes the communities most at risk (climate justice) and those that ensure a positive impact on the environment (environmental justice). These tensions must be carefully considered as climate mitigation and adaptation policies are developed. Climate justice and environmental justice do not need to be in competition, but the mechanism of how policies are implemented will need to be carefully considered if they are to be complementary. Most climate warming is caused by the rapid rise in the concentrations of greenhouse gases, carbon dioxide (CO_2) and methane, in the atmosphere. Greenhouse gases trap heat radiating off land in an atmospheric layer around the Earth. Climate mitigation approaches work to reduce the concentration of greenhouse gases in the atmosphere, either by stopping the emission of these gases into the atmosphere from human-caused sources or by taking greenhouse gases out of the atmosphere. Climate adaption approaches address the changes needed to ensure resilience of our society and infrastructure to future climate impacts.

In most cases, decarbonization approaches that reduce the combustion of fossil fuels will dramatically improve air quality. In this chapter, I will discuss some examples of historical climate-related policies where the unintended consequences negatively affected air quality, and where air quality policies were not beneficial to climate. While, thankfully, these cases do not occur frequently, their impact can be life-changing for the communities affected. I will also discuss some climate mitigation and adaption approaches, two of which have been proposed for New York City (NYC) that, even within a climate justice framework, *could* cause air quality problems. By the end of this chapter, I hope to have convinced you that the development of comprehensive climate policies requires the integration of both climate justice and environmental justice concerns, rather than prioritizing one impact over the other.

LINKS BETWEEN CLIMATE WARMING AND AIR QUALITY

The combustion of fossil fuels has driven the rapid increase in global CO_2 concentrations in the atmosphere since the 1850s.[1] Fossil fuel combustion also produces air pollutants that are toxic to humans, animals, and plants. For example, high temperature combustion can split nitrogen molecules in air and form nitric oxide (NO) that is rapidly oxidized to nitrogen dioxide (NO_2), which together are known as nitrogen oxide or NOx. In the daylight atmosphere, NOx can react with other organic compounds in the atmosphere to create ozone (O_3) and can also be a source of particulate matter (PM). Toxic air pollutants, including NOx, ozone, and particulate matter, cause serious health problems, including heart attacks and strokes: In 2019 alone, air pollution was thought to have caused the premature deaths of 6.7 million people worldwide.[2]

The amount and type of NOx and other air pollutants co-emitted with CO2 during combustion depend on the fuel type. Natural gas is the cleanest of the fossil fuels but can be difficult to burn to completion, so boiler exhaust can contain unburned methane, carbon monoxide (CO), formaldehyde, etc. The combustion of liquid fossil fuels such as home heating oil, petroleum, and diesel also releases various volatile organic compounds. Oil and diesel, depending on the quality/sulfur content of the fuel, also release sulfur dioxide (SO2), which can be oxidized in the atmosphere to create acid rain and particulate matter. Coal is the dirtiest fossil fuel, releasing vast amounts of sulfur, chlorine, mercury, heavy metals, particulate matter of various sizes, and soot (black carbon), all of which degrade air quality downwind. As we will see, various Environmental Protection Agencies (EPAs) have tried to remove the worst of the pollutants from the combustion exhaust over the twentieth century, with varying results. Most coal power plants in the United States now have scrubbers to reduce NOx, sulfur, and mercury emissions, which has had some surprising global climate impacts that are discussed in the next section. Cars and trucks are now required to have catalytic converters to reduce emissions of NOx and particulate matter from each vehicle, which was used to justify climate policy decisions in Europe.[3]

POLICIES TO IMPROVE AIR QUALITY THAT INCREASED CLIMATE WARMING

In general, global surface air temperatures have steadily increased since 1850, except during the 1940s–1970s. With our massive increase in coal combustion beginning with World War II, concentrations of sulfur gas in the atmosphere increased rapidly. Sulfur dioxide gas reacts in the atmosphere to eventually form sulfur-based particles (also called particulate matter or aerosols), which can reflect sunlight back to space. This leveling off of the global temperature has been labeled "global dimming" and has often been described as accidental geoengineering.

However, these same climate cooling sulfur particles have a massive negative impact on humans, plants, and animals. High concentrations of particulate matter in the air are toxic to human health. The London Fog event of December 1952, where a stagnant weather system trapped toxic coal emissions in a dense smog over London for days on end, was estimated to have caused 12,000 additional premature deaths in the weeks after the event.[4] The

London Fog event eventually led to political discussions about the dangers of air pollutants and to the UK Clean Air Act of 1956. The sulfur gas emitted from coal combustion also causes acid rain that damages forests, farms, and soil in areas far downwind of where it is emitted. By the early 1970s, most countries around the world recognized the dangers of unregulated sulfur emissions, and chemical scrubbers in coal and oil-fired power plants were introduced to reduce the emission of sulfur. But phasing out such emissions has been a slow process: For example, NYC finally banned the use of high-sulfur bunker fuel (No. 6 distillate fuel oil) for home residential heating in 2015. Since then, the sulfur gas concentrations in the city have been reduced dramatically (almost below detection levels), and the wintertime air quality is much cleaner than at any previous time in modern memory.

One recent air quality policy to reduce sulfur emissions has also had an unprecedented impact on climate.[5] Emissions of sulfur gas from the exhaust of ships leads to the creation of ship tracks: bright clouds made up of small water droplets that satellites can see over the ocean for a day or two after a ship has passed by. These clouds are generated in a similar way to the proposed geoengineering approach called marine cloud brightening, and they reflect sunlight back to space in a similar way to the global dimming effect of the 1950s. For ship tracks, the bright clouds stop the sunlight reaching the dark ocean underneath. In 2020 the International Maritime Organization banned the burning of high-sulfur bunker fuel in ships due to its toxic air quality impacts.[6] By 2023 the ship tracks usually seen from space had mostly disappeared.[7] While removing the last large source of toxic sulfur gas emissions is good for air quality, especially for coastal communities, the lack of ship track clouds may also lead to more heat being absorbed by the dark oceans.[8] It has also been suggested as a driver of the record-breaking global temperatures of 2023.[9] The impact of this change in cloud cover on climate warming is not yet known but continues to be evaluated by scientists.

CLIMATE POLICIES THAT DEGRADED AIR QUALITY

The European Union (EU) has introduced a number of climate policies that have been or are likely to be detrimental to air quality in Europe. These policies were focused on vehicle emissions and wood burning.

- **Vehicles** running on fossil fuels, petrol and diesel, are a large source of both greenhouse gases and air pollutants. You can generally travel farther on a gallon of diesel than on a gallon of petrol. Over the past thirty years, the EU has used this fact to justify promoting the purchase of diesel vehicles in order to meet their Kyoto Protocol climate commitments. However, these climate mitigation policies did not consider the air-quality implications: The higher temperature burn of a diesel engine produces more NOx and other air pollutants than a petrol engine. Both engine types are supposed to use catalytic converters to reduce these emissions, but observations of air pollutants around Europe had long suggested that emissions from vehicles were much larger than reported by the car companies. Then, in 2014, a series of long-distance road tests across the United States confirmed that the NOx emissions from some diesel engines were thirty to forty times greater than reported.[10] Shockingly, some car companies had illegally introduced software to detect emissions test conditions and to only use the air pollution control measures during the emissions test. Outside of the test, the vehicles would bypass the catalytic converter to improve performance. Known as "dieselgate," eventually the scandal extended to nearly eleven million vehicles around the world and resulted in Volkswagen paying over USD$30 billion in fines and settlements in various countries.[11] But the real impact of this climate policy was on air quality and health—diesel-related pollutant emissions in excess of the regulations massively increased ozone and PM concentrations that contributed to approximately 38,000 premature deaths globally in 2015.[12]
- **Wood Burning** has been framed as a carbon-neutral source of energy by the EU, based on the idea that the CO_2 emitted during such burning will be reabsorbed by more trees in the future. Biomass (firewood, etc.) was 60 percent of the EU's renewable energy budget in 2023. But by burning this reservoir of carbon, we are releasing CO_2 that was captured thirty to fifty years ago back into the atmosphere. Similar to wildfires, the combustion of wood and solid biofuels results in very high concentrations of particulate matter as well as hundreds of toxic pollutants. The EU acceptance of

wood burning as carbon neutral has promoted the use of wood stoves worldwide, most often without any filtration of particulate matter from the exhaust. Toxic wintertime pollution events regularly occur in cities and towns around the world, especially in Europe (e.g., Paris), but also in cold cities like Fairbanks, Alaska, and Ulaanbaatar, Mongolia.[13]

CLIMATE MITIGATION AND ADAPTATION IMPACTS ON AIR QUALITY

Most of the proposed climate mitigation and adaptation policies have been examined through a climate justice lens, but they are not often assessed for their impact on air quality. Cities are a key area for the implementation of both climate mitigation and adaption approaches—over 70 percent of fossil fuel–driven CO_2 emissions occur in cities.[14] In this section I discuss examples of policies that need to be implemented in very specific ways to ensure minimal air quality damage.

Alternative Fuels: Hydrogen

A carbon-free fuel source is urgently needed for processes that require high-temperature combustion, such as steel manufacturing. Hydrogen has been proposed as a carbon-free alternative. In late 2023 the US government invested USD$7 billion in hydrogen hubs to fast-track commercialization.[15] Hydrogen can be produced from the electrolysis of water, and the combustion of pure hydrogen is carbon free. Hydrogen combustion produces water vapor in pure oxygen, but in air, which is 78 percent nitrogen, the high combustion temperature of hydrogen will produce more highly toxic NOx than the combustion of natural gas or oil, which burn at cooler temperatures.

The source and storage of hydrogen have generated much discussion.[16] As a small, explosive molecule, hydrogen is difficult to store or transport in gas form. The electrolysis of water is very energy intensive, which has limited the use of renewable energy in hydrogen production so far.

Currently most hydrogen is inefficiently generated from methane (CH_4) in natural gas, in a process that releases CO_2 and other air pollutants to the atmosphere. Methane is a strong greenhouse gas, with a warming impact over eighty times that of CO_2 over twenty years, and additional in leaks of natural gas would also increase climate warming. Ammonia (NH_3), a common agricultural fertilizer, has also been proposed as a source and carrier gas for hydrogen. Most ammonia is produced industrially through the high temperature Haber-Bosch process using hydrogen released from natural gas. Ammonia can thermally decompose to hydrogen at high temperatures. Most of the hydrogen or ammonia production steps are highly inefficient, which could lead to large additional NOx emissions from fossil fuel combustion alongside the already larger NOx production from hydrogen combustion. Care must be taken to minimize the fossil fuel combustion and inevitable leaks of natural gas or ammonia, as well as to reduce emissions of air pollutants. Note too that, in cold atmospheric conditions, ammonia can form particulate matter that leads to air pollution events separate from any combustion process.

Hydrogen is not a greenhouse gas itself, but leaked hydrogen can act as an indirect greenhouse gas. Any leaked hydrogen will react with the hydroxyl radical, OH, in the atmosphere to form water vapor. OH is the oxidant that destroys 90 percent of methane and is key to breaking down almost all other air pollutants in the atmosphere. By reducing the available concentrations of OH in the atmosphere, hydrogen will lead to a longer lifetime of methane in the atmosphere, which means that the warming potential of every methane molecule will be increased. The few studies that have examined the climate impacts of hydrogen estimate the global warming potential as approximately eleven times that of CO_2 and recommend minimizing hydrogen leaks.[17]

Improved Energy Efficiency of Buildings Can Affect Air Quality

Energy consumption for cooling and heating of buildings is a huge fraction of total urban energy use and CO_2 emissions. Cooling is the largest power demand of a building, and approximately 20 percent of total electricity use worldwide has been attributed to such cooling.[18] Many old buildings are "leaky"—they exchange air and heat with the outside atmosphere at a rapid rate. With more extreme heat waves expected from climate warming, the cooling of buildings, already the most intense energy use in such structures, will require even more energy. Energy use in households increases by about 40 percent when they adopt air conditioning, so the energy needed for the cooling of buildings is expected to grow rapidly with warming temperatures.[19] Socioeconomic factors often dictate when older buildings are refurbished, which often means poorer communities will see increased energy costs. To ensure a climate just transition, energy efficiency of all existing buildings must be improved so that the associated greenhouse gas emissions from cities do not continue to rise alongside the temperatures. Electrifying all the existing combustion sources in cities and providing electricity from centralized fossil fuel power plants would reduce emissions of greenhouse gases across the city to a single point and dramatically improve air quality at the same time. These fossil fuel power plants would provide far fewer CO_2 emission locations that could then undergo CO_2 removal (known as direct air capture) from the exhaust, thus, providing a short-term climate and environmental solution while large-scale, carbon-free energy sources are found.

However, improving energy efficiency of buildings could also exacerbate one of the most prevalent environmental justice problems in many communities: indoor air quality. Sealing up buildings to reduce their energy use without adequate and sometimes expensive ventilation could mean that the air inside these buildings would rapidly become toxic. Most people spend over 90 percent of their day indoors, and, while indoor air quality became a discussion point in 2020 due to the COVID-19 pandemic, it is still a vastly understudied research area. Unlike the various national air quality standards for outdoor air, there are few indoor air quality standards. "Sick-building syndrome" has become the catch-all term for health symptoms that people get in particular buildings, and it is caused by unhealthy indoor air quality. Radon, a known carcinogen, is one of the few gases regularly checked for in most buildings. It is a radioactive decay product of uranium that occurs naturally, mostly in granite soils. It typically moves up into a building through the foundation, where it can get trapped inside, especially in well-sealed, energy-efficient buildings. Radon cannot be removed by air filters and must be ventilated. Most radon reduction plans are designed to seal off the basement to ensure that the gas cannot enter the building, and they are usually quite effective.

But what happens when toxic gases are emitted inside the building and not just infiltrating from outside? The organic compounds in solvents, adhesives, and glues used in building materials can evaporate out of the materials and accumulate in the indoor air if they are not properly ventilated. For example, a cluster of cancers on the North Carolina State University campus in late 2023 has been linked to high concentrations of Polychlorinated biphenyls (PCBs) in the air inside a building on campus.[20] PCBs are highly carcinogenic chemical compounds that were extensively used in industrial and consumer products until the 1980s but, due to their long-lasting nature, are still in use today.

Some buildings also have indoor combustion to ventilate. Cooking food produces particulate matter that could be filtered away by a carbon or charcoal filter over the stove/oven.[21] However, the combustion of natural gas or propane produces toxic pollutants (NOx, formaldehyde, benzene, etc.), which will not be removed by the carbon filter, and this exhaust must be ventilated outside. Stove-top flames are often the most obvious source of combustion, but ovens cause a much larger buildup of air pollutants due to the long duration required for preheating and longer cooking time. Commercial kitchens are generally required by law to mechanically ventilate their exhaust, but many home kitchens are not subject to the same requirements. In NYC, any kitchen with a window is considered legally ventilated, even if it is difficult to open the window or to vent the kitchen air through that window. Recent studies have shown that gas ovens can raise indoor concentrations of NOx and benzene, a known carcinogen, to levels of concern by EPA standards.[22] Alongside the energy wasted through open windows, outdoor pollution events (wildfires, ozone exceedance events, etc.) and heat extremes are occurring increasingly regularly, so opening windows to ventilate the kitchen becomes a problem. There must be a complementary discussions of energy efficiency and ventilation at all stages of policy development or toxic levels of unventilated indoor air could become more prevalent as buildings become better sealed and more energy efficient.

Nature-Based Climate Solutions with Impacts on Air Quality in Cities

Across nearly all parts of the globe, large investments are being made in nature-based climate solutions (NbCS) to reduce CO_2 in the atmosphere. Despite these investments, few NbCS approaches have been evaluated to quantify their net CO_2 uptake, and even fewer have been evaluated for their impact on air quality.[23]

One of the most popular nature-based climate solutions is "urban greening," or planting trees in cities: It improves urban livability in many different ways, including by reducing the urban heat island effect. Extreme temperatures are detrimental to human health, and air temperatures are even higher in cities where the concrete buildings and pavements magnify temperatures. Urban greening has been particularly promoted as a climate-just adaptation with some additional climate mitigation: Trees provide shade and transpire water vapor during the day as they absorb CO_2 during photosynthesis, thus reducing the temperatures and storing carbon at the same time. Trees can absorb a lot of CO_2 during the daytime, but some CO_2 is respired back out again at night. The CO_2 stored within the tree, known as the net CO_2 uptake, allows the tree to grow. Trees in cities grow in hotter, more polluted, and sometimes more water-limited conditions than rural forests. Despite this, urban vegetation has been found to grow several times faster and respire at lower rates than predicted in some regions like the northeastern United States.[24] But the high stress conditions around street trees (vehicle pollution, mechanical damage, shading from tall buildings, excessive heat, small root volumes, etc.) eventually have a detrimental effect and result in such trees having a much shorter lifespan than a forest tree.[25] For tropical cities with year-round green vegetation, it is expected that vegetation will take up even greater amounts of CO_2, but little data currently exists to evaluate that expectation. Most urban-greening approaches favor the planting of climate-resilient trees—those species that will survive the hot and polluted conditions experienced in cities. Oak species are usually on the top of the lists of such tree species for urban greening in US cities.[26]

The NYC Parks Department has enthusiastically embraced urban greening as an adaptation to the urban heat island effect. The metro area is the third largest urban source of greenhouse gases in the world and the largest air pollution source region in the United States, but thanks to some favorable winds off the ocean, NYC only has the second worst summertime air quality in the country (after Los Angeles). About 22 percent of the area of NYC is covered in trees, and the city maintains a database of all the trees on the street and many in city parks.[27] The hottest and most polluted parts of the city are often the neighborhoods with the highest poverty rates (e.g., the South Bronx), and these areas are top of the list for urban greening. It has been estimated that trees in cities take up 20–40 percent of the afternoon CO_2 emissions released from fossil fuel combustion in peak summer.[28] While this is a large instantaneous uptake, the growing season is only over the summer, so the overall impact of the urban trees on the total CO_2 budget for the city is currently quite small. It will become more important as building efficiency improves, fossil fuel combustion decreases, and more trees are planted.

However, the type of tree species planted is important for air quality, especially in already polluted cities like New York. Isoprene is a volatile organic compound that is emitted from tree species like oak and sweetgum when the trees are stressed by the heat. The emission of isoprene from vegetation has a strong light and temperature dependence, with most isoprene emitted during the daytime when air temperatures approach 35°C, but decreasing rapidly at hotter temperatures. In the atmosphere, isoprene is oxidized by OH and goes on to react with NOx emitted from the combustion of fossil fuels to produce ozone. In NYC about half of the trees in parks and about 20 percent of the street trees are isoprene-producing oak and sweetgum, with the remainder being non-isoprene-producing species (London plane tree, maple, linden, cherry, etc.). A recent study found that urban greening strategies that favor planting oak trees could double future isoprene emissions in NYC, resulting in an increase in peak ozone on hot days.[29] To avoid more severe ozone exceedance events that adversely affect human and plant health, the city would have to reduce NOx emissions by a factor of 5, which is vastly more than the NOx reduction observed during the citywide lockdowns of the COVID-19 pandemic in 2020.[30]

FUTURE POLICY RECOMMENDATIONS

To ensure that climate justice and environmental justice are complementary and not in competition, we need to carefully consider the mechanisms of how climate policies are implemented. There are many examples of climate-just mitigation and adaptation policies that would also be environmentally just that have not been discussed here. Any process that reduces the combustion of fossil fuels will lead to better air quality. For example, electric vehicles (EVs) would vastly improve air quality, and the magnitude of the reduction in greenhouse gases emissions would depend on the energy source. But there are also environmental concerns about the batteries used in EVs (e.g., lithium mining). The dangers of diesel emissions and wood burning have long been highlighted in the air quality and environmental justice communities. Ventilation to maintain healthy indoor air quality will become more important for all communities with the increased energy efficiency of buildings. The massive human and economic cost of air quality (e.g., healthcare costs after a pollution event) should be accounted for when considering the economic cost of burning fossil fuels and wood. The few cautionary examples described in this chapter should highlight why air quality and environmental justice experts need to be consulted alongside climate justice advocates when climate policies for mitigation and adaptation are being developed. Future climate policies need to be developed with a full consideration of their impact on environmental justice or we may again see the unnecessary deaths of tens of thousands of people. Indeed, calling for improved air quality could be a more tangible and urgent motivator to phase out fossil fuel and wood combustion, which is the most important step in any effective climate mitigation plan.

Notes

1. Masson-Delmotte et al., "IPCC, 2021: Summary for Policymakers."
2. Fuller et al., "Pollution and Health."
3. For more detailed information, see Commane and Schiferl, "Climate Mitigation Policies for Cities."
4. Bell et al., "A Retrospective Assessment of Mortality."
5. Voosen, "Ship Fuel Rules Have Altered Clouds."

6. International Maritime Organization, "Amendments to MARPOL Annex VI."
7. Watson-Parris et al., "Shipping Regulations Lead to Large Reduction."
8. Diamond, "Detection of Large-Scale Cloud Microphysical Changes."
9. Goessling et al., "Recent Global Temperature Surge."
10. Thompson et al., "In-Use Emissions Testing," 133.
11. Kell, "From Emissions Cheater to Climate Leader."
12. Anenberg et al., "Impacts and Mitigation."
13. Favez et al., "Evidence for a Significant Contribution"; Ward et al., "Source Apportionment of PM2.5"; Guttikunda et al., "Particulate Pollution in Ulaanbaatar."
14. Intergovernmental Panel on Climate Change (IPCC), ed., "Technical Summary."
15. Biden-Harris administration, "Biden-Harris Administration Announces."
16. Murdoch et al., "Pathways to Commercial Liftoff."
17. Sand, "A Multi-Model Assessment"; Paulot et al., "Reanalysis of NOAA H_2 Observations."
18. IEA Paris, "The Future of Cooling."
19. Randazzo et al., "Air Conditioning and Electricity Expenditure."
20. Vespa, "NC State Found PCBs."
21. Farmer et al., "Overview of HOMEChem."
22. We Act for Environmental Justice, "Out of Gas, in with Justice"; Kashtan et al., "Gas and Propane Combustion from Stove."
23. Novick et al., "We Need a Solid Scientific Basis"; Buma et al., "Expert Review of the Science."
24. Hardiman et al., "Accounting for Urban Biogenic Fluxes."
25. Smith et al., "Live Fast, Die Young."
26. Roloff et al. "The Climate-Species-Matrix"; Cannon and Petit, "The Oak Syngameon."
27. NYC Parks Department, "NYC Parks Tree Map."
28. Wei et al., "High Resolution Modeling of Vegetation."
29. Wei et al., "Modeling of Summertime Biogenic Isoprene Emissions."
30. Tzortziou et al., "Declines and Peaks in NO_2 Pollution."

BIBLIOGRAPHY

Anenberg, Susan C., Joshua Miller, Ray Minjares et al. "Impacts and Mitigation of Excess Diesel-Related NOx Emissions in 11 Major Vehicle Markets." *Nature* 545, no. 7655 (May 2017): 467–71. https://doi.org/10.1038/nature22086.

Bell, Michelle L., Devra L. Davis, and Tony Fletcher. "A Retrospective Assessment of Mortality from the London Smog Episode of 1952: The Role of Influenza and Pollution." *Environmental Health Perspectives* 112, no. 1 (January 2004): 6–8. https://doi.org/10.1289/ehp.6539.

Biden-Harris administration. "Biden-Harris Administration Announces Regional Clean Hydrogen Hubs to Drive Clean Manufacturing and Jobs." The White House,

October 13, 2023. https://www.whitehouse.gov/briefing-room/statements-releases/2023 /10/13/biden-harris-administration-announces-regional-clean-hydrogen-hubs-to -drive-clean-manufacturing-and-jobs/.

Buma, B., D. R. Gordon, K. M. Kleisner et al. "Expert Review of the Science Underlying Nature-Based Climate Solutions." *Nature Climate Change* 14, no. 4 (April 2024): 402– 6. https://doi.org/10.1038/s41558-024-01960-0.

Cannon, Charles H., and Rémy J. Petit. "The Oak Syngameon: More than the Sum of Its Parts." *New Phytologist* 226, no. 4 (May 2020): 978–83. https://doi.org/10.1111/nph.16091.

Commane, Róisín, and Luke D. Schiferl. "Climate Mitigation Policies for Cities Must Consider Air Quality Impacts." *Chem* 8, no. 4 (April 2022): 910–23. https://doi.org/10.1016 /j.chempr.2022.02.006.

Diamond, Michael S. "Detection of Large-Scale Cloud Microphysical Changes Within a Major Shipping Corridor After Implementation of the International Maritime Organization 2020 Fuel Sulfur Regulations." *Atmospheric Chemistry and Physics* 23, no. 14 (July 25, 2023): 8259–69. https://doi.org/10.5194/acp-23-8259-2023.

Farmer, D. K., M. E. Vance, J.P.D. Abbatt et al. "Overview of HOMEChem: House Observations of Microbial and Environmental Chemistry." *Environmental Science: Processes & Impacts* 21, no. 8 (2019): 1280–1300. https://doi.org/10.1039/C9EM00228F.

Favez, Olivier, Hélène Cachier, Jean Sciare, Roland Sarda-Estève, and Laurent Martinon. "Evidence for a Significant Contribution of Wood Burning Aerosols to PM2.5 during the Winter Season in Paris, France." *Atmospheric Environment* 43, no. 22–23 (July 2009): 3640–44. https://doi.org/10.1016/j.atmosenv.2009.04.035.

Fuller, Richard, Philip J. Landrigan, Kalpana Balakrishnan et al. "Pollution and Health: A Progress Update." *Lancet Planetary Health* 6, no. 6 (June 2022): e535–47. https://doi .org/10.1016/S2542-5196(22)00090-0.

Goessling, Helge F., Thomas Rackow, and Thomas Jung. "Recent Global Temperature Surge Intensified by Record-Low Planetary Albedo." *Science* 387, no. 6729 (January 3, 2025): 68–73. https://doi.org/10.1126/science.adq7280.

Guttikunda, Sarath K., Sereeter Lodoysamba, Baldorj Bulgansaikhan, and Batdorj Dashdondog. "Particulate Pollution in Ulaanbaatar, Mongolia." *Air Quality, Atmosphere & Health* 6, no. 3 (September 2013): 589–601. https://doi.org/10.1007/s11869-013-0198-7.

Hardiman, Brady S., Jonathan A. Wang, Lucy R. Hutyra, Conor K. Gately, Jackie M. Getson, and Mark A. Friedl. "Accounting for Urban Biogenic Fluxes in Regional Carbon Budgets." *Science of the Total Environment* 592 (August 2017): 366–72. https://doi.org /10.1016/j.scitotenv.2017.03.028.

IEA Paris. "The Future of Cooling." 2018. https://www.iea.org/reports/the-future-of -cooling.

Intergovernmental Panel on Climate Change (IPCC), ed. "Technical Summary." In *Climate Change 2021—The Physical Science Basis: Working Group I Contribution to the Sixth Assessment Report of the Intergovernmental Panel on Climate Change*, 35–144. Cambridge: Cambridge University Press, 2023. https://doi.org/10.1017/9781009157896.002.

International Maritime Organization. "Amendments to MARPOL Annex VI." 2020. https: //wwwcdn.imo.org/localresources/en/KnowledgeCentre/IndexofIMOResolutions /MEPCDocuments/MEPC.305%2873%29.pdf.

Kashtan, Yannai S., Metta Nicholson, Colin Finnegan et al. "Gas and Propane Combustion from Stoves Emits Benzene and Increases Indoor Air Pollution." *Environmental Science & Technology* 57, no. 26 (July 4, 2023): 9653–63. https://doi.org/10.1021/acs.est.2c09289.

Kell, Georg. "From Emissions Cheater to Climate Leader: VW's Journey from Dieselgate to Embracing E-Mobility." *Forbes Magazine*, December 5, 2022. https://www.forbes.com/sites/georgkell/2022/12/05/from-emissions-cheater-to-climate-leader-vws-journey-from-dieselgate-to-embracing-e-mobility/#.

Masson-Delmotte, V., P. Zhai, A. Pirani et al. "IPCC, 2021: Summary for Policymakers." In *Climate Change 2021: The Physical Science Basis. Contribution of Working Group I to the Sixth Assessment Report of the Intergovernmental Panel on Climate Change*. Cambridge: Cambridge University Press, 2021. https://www.ipcc.ch/report/ar6/wg1/downloads/report/IPCC_AR6_WGI_Full_Report.pdf.

Murdoch, Hannah, Jason Munster, Sunita Satyapal, Neha Rustagi, Amgad Elgowainy, and Michael Penev. "Pathways to Commercial Liftoff: Clean Hydrogen." US Department of Energy, March 20, 2023. https://liftoff.energy.gov/wp-content/uploads/2023/03/20230320-Liftoff-Clean-H2-vPUB.pdf.

Novick, Kimberly A., Trevor F. Keenan, William R. L. Anderegg et al. "We Need a Solid Scientific Basis for Nature-Based Climate Solutions in the United States." *Proceedings of the National Academy of Sciences* 121, no. 14 (April 2, 2024): e2318505121. https://doi.org/10.1073/pnas.2318505121.

NYC Parks Department. "NYC Parks Tree Map." n.d. https://tree-map.nycgovparks.org/tree-map.

Paulot, Fabien, Gabrielle Pétron, Andrew M. Crotwell, and Matteo B. Bertagni. "Reanalysis of NOAA H_2 Observations: Implications for the H_2 Budget." *Atmospheric Chemistry and Physics* 24, no. 7 (April 9, 2024): 4217–29. https://doi.org/10.5194/acp-24-4217-2024.

Randazzo, Teresa, Enrica De Cian, and Malcolm N. Mistry. "Air Conditioning and Electricity Expenditure: The Role of Climate in Temperate Countries." *Economic Modelling* 90 (August 2020): 273–87. https://doi.org/10.1016/j.econmod.2020.05.001.

Roloff, Andreas, Sandra Korn, and Sten Gillner. "The Climate-Species-Matrix to Select Tree Species for Urban Habitats Considering Climate Change." *Urban Forestry & Urban Greening* 8, no. 4 (January 2009): 295–308. https://doi.org/10.1016/j.ufug.2009.08.002.

Sand, Maria, Ragnhild Bieltvedt Skeie, Marit Sandstad et al. "A Multi-Model Assessment of the Global Warming Potential of Hydrogen." *Communications Earth & Environment* 4, no. 1 (June 7, 2023): 203. https://doi.org/10.1038/s43247-023-00857-8.

Smith, Ian A., Victoria K. Dearborn, and Lucy R. Hutyra. "Live Fast, Die Young: Accelerated Growth, Mortality, and Turnover in Street Trees," ed. Dafeng Hui. *PLOS ONE* 14, no. 5 (May 8, 2019): e0215846. https://doi.org/10.1371/journal.pone.0215846.

Thompson, Gregory J., Daniel K. Carder, Marc C. Besch, Arvind Thiruvengadam, and Hemanth K. Kappanna. "In-Use Emissions Testing of Light-Duty Diesel Vehicles in the United States." International Council on Clean Transportation, 2014.

Tzortziou, Maria, Charlotte F. Kwong, Daniel Goldberg et al. "Declines and Peaks in NO_2 Pollution During the Multiple Waves of the COVID-19 Pandemic in the New York

Metropolitan Area." *Atmospheric Chemistry and Physics* 22, no. 4 (February 22, 2022): 2399–2417. https://doi.org/10.5194/acp-22-2399-2022.

Vespa, Emily. "NC State Found PCBs in 2018 Testing of Poe Hall, D. H. Hill Jr. Library, New Records Show." Technicianonline.com," March 7, 2024. https://www.technician online.com/news/nc-state-found-pcbs-in-2018-testing-of-poe-hall-d-h-hill-jr-library /article_d09785d2-dcfe-11ee-9534-8f6e58cca055.html.

Voosen, Paul. "Ship Fuel Rules Have Altered Clouds and Warmed Waters." *Science* 381, no. 6657 (August 4, 2023): 467–68. https://doi.org/10.1126/science.adk0831.

Ward, Tony, Barbara Trost, Jim Conner, James Flanagan, and R.K.M. Jayanty. "Source Apportionment of PM2.5 in a Subarctic Airshed—Fairbanks, Alaska." *Aerosol and Air Quality Research* 12, no. 4 (2012): 536–43. https://doi.org/10.4209/aaqr.2011.11.0208.

Watson-Parris, Duncan, Matthew W. Christensen, Angus Laurenson, Daniel Clewley, Edward Gryspeerdt, and Philip Stier. "Shipping Regulations Lead to Large Reduction in Cloud Perturbations." *Proceedings of the National Academy of Sciences* 119, no. 41 (October 11, 2022): e2206885119. https://doi.org/10.1073/pnas.2206885119.

We Act for Environmental Justice. "Out of Gas, in with Justice," 2023. https://www.weact .org/wp-content/uploads/2023/02/Out-of-Gas-Report-FINAL.pdf.

Wei, Dandan, Cong Cao, Karambelas Alexandra, John Mak, Andrew Reinmann, and Róisín Commane. "High-Resolution Modeling of Summertime Biogenic Isoprene Emissions in New York City." *Environmental Science & Technology.* 2024.

Wei, Dandan, Andrew Reinmann, Luke D. Schiferl, and Róisín Commane. "High Resolution Modeling of Vegetation Reveals Large Summertime Biogenic CO_2 Fluxes in New York City." *Environmental Research Letters* 17, no. 12 (December 1, 2022): 124031. https:// doi.org/10.1088/1748-9326/aca68f.

Climate Change as a Harm Multiplier

Exploring Sociology, Migration, and Environmental Health

6

Sociological Perspectives on Climate Justice

JENNIFER E. GIVENS AND MUFTI NADIMUL QUAMAR AHMED

limate activist Greta Thunberg's book about the climate crisis begins, "To solve this problem, we need to understand it," but this is no small task, because the problem is actually "a series of interconnected problems."[1] Sociologists agree that the climate crisis is a complex *social* problem, and to address it we need to both accurately identify and address multiple, connected, anthropogenic drivers and attend to inequitable impacts. A multidisciplinary approach that incorporates sociological insights is required to come up with effective and just ways forward that remedy rather than exacerbate inequalities and suffering.[2] Without multidisciplinary understanding, there is a risk that effort and resources will be devoted to perceived solutions that are unlikely to succeed.

This chapter provides an overview of sociological contributions to a better understanding of climate justice, with the aim of fostering greater collaboration across disciplines moving forward. Within the discipline of sociology, there is a rich tradition of scholarship related to inequality, equity, and justice. Sociologists incorporate these concepts when examining the social drivers, impacts, and potential ways to address climate change. Core areas of focus within sociology include inequality, power, groups, norms, social conflict, social cohesion, and social change. These are all central issues in climate justice. From causes to consequences to potential ways to respond, inequalities are inherent. In the context of climate change, "sociology is one of the major disciplines to foster understanding and protection of the livelihoods of local people."[3] In addition to important individual and community considerations, sociological research also examines social *systems* and "the ways that *social and structural inequality* play foundational roles in shaping environmental harms, environmental injustices, and potential solutions."[4] In reviewing these sociological contributions, we also cover some work from related disciplines inherently intertwined in sociological approaches, which provides additional support for our argument for the importance of disciplinary cross-pollination and collaboration to address the climate crisis and work toward climate justice.

SOCIOLOGY AND ENVIRONMENTAL SOCIOLOGY

Sociology is a social science that utilizes scientific methods to gain a better understanding of society as the object of study. The scope of society can be at many scales, from individuals in local groups to global society. Using both quantitative and qualitative methods, sociologists examine patterns and cases related to social phenomena such as social conflict, social cohesion, and social change.

Studies of climate justice often fall within the realm of environmental sociology, which focuses on relationships between environment and society. Environmental sociology is one of many subdisciplines within sociology, and aspects of climate justice overlap with many other areas of sociological research such as development, global political economy, community, rural sociology, urban sociology, race, gender, class, and collective behavior and social movements.

ENVIRONMENTAL JUSTICE

Sociologists claim some of the roots of environmental justice scholarship, especially the early work by sociologist Robert Bullard documenting racial disparities in exposures to toxic waste sites.[5] This work inspired the expansion of a broad literature within sociology that explores inequalities in exposure to environmental harms that certain communities and groups face that are shaped by race, class, gender, age, sexuality, queer identities, disability status, Indigeneity, ethnicity, nationality/citizenship, migration status, other social factors, and intersections between these factors.[6] The relationship between environmental harms and some of these inequalities have been examined more fully than others, and ongoing research explores these issues in different contexts and as they change, or not, over time.

Work in environmental justice documents disparities and increasingly explores intersectionality, attends to underlying drivers of inequality, and examines social movements and other efforts by entities such as communities and the state to oppose, or not, disproportionate exposures to environmental harms.[7] Scholars in sociology and activists studying environmental justice often discuss different types of justice, such as distributive justice, which looks at the fairness of the distribution of environmental goods and harms; procedural justice, which assesses equitable access to processes; and recognition justice, which recognizes group differences and attends to issues of dignity, livelihoods, and capabilities.[8] Increasingly, environmental justice scholarship in sociology incorporates considerations of nonhuman species and draws on or incorporates Indigenous understandings, including reciprocity, responsibility, and interconnectedness.[9] The role of governments in environmental justice is a contested topic. David Pellow contributes important work on critical environmental justice that expands the field to include political issues not often associated with environmental justice, and he puts forth a critical view of the state as a complicit source of environmental injustice rather than as an effective component of and target for environmental justice efforts.[10]

While the roots of environmental justice scholarship are usually traced to Bullard's work in the United States, other scholars studying regions outside the country draw attention to the global "environmentalism of the poor" and global perspectives on environmental justice.[11] The Environmental Justice Atlas (EJAtlas) is a multidisciplinary, dynamic, and growing inventory of environmental justice and climate justice conflicts around the world.[12] Joan Martínez-Alier et al. trace the increasing number of such conflicts around the world, cataloged in the EJAtlas, to the changing nature of the economy, and they link these to various concepts such as climate justice, climate debt, and ecologically unequal exchange, each of which we describe further below.[13]

CLIMATE JUSTICE

Climate justice "is the concern that the causes and consequences of climate change and the impacts of efforts to reduce the magnitude of climate change and adapt to it are inequitably distributed."[14] Sociologists focus on issues of inequality, power, groups, and norms, and environmental justice to position the discipline to provide useful insights on climate justice. As with environmental justice, climate justice scholars study how multiple aspects of inequality related to race, class, gender, age, sexuality, queer identities, disability status, Indigeneity, ethnicity, nationality/citizenship, migration status, others social factors, and intersections between these factors shape climate inequalities.[15] An example of sociologists examining climate justice through an intersectional lens includes studying gendered vulnerability and resilience in US Indigenous communities.[16] Increasingly sociologists examine human-animal relationships, nonhuman species, and multispecies justice as important components of climate justice.[17] Multispecies justice conceptualizations incorporate social justice and often build on Indigenous cultural systems based on reciprocal relationships with the natural environment, including plants, animals, and ecosystems.[18] Proponents of this approach argue that climate justice must take into consideration nonhuman/more-than-human species in order to truly be just, to challenge the problematic thinking that characterizes humans as separate from other species, and thus to more effectively address the climate crisis (for more discussion, see chapter 10 of this volume).[19]

Climate justice scholarship in sociology both builds on insights from environmental justice scholarship and is unique in that much sociological work on climate justice applies to the climate crisis systemic and structural political economic perspectives drawn from broader sociological and environmental sociology theory. In other words, climate justice scholars examine topics ranging from studies of communities around the world facing environmental justice issues related to climate change, including threats to livelihoods, to research on global climate negotiations, uneven and inequitable international flows of resources, and international inequalities in drivers and impacts of climate change.[20] Likewise, while environmental justice movements tend to grow from local, community-focused cases at local scales, climate justice movements span scales from the local to the global, from local decarbonization movements to the global climate youth movement sparked by Greta Thunberg and the Fridays for Future school strikes. Climate change is a global issue, and to be truly equitable, climate justice must also be global. Sociological research makes especially strong contributions that demonstrate the relationships between global inequalities and climate change: therefore, in the next sections we focus on sociological research that addresses the global scale of this issue.

Sociologists conducted early foundational research on climate justice, documenting global political economic inequalities in contributions to global climate change compared to exposure to its impacts, highlighting the inequalities and injustice in these relationships. This work is key to the now accepted understanding that those who contributed the least to climate change are likely to face the harshest climate change outcomes, an idea central to the concept of climate justice.[21] In their book, *A Climate of Injustice: Global Inequality, North-South Politics, and Climate Policy,* Timmons Roberts and Bradley Parks empirically examine how global inequality negatively affects trust and cooperation between nation-states on global climate agreements. They document early efforts for climate justice, noting, "The issue of reconciling social justice with environmental protection has surfaced at every major international meeting since the first environment and development conference at Stockholm in 1972."[22] The authors make the case that "responses to climate change are wound up with other social and economic issues facing nations and are fundamentally about inequality and injustice by bringing theorizing about international environmental politics and global political economy to bear on empirical analyses of climate inequality and international cooperation."[23] In later work, Roberts and Parks continue to document competing demands at global climate negotiations for emissions reductions versus development assistance to address inequalities in human well-being.[24] They also review the early literature on ecological debt and ecologically unequal exchange, two literatures that are important to sociological perspectives on climate justice.

CLIMATE DEBT

Climate debt demonstrates the importance of a multidisciplinary approach because this concept spans disciplines such as sociology, environmental politics, ecological economics, and human ecology. Rikard Warlenius defines climate debt as "the idea that climate change is caused by rich people while mainly harming people that are poor, and therefore, the former should take the burden of mitigation and adaptation costs" and says this "is at the very core of climate justice."[25] The concept of climate debt draws on the international scale of "disparities of responsibility and impact" in the climate crisis.[26] Put another way, climate debt is the idea that developed/Global North countries owe a debt to less developed countries of the Global South for their contributions to climate change in terms of emissions, but also because of the environmental and social damage embodied in uneven energy intensive trade and material-intensive consumption more broadly.[27] Warlenius makes distinctions between earlier literature on ecological debt, which called for international development loan forgiveness, and climate debt, which is a violation of more fundamental rights and supports calls for climate justice involving restorative justice, where what has been taken will be returned as an effort toward decolonization. He goes on to make the point that while on the surface this is basic ethics, asking those who made the mess to clean it up, it is actually a radical notion because of the "magnitude of the mess and the skewness of responsibility."[28] This understanding helps explain both the pushback at the international scale in terms of cooperation and the inability to make sufficient progress at reducing climate change causing emissions, and yet this weakness is also a strength in that it calls for "radical redistributive measures with a strong legitimacy."[29] In the next section we discuss sociological perspectives on obstruction and lack of progress on climate change, and reform versus system change

ECOLOGICAL UNEQUAL EXCHANGE AND ENVIRONMENTAL LOAD DISPLACEMENT

Ecologically unequal exchange is a global political economic environmental theory that posits that inequitable trade relationships between nations shape the uneven distribution of environmental harms and human well-being. A shorthand way of thinking of this is that more developed countries use less developed countries as a tap for resources and a sink for wastes.[30] Ecologically unequal exchange is also multidisciplinary, with roots in environmental sociology, ecological economics, and world-systems theory. An important tenet of world systems theory that relates to climate justice is that national development needs to be contextualized with an understanding of a nation-state's integration into a global system characterized by inequality in terms of economic and military power.[31] Trade relations are ecologically unbalanced and unfair because poorer nations export undervalued raw materials and products and the social and environmental costs of the extraction and production are not accounted for in the pricing. In other words, poorer countries are sacrificing environmental and social well-being for perceived national economic development.[32] While this perspective focuses on the national scale, comparable inequalities exist at subnational scales.

Empirical evidence of ecologically unequal exchange calls into question development recommendations from entities like the World Bank and World Trade Organization by demonstrating that trade is not necessarily improving development and human well-being and is negatively affecting the environment in countries of the Global South engaged in export-oriented development and trade with the Global North. This uneven trade from the Global South to the Global North, termed the vertical flow of exports, may improve GDP in developing nations, but GDP growth does not accurately capture societal development or well-being.[33] It also calls into question the idea that wealthier countries are dematerializing; rather, empirical evidence demonstrates they are shifting the burden of their consumption elsewhere via trade, in a mechanism known as environmental load displacement.[34]

The consumption/degradation paradox that results from these structural relationships, wherein some nations, locations, and groups are removed from the impacts of their actions, can relegate calls for climate justice to those who are most affected, while others who are less affected do not have increased concern or an urgency to act.[35] Informed by a global perspective of ecologically unequal exchange, examinations of teleconnections, environmental and socioeconomic interactions across space and time and across the extraction, production, consumption, and disposal commodity chain, each stage of which produces greenhouse gas emissions and other environmental harms, illuminate how the unequal global political economic system continues to structure inequality, the climate crisis, and climate injustice.

INEQUALITIES IN RESPONSIBILITY AND WELL-BEING

Scholars highlight the massive inequalities in responsibility for climate change. "With only four percent of the world's population, the US is responsible for over 20 percent of all global emissions. That can be compared to 136 developing countries that together are only responsible for 24 percent of global emissions."[36] "The fact that 3 billion people use less energy on an annual per capita basis than a standard American refrigerator gives you an idea of how far away from global equity and climate justice we currently are."[37] These memorable examples point to vast inequity in responsibility for climate change.

Sociologists have also examined how these emissions have corresponded to development that has a relationship with the creation of human well-being, which also influences unequal effects of climate change across different groups and national populations. Sociological literature on the carbon intensity of well-being quantifies how carbon intensely nation-states and other entities produce well-being for their citizens. This measure quantifies the relationships between human well-being and the stress humans, in the pursuit of well-being, often put on the environment.[38] It highlights the disproportionate contribution some nations and groups have made to climate change, and the disparities in well-being outcomes, another component of climate justice.[39] The carbon intensity of well-being is a measure that helps scholars document, conceptualize, and examine global inequalities in drivers and outcomes, informing both climate mitigation and adaptation efforts and providing empirical evidence for climate justice efforts.[40] Research on both production- and consumption-based carbon intensity of well-being also enables examinations of ecologically unequal exchange and environmental load displacement.[41] The indicator is applicable across scales when data are available.[42]

Varying levels of development and carbon intensity of well-being are important to examine globally. Renowned climate scholar Saleemul Huq elaborates on this in the context of his home country, Bangladesh, providing an important perspective from the Global South. "We need to think about global injustice now. The manifest injustice whereby polluters—largely rich people around the world—are hurting poor people. This disparity in consequences is a global phenomenon."[43] Huq provides an important perspective on climate adaptation and resilience, key to better understandings of climate justice:

> The communities impacted by environmental degradation and climate change are overwhelmingly poor people of colour, even in rich countries such as the United States. . . . Bangladesh, my home, is facing a slow-moving disaster as sea-level rise threatens the coasts, which may lead to the displacement of millions of people. . . . But the story of Bangladesh is not a story of victims. It is a story of heroes, a story of the future of the planet. The rest of the planet is going to face tomorrow what we are facing today, and the rest of the planet is going to have to come and learn from us how to deal with this problem. . . . What is truly important in times of crisis is social cohesion—people helping each other—and we have that in droves in Bangladesh.[44]

Huq's words illustrate multiple things, including the importance of analysis of quantitative global and national patterns, attention to local contexts embedded in the global system, qualitative examinations of specific cultural contexts, and the importance of attending to and elevating diverse voices from the Global South for a better understanding of the complexities of climate justice and a more just approach. Specifically, Huq suggests that scholars across disciplines can examine Bangladesh for insights into how social cohesiveness and social networks, also known as social capital, play key roles in fostering resilience in crisis circumstances, particularly in the context of the climate catastrophe.

SOCIAL DRIVERS OF THE CLIMATE CRISIS

Understanding the drivers of climate change, including inequalities inherent in those drivers and inequality as a driver itself, is imperative if we want to work toward climate justice and effectively address climate change. "Understanding why individuals, organizations, and nations take actions that influence the climate is in some sense prior to all other issues related to climate change."[45] It is vital to accurately understand drivers in all their complexity in order to design effective ways to address them, and to do this, a multidisciplinary approach is needed. Otherwise, there is a risk that great amounts of effort and resources will be devoted to perceived solutions that, due to a lack of interdisciplinary understanding, misunderstand the true drivers of the issues. These will only delay effective actions, leading us further into climate crisis.

Anthropogenic drivers are the human actions, social systems, and other social factors that create and shape the production of climate change–causing emissions.[46] Within environmental sociology, research on ecologically unequal exchange is part of the literature on anthropogenic drivers; other drivers include economic growth, affluence, income and wealth inequality, population, urbanization, corporations, and militarization.[47]

There are key debates around some of these drivers. One central debate concerns economic growth; sociological research on this topic helps us examine whether it is possible for an economy to have continued growth without putting sustainability and survival at risk by surpassing carrying capacity versus the need for degrowth.[48] Another characterization of this important question is whether continued development is possible or if it will necessarily result in unsustainable environmental degradation.[49] A related debate is whether increasing affluence is associated with increased emissions, which the preponderance of empirical evidence suggests is the case.[50] This research builds on early work examining how environmental impact is shaped by the combination of population, affluence, and technology.[51] In a key example of later work examining and finding support for the detrimental environmental effects of affluence and economic growth on increasing emissions, Andrew Jorgenson and Brett Clark analyze the relationship between GDP per capita and total emissions, per capita emissions, and emissions per unit of GDP from 1960 to 2005 across developed and developing nations.[52]

Research finds inequality itself can be a climate change driver, as are corporations, and components of militarization.[53] In terms of inequality, across multiple scales and contexts, higher inequality is associated with higher emissions, indicating inequality reduction may be a sustainability strategy.[54] Mechanisms by which inequality may drive climate change include consumption dynamics, use of power and influence by those that benefit from the current system, and capitalist system dynamics.[55] For example, some sociologists argue that the key driver of the climate crisis and climate injustice is the capitalist system itself, with the continuous pursuit of profit and with inequality an inherent component of this political-economic system.[56] In terms of corporate emissions, these are also characterized by disproportionality, where a comparatively small number of corporations contribute a disproportionate amount to the climate crisis; this too suggests a sustainability strategy: targeting the super polluters.[57] In terms of militarization, factors such as the size and capital intensiveness of nations' militaries are important for understanding carbon emissions, because militaries are carbon intense even when not engaged in active conflict, for example with the ongoing production, maintenance, and use of carbon-intensive planes, ships, vehicles, and other military materials and supplies and other routine activities.[58] A sociological perspective reveals the multiple ways in which military conflict, in addition to being catastrophic and inhumane, is not good for the climate; several recent news articles have reported copious carbon emissions contributed by current conflicts, such as Israel's war in Gaza and Russia's war in Ukraine. Thus, reducing militarization and active conflict could be both a justice and a sustainability strategy (for more discussion of this topic, see chapter 3 in this volume).

The literature on drivers makes a vital contribution to climate justice by enhancing our understanding of the complex social drivers of climate change, inequalities within those drivers, and potential strategies to address the climate crisis and climate justice. Since the climate crisis is anthropogenic, or human caused, we will need ways to address the crisis and its injustices that rely on accurate understandings of both social drivers and social constraints that may limit potential solutions.

SOCIAL CONSTRAINTS AND OBSTRUCTION

Without multidisciplinary understanding, there is also a risk that effort and resources will be devoted to perceived solutions that are unlikely to succeed because of structural factors and other forces of obstruction. Some proposed climate solutions lack an understanding of the realities of context, especially the context of the current global political economic system and related political economic motives at smaller scales. This highlights the importance of a sociological understanding of the drivers of climate change, especially the structural and systemic drivers inherent in our current global system, characterized by both inequality and dominant expectations for continuous economic growth. Timmons Roberts et al. write, "Understanding social and political barriers to switching from high-carbon to lower-carbon modes of production and consumption, and ways to overcome them, will be fundamental." They go on to say it is "feasible to achieve low carbon emissions and high human well-being for all nations. But the barriers are substantial, and include addressing existing vested interests of economic sectors, technological lock-in, assumptions embedded in culture, and political structures. Unfortunately, these areas are currently the weakest link in the existing research and policy chain."[59]

Within the sociological literature, there is also work illuminating active efforts to obscure realities about climate change in order to maintain the current system, which, however socially damaging and ecologically unsustainable overall, benefits some groups in the short term.[60] For example, Justin Farrell notes that much study of climate denial has focused on individual level beliefs, whereas sociological literature demonstrates the importance of the "institutional and corporate structure of the climate change counter-movement." Farrell utilizes "machine-learning text analysis to show the movement's influence in the news media and bureaucratic politics."[61] Brulle's research demonstrates the importance of funding, lobbying, and coalitions to climate change countermovement organizations.[62] William Lamb et al. examine the discourses of climate delay that justify inaction.[63] Edwards et al. call for more attention to dynamics of obstruction in the Global South.[64] These scholars utilize a sociological, structural approach to illuminate efforts to obstruct progress on climate change, with clear justice implications; there is also work on efforts aiming to counteract such obstruction.[65]

REFORM VERSUS TRANSFORM

Climate justice solutions involve the concept of just transitions. This concept has roots in the US and international labor movements and includes movement toward a low carbon economy, climate resilience, and a system that is equitable and inclusive. In the most effective versions of this perspective, addressing justice is also a way to address key drivers of climate change. Along these lines, David Ciplet utilizes sociological theoretical perspectives about power and social change, including insights from Antonio Gramsci and Karl Polanyi, to analyze the potential of calls for just transitions to be truly transformative "in shifting the political economic structures that cause, sustain, and deepen injustices."[66] Ciplet develops a typology ranging from those efforts that maintain the status quo to those that could be transformative, but he notes that transformation potential depends on building diverse coalitions that are difficult to form and sustain, especially in the face of structural obstacles and opposition from dominant hegemonic coalitions that preserve the status quo. Along these lines, Theo LeQuesne examines one social movement's potential for leading to just transition, specifically looking at an Indigenous-led environmental justice movement in the United States.[67] These are two examples that examine potential for transformation in various contexts in the United States.

At the international negotiations scale, the establishment of a loss and damage fund at the United Nations Framework on Climate Change Conference of Parties (COP 27) aims to address international inequities by providing financial assistance to countries most affected by or at risk of being affected by climate change. Danielle Falzon and Pinar Batur see the possibility of this mechanism for holding nations accountable for their destructive pasts, including racist colonialism, if appropriately implemented and funded. They argue that damage funding and climate finance will need to carefully navigate assessments of vulnerability while recognizing the tremendous debt that is owed.[68]

There are competing perspectives on the potential for transformative change within the current capitalist political economic system. Research by scholars such as David Pellow and Jill Harrison call into question the ability of the state under the current system to be an ally in environmental justice efforts, and rather see the state as complicit in environmental injustice or ineffective at moving toward justice.[69] David Purucker, on the other hand, pushes back on this characterization, coming up with an alternative theoretical perspective on the state that "allows for the possibility of action both against and within states."[70] Scholars such as John Foster see the capitalist economic system as the key driver of climate change and climate injustice, and thus system transformation as the only option to address the climate crisis and climate injustice.[71] Other scholars (e.g., Tschakert) point to cultural reforms, or transformations, such as multispecies justice approaches that would require systemic cultural change.[72] Regardless, because of its disciplinary focus on inequality and its tradition of examining social structure, sociological research is well suited to identifying alternatives to our modern inequitable and unsustainable world.[73] The discipline has the theoretical tools and methods to question and examine potential for system reform versus system change.[74]

CONCLUSION: MULTIDISCIPLINARY APPROACHES

There is a growing sociology of climate change. A number of sociologists call on other sociologists to make climate change more central to the discipline of sociology as this crisis is becoming an increasingly important component of the study of society.[75] Such scholars point to the "sociological naivety of IPCC assessments and many of the policies they subsequently inform" and how the consensus process of the IPCC struggles to incorporate controversial topics.[76] Likewise, scholars note that international climate talks have avoided addressing issues of equity, and they argue for the incorporation of equity and climate justice into climate negotiations, for which the discipline of sociology is particularly well suited.[77]

This chapter is one attempt to demonstrate some of what the sociological perspective has to offer to work on climate justice, especially in order to emphasize what sociology has to contribute to multidisciplinary approaches. As climate change is a global problem characterized by global inequality, we focus on some of the global approaches in sociology on climate justice. As noted above, this sociological work also relies on research from outside of the social sciences. For example, carbon emissions data are created by multidisciplinary efforts; these data are necessary to examine relationships between environment and society, also called socio-ecological relationships, socio-ecological systems, coupled human and natural systems, or related terms.[78]

Sociology is well situated with its emphasis on inequality to contribute to literature on climate justice. Social change and social cohesion, concepts sociologists often examine, are crucial for concerted efforts to combat climate change; however, the importance of interdisciplinary and multidisciplinary work cannot be overstated. Climate change and climate justice are complex, interrelated problems that will not be adequately addressed with simple solutions from one discipline. In addition, within sociology there are multiple approaches, from qualitative research to work that engages community voices in coproduction of knowledge, that should also be incorporated to address issues of climate justice moving forward, that were outside the scope of this review.[79] Furthermore, outside of academia there are activists and frontline communities engaged in environmental and climate justice work. Engaging with various communities, methods, and disciplines offers ways to better incorporate voices, perceptions, and needs of diverse communities in the face of climate change. "To realize its potential to contribute to the societal discourse on climate change, sociology must continue to be theoretically integrated, to engage with other disciplines, and to remain concerned with issues related to environmental and climate inequalities."[80]

In her book, Greta Thunberg writes, "The transformation we need . . . may not be politically possible today. But we are the ones who determine what will be politically possible tomorrow."[81] She argues for individual change, but also individual change in groups, acting collectively, changing norms, addressing inequalities, and speaking truth to power to work toward system change. This directly aligns with sociology's focus on inequality, power, groups, and norms. It is up to sociologists to focus on climate change and to reach across disciplines to contribute a sociological perspective, and for other disciplines to reach back in cooperation to address the climate crisis and work toward climate justice now.

Notes

1. Thunberg, *The Climate Book*, 2–3.
2. Dietz et al., "Climate Change and Society."
3. Hossen, "Decolonizing Sociology for Social Justice in Bangladesh."
4. Caniglia et al., "Introduction: A Twenty-First Century Public Environmental Sociology," 4.
5. Bullard, "Solid Waste Sites"; Bullard, *Dumping in Dixie*.
6. Dietz et al., "Climate Change and Society."
7. Mohai et al., "Environmental Justice"; Agyeman et al., "Trends and Directions in Environmental Justice"; Pellow, *Resisting Global Toxics*; Malin and Ryder, "Environmental Justice Scholarship"; Harrison, *From the Inside Out*; Malin et al., "Sites of Resistance, Acceptance, and Quiescence."
8. Schlosberg, "Theorising Environmental Justice. See also chapter 2 in this volume.
9. Kimmerer, *Braiding Sweetgrass: Indigenous Wisdom*; Schlosberg, "Theorising Environmental Justice"; Whyte, "Indigenous Environmental Justice."
10. Pellow, *What Is Critical Environmental Justice?* See also Purucker, "Critical Environmental Justice and the State."
11. See, for example, Martínez-Alier et al., "Is There a Global Environmental Justice Movement?"; Anguelovski and Martínez-Alier, "The 'Environmentalism of the Poor' Revisited"; Pellow, *Resisting Global Toxics*; Temper et al., "Mapping the

Frontiers and Front Lines"; Temper et al., "The Global Environmental Justice Atlas (EJAtlas)"; Givens et al., "Ecologically Unequal Exchange."

12. Environmental Justice Atlas (EJAtlas), https://ejatlas.org.
13. Martínez-Alier et al., "Is There a Global Environmental Justice Movement?"
14. Dietz et al., "Climate Change and Society," 144.
15. Dietz et al.
16. Vinyeta et al., *Climate Change Through an Intersectional Lens*.
17. Dietz et al., "Climate Change and Society"; Guenther, "An Invitation to Bring Animals"; Taylor et al., "A Sociology of Multi-Species Relations."
18. Vinyeta et al., *Climate Change Through an Intersectional Lens*.
19. Tschakert et al., "Multispecies Justice," e699.
20. Mohai et al., "Environmental Justice."
21. Roberts and Parks, *A Climate of Injustice*; Roberts and Parks, "Ecologically Unequal Exchange."
22. Roberts and Parks, *A Climate of Injustice*, 2.
23. Roberts and Parks, "Ecologically Unequal Exchange," 4–5.
24. Roberts and Parks.
25. Warlenius, "Decolonizing the Atmosphere," 132.
26. Roberts and Parks, "Ecologically Unequal Exchange," 393.
27. Roberts and Parks; Muradian et al., "Embodied Pollution in Trade"; Martinez-Alier, *The Environmentalism of the Poor*.
28. Warlenius, "Decolonizing the Atmosphere," 151.
29. Warlenius.
30. See Givens et al., "Ecologically Unequal Exchange"; Givens and Huang, "Ecologically Unequal Exchange and Environmental Load Displacement"; Givens and Huang, "Ecologically Unequal Exchange"; Hornborg and Martinez-Alier, "Introduction: Ecologically Unequal Exchange"; Jorgenson, "The Sociology of Ecologically Unequal Exchange"; Frey et al., eds, *Ecologically Unequal Exchange*.
31. Bunker, "Modes of Extraction, Unequal Exchange"; Bunker, *Underdeveloping the Amazon*; Wallerstein, *The Modern World-System I*; Chase-Dunn, *Global Formation: Structures of the World-Economy*; Chase-Dunn and Grimes, "World-Systems Analysis."
32. Hornborg, "Towards an Ecological Theory of Unequal Exchange"; Hornborg, "Zero-Sum World"; Jorgenson, "Unequal Ecological Exchange and Environmental Degradation"; Jorgenson, "The Sociology of Ecologically Unequal Exchange"; Muradian and Martinez-Alier, "South-North Materials Flow"; Muradian and Martinez-Alier, "Trade and the Environment"; Muradian et al., "Embodied Pollution in Trade"; Rice, "Ecological Unequal Exchange: International Trade"; Rice, "Material Consumption and Social Well-Being"; Roberts and Parks, "Ecologically Unequal Exchange."
33. Jorgenson, "The Sociology of Unequal Exchange"; Jorgenson, "Carbon Dioxide Emissions"; Jorgenson, "The Sociology of Ecologically Unequal Exchange"; Roberts and Parks, "Ecologically Unequal Exchange."

34. Hornborg, "Towards an Ecological Theory of Unequal Exchange"; Hornborg, "Zero-Sum World"; Muradian and Martinez-Alier, "South-North Materials Flow"; Muradian and Martinez-Alier, "Trade and the Environment"; Muradian et al., "Embodied Pollution in Trade"; for reviews, see Givens et al., "Ecologically Unequal Exchange: A Theory of Global Environmental Injustice"; Givens and Huang, "Ecologically Unequal Exchange and Environmental Load Displacement"; Givens and Huang, "Ecologically Unequal Exchange"; Jorgenson, "The Sociology of Ecologically Unequal Exchange."

35. Jorgenson, "Consumption and Environmental Degradaion"; Jorgenson et al., "Ecologically Unequal Exchange and the Resource Consumption"; Givens and Jorgenson, "The Effects of Affluence"; Muradian and Martinez-Alier, "Trade and the Environment."

36. Roberts and Parks, "A Climate of Injustice"; Roberts and Parks, "Ecologically Unequal Exchange."

37. Thunberg, *The Climate Book*, 154.

38. Dietz and Jorgenson, "Towards a New View of Sustainable Development"; Jorgenson, "Economic Development and Carbon Intensity"; Jorgenson, "Inequality and Carbon Intensity."

39. Lamb et al., "Transitions in Pathways of Human Development"; Roberts et al., "Four Agendas for Research and Policy," e3. For a review, see Givens and Briscoe, "Carbon Intensity of Well-Being."

40. Roberts et al., "Four Agendas for Research and Policy."

41. Jorgenson and Givens, "The Changing Effect of Economic Development"; Givens, "Ecologically Unequal Exchange"; Givens, "World Society, World Polity."

42. Jorgenson et al., "Inequality, Poverty, and Carbon Intensity"; Briscoe et al., "Intersectional Indicators."

43. Huq, "Life at 1.1 Degrees C," 161.

44. Huq, 159–60.

45. Dietz et al., "Climate Change and Society," 137.

46. Dietz et al.; Jorgenson et al., "Social Science Perspectives on Drivers," e554; Jorgenson et al., "Advances in Research on Anthropogenic Drivers."

47. For a multidisciplinary review, see Jorgenson et al., "Social Science Perspectives on Drivers."

48. Jorgenson et al., "Social Science Perspectives on Drivers"; Kallis, "In Defence of Degrowth"; Kallis et al., "Research on Degrowth."

49. For a review of opposing theoretical perspectives and a new theoretical proposition, see Fisher and Jorgenson, "Ending the Stalemate."

50. Dietz et al., "Climate Change and Society."

51. For example, see York et al., "STIRPAT, IPAT and ImPACT."

52. Jorgenson and Clark, "Are the Economy and the Environment Decoupling?"

53. Jorgenson, "Social Science Perspectives"; Jorgenson et al., "Advances in Research on Anthropogenic Drivers."

54. Jorgenson, "Social Science Perspectives"; Givens et al., "Inequality, Emissions, and Human Well-Being."

55. Dietz et al., "Climate Change and Society"; Foster, "Marx's Theory of Metabolic Rift"; Piketty, *Capital in the Twenty-First Century*; Schor and White, *Plenitude: The New Economics of True Wealth*; Schor, "Climate, Inequality, and the Need for Reframing"; for a review, see Givens et al., "Inequality, Emissions, and Human Well-Being."

56. Foster, *Capitalism in the Anthropocene*; Foster et al., *The Ecological Rift*.

57. Grant et al., *Super Polluters*.

58. Hooks et al., "Recasting the Treadmills of Production"; Jorgenson et al., "Guns Versus Climate."

59. Roberts et al., "Four Agendas for Research and Policy."

60. Brulle, "Institutionalizing Delay"; Brulle, "Networks of Opposition," 603–4; Farrell, "Corporate Funding and Ideological Polarization"; Farrell, "Network Structure and Influence"; Roberts et al., "Informing Strategic Climate Action."

61. Farrell, "Network Structure and Influence," 370.

62. Brulle, "Institutionalizing Delay"; Brulle, "Networks of Opposition."

63. Lamb et al., "Transitions in Pathways of Human Development."

64. Edwards et al., "Climate Obstruction in the Global South."

65. Roberts et al., "Informing Strategic Climate Action."

66. Ciplet, "Transition Coalitions," 315.

67. LeQuesne, "Petro-Hegemony and the Matrix of Resistance." LeQuesne draws insights from the Standing Rock Water Protectors' uprising against the Dakota Access Pipeline, Patricia Hill Collins's insights on intersectionality and the matrix of domination, and Gramsci's theory of hegemony to understand petro-hegemony and the potential for a matrix of resistance that incorporates intersectionality, collective liberation, and a balance between universalism and particularism to counter petro-capitalism's hegemony.

68. Falzon and Batur, "Lost and Damaged." See also Roberts, "Calculating What We Owe"; Roberts et al., "Rebooting a Failed Promise"; Robinson et al., "Vulnerability-Based Allocations."

69. Pellow, *What Is Critical Environmental Justice?*; Harrison, *From the Inside Out*.

70. David Purucker, "Critical Environmental Justice and the State," 176.

71. Foster, *Capitalism in the Anthropocene*; Foster, "Marx's Theory of Metabolic Rift"; Foster et al., *The Ecological Rift*.

72. Tschakert et al., "Multispecies Justice."

73. Klinenberg et al., "Sociology and the Climate Crisis."

74. Dietz et al., "Climate Change and Society."

75. Klinenberg et al., "Sociology and the Climate Crisis."

76. Lockie, "Mainstreaming Climate Change Sociology," 1.

77. Falzon et al., "Sociology and Climate Change"; Sonja Klinsky et al., "Why Equity Is Fundamental."

78. See, for example, Caniglia and Mayer, "Socio-Ecological Systems"; Liu et al., "Complexity of Coupled Human and Natural Systems."
79. For examples, see Lockie, "Mainstreaming Climate Change Sociology."
80. Dietz et al., "Climate Change and Society"; see also chapter 15 in this volume.
81. Thunberg, *The Climate Book*, 421.

BIBLIOGRAPHY

Agyeman, Julian, David Schlosberg, Luke Craven, and Caitlin Matthews. "Trends and Directions in Environmental Justice: From Inequity to Everyday Life, Community, and Just Sustainabilities." *Annual Review of Environment and Resources* 41 (2016): 321–40.
Anguelovski, Isabelle, and Joan Martínez-Alier. "The 'Environmentalism of the Poor' Revisited: Territory and Place in Disconnected Glocal Struggles." *Ecological Economics* 102 (2014): 167–76.
Berger, Roni. "Now I See It, Now I Don't: Researcher's Position and Reflexivity in Qualitative Research." *Qualitative Research* 15, no. 2 (2015): 219–34.
Bourke, Brian. "Positionality: Reflecting on the Research Process." *The Qualitative Report* 19, no. 33 (2014): 1–9.
Briscoe, Michael D., Jennifer E. Givens, and Madeleine Alder. "Intersectional Indicators: A Race and Sex-Specific Analysis of the Carbon Intensity of Well-Being in the United States, 1998–2009." *Social Indicators Research* 155 (2021): 97–116.
Brulle, Robert J. "Institutionalizing Delay: Foundation Funding and the Creation of US Climate Change Counter-Movement Organizations." *Climatic Change* 122 (2014): 681–94.
Brulle, Robert J. "Networks of Opposition: A Structural Analysis of US Climate Change Countermovement Coalitions 1989–2015." *Sociological Inquiry* 91, no. 3 (2021): 603–24.
Bullard, Robert D. *Dumping in Dixie: Race, Class, and Environmental Quality.* Boulder: Westview Press, 1990.
Bullard, Robert D. "Solid Waste Sites and the Black Houston Community." *Sociological Inquiry* 53, no. 2–3 (1983): 273–88.
Bunker, Stephen G. "Modes of Extraction, Unequal Exchange, and the Progressive Underdevelopment of an Extreme Periphery: The Brazilian Amazon, 1600–1980." *American Journal of Sociology* 89, no. 5 (1984): 1017–64.
Bunker, Stephen G. *Underdeveloping the Amazon: Extraction, Unequal Exchange, and the Failure of the Modern State.* Chicago: University of Chicago Press, 1985.
Caniglia, Beth Schaefer, Andrew Jorgenson, Stephanie A. Malin, Lori Peek, and David N. Pellow. "Introduction: A Twenty-First Century Public Environmental Sociology." *Handbook of Environmental Sociology* (2021): 1–11.
Caniglia, Beth Schaefer, and Brian Mayer. "Socio-Ecological Systems." *Handbook of Environmental Sociology* (2021): 517–36.
Chase-Dunn, Christopher. *Global Formation: Structures of the World-Economy.* Rowman & Littlefield, 1998.

Chase-Dunn, Christopher, and Peter Grimes. "World-Systems Analysis." *Annual Review of Sociology* 21, no. 1 (1995): 387–417.

Ciplet, David. "Transition Coalitions: Toward a Theory of Transformative Just Transitions." *Environmental Sociology* 8, no. 3 (2022): 315–30.

Dietz, Thomas, and Andrew K. Jorgenson. "Towards a New View of Sustainable Development: Human Well-Being and Environmental Stress." *Environmental Research Letters* 9, no. 3 (2014): 031001.

Dietz, Thomas, Rachael L. Shwom, and Cameron T. Whitley. "Climate Change and Society." *Annual Review of Sociology* 46 (2020): 135–58.

Edwards, Guy, Paul K. Gellert, Omar Faruque et al. "Climate Obstruction in the Global South: Future Research Trajectories." *PLOS Climate* 2, no. 7 (2023): e0000241.

Falzon, Danielle, and Pinar Batur. "Lost and Damaged: Environmental Racism, Climate Justice, and Conflict in the Pacific." *Handbook of the Sociology of Racial and Ethnic Relations* (2018): 401–12.

Falzon, Danielle, J. Timmons Roberts, and Robert J. Brulle. "Sociology and Climate Change: A Review and Research Agenda." *Handbook of Environmental Sociology* (2021): 189–217.

Farrell, Justin. "Corporate Funding and Ideological Polarization about Climate Change." *Proceedings of the National Academy of Sciences* 113, no. 1 (2016): 92–97.

Farrell, Justin. "Network Structure and Influence of the Climate Change Counter-Movement." *Nature Climate Change* 6, no. 4 (2016): 370–74.

Fisher, Dana R., and Andrew K. Jorgenson. "Ending the Stalemate: Toward a Theory of Anthro-Shift." *Sociological Theory* 37, no. 4 (2019): 342–62.

Foster, John Bellamy. *Capitalism in the Anthropocene: Ecological Ruin or Ecological Revolution*. New York: NYU Press, 2022.

Foster, John Bellamy. "Marx's Theory of Metabolic Rift: Classical Foundations for Environmental Sociology." *American Journal of Sociology* 105, no. 2 (1999): 366–405.

Foster, John Bellamy, Brett Clark, and Richard York. *The Ecological Rift: Capitalism's War on the Earth*. New York: NYU Press, 2011.

Frey, R. Scott, Paul K. Gellert, and Harry F. Dahms, eds. *Ecologically Unequal Exchange: Environmental Injustice in Comparative and Historical Perspective*. Springer, 2018.

Givens, Jennifer E. "Ecologically Unequal Exchange and the Carbon Intensity of Well-Being, 1990–2011." *Environmental Sociology* 4, no. 3 (2018): 311–24.

Givens, Jennifer E. "World Society, World Polity, and the Carbon Intensity of Well-Being, 1990–2011." *Sociology of Development* 3, no. 4 (2017): 403–35.

Givens, Jennifer E., and Michael Briscoe. "Carbon Intensity of Well-Being." *Elgar Encyclopedia of Environmental Sociology* (2024).

Givens, Jennifer E., and Xiaorui Huang. "Ecologically Unequal Exchange and Environmental Load Displacement: Global Perspectives on Structural Inequalities and the Environment." *Handbook of Environmental Sociology* (2021): 53–70.

Givens, Jennifer E., and Xiaorui Huang. "Ecologically Unequal Exchange." *Elgar Encyclopedia of Environmental Sociology* (2024).

Givens, Jennifer E., Xiaorui Huang, and Andrew K. Jorgenson. "Ecologically Unequal Exchange: A Theory of Global Environmental Injustice." *Sociology Compass* 13, no. 5 (2019): e12693.

Givens, Jennifer E., and Andrew K. Jorgenson. "The Effects of Affluence, Economic Development, and Environmental Degradation on Environmental Concern: A Multilevel Analysis." *Organization & Environment* 24, no. 1 (2011): 74–91.

Givens, Jennifer E., Orla M. Kelly, and Andrew K. Jorgenson. "Inequality, Emissions, and Human Well-Being." *Handbook on Inequality and the Environment.* Edward Elgar Publishing, 2023, 308–24.

Grant, Don, Andrew Jorgenson, and Wesley Longhofer. *Super Polluters: Tackling the World's Largest Sites of Climate-Disrupting Emissions.* New York: Columbia University Press, 2020.

Guenther, Katja M. "An Invitation to Bring Animals Into Feminist and Queer Sociology." *Sociology Compass* 18, no. 4 (2024): e13198.

Harrison, Jill Lindsey. *From the Inside Out: The Fight for Environmental Justice Within Government Agencies.* Cambridge: MIT Press, 2019.

Hooks, Gregory, Michael Lengefeld, and Chad L. Smith. "Recasting the Treadmills of Production and Destruction: New Theoretical Directions." *Sociology of Development* 7, no. 1 (2021): 52–76.

Hornborg, Alf. "Towards an Ecological Theory of Unequal Exchange: Articulating World System Theory and Ecological Economics." *Ecological Economics* 25, no. 1 (1998): 127–36.

Hornborg, Alf. "Zero-Sum World: Challenges in Conceptualizing Environmental Load Displacement and Ecologically Unequal Exchange in the World-System." *International Journal of Comparative Sociology* 50, no. 3–4 (2009): 237–62.

Hornborg, Alf, and Joan Martinez-Alier. "Introduction: Ecologically Unequal Exchange and Ecological Debt." *Journal of Political Ecology* 23, no. 1 (2016): 328–33.

Hossen, M. Anwar. "Decolonizing Sociology for Social Justice in Bangladesh: Delta Scholarship Matters." *Critical Sociology* 49, no. 3 (2023): 545–61.

Huq, Saleemul. "Life at 1.1 Degrees C." In *The Climate Book: The Facts and the Solutions,* by Greta Thunberg. New York: Penguin, 2023, 158–61.

Jafar, Anisa JN. "What Is Positionality and Should It Be Expressed in Quantitative Studies?." *Emergency Medicine Journal* 35, no. 5 (2018): 323–24.

Japan Times. "Ukraine War Responsible for 150 million Tons of CO2 Emissions." December 5, 2023. https://www.japantimes.co.jp/environment/2023/12/05/climate-change/ukraine-war-co2-emissions/.

Jorgenson, Andrew K. "Carbon Dioxide Emissions in Central and Eastern European Nations, 1992–2005: A Test of Ecologically Unequal Exchange Theory." *Human Ecology Review* (2011): 105–14.

Jorgenson, Andrew K. "Consumption and Environmental Degradation: A Cross-National Analysis of the Ecological Footprint." *Social Problems* 50, no. 3 (2003): 374–94.

Jorgenson, Andrew K. "Economic Development and the Carbon Intensity of Human Well-Being." *Nature Climate Change* 4, no. 3 (2014): 186–89.

Jorgenson, Andrew K. "Inequality and the Carbon Intensity of Human Well-Being." *Journal of Environmental Studies and Sciences* 5 (2015): 277–82.

Jorgenson, Andrew K. "The Sociology of Ecologically Unequal Exchange and Carbon Dioxide Emissions, 1960–2005." *Social Science Research* 41, no. 2 (2012): 242–52.

Jorgenson, Andrew K. "The Sociology of Ecologically Unequal Exchange, Foreign Investment Dependence and Environmental Load Displacement: Summary of the Literature and Implications for Sustainability." *Journal of Political Ecology* 23, no. 1 (2016): 334–49.

Jorgenson, Andrew K. "The Sociology of Unequal Exchange in Ecological Context: A Panel Study of Lower-Income Countries, 1975–2000." *Sociological Forum* 24, no. 1 (2009): 22–46.

Jorgenson, Andrew K. "Unequal Ecological Exchange and Environmental Degradation: A Theoretical Proposition and Cross-National Study of Deforestation, 1990–2000." *Rural Sociology* 71, no. 4 (2006): 685–712.

Jorgenson, Andrew K., Kelly Austin, and Christopher Dick. "Ecologically Unequal Exchange and the Resource Consumption/Environmental Degradation Paradox: A Panel Study of Less-Developed Countries, 1970–2000." *International Journal of Comparative Sociology* 50, no. 3–4 (2009): 263–84.

Jorgenson, Andrew K., and Brett Clark. "Are the Economy and the Environment Decoupling? A Comparative International Study, 1960–2005." *American Journal of Sociology* 118, no. 1 (2012): 1–44.

Jorgenson, Andrew K., Brett Clark, Ryan P. Thombs et al. "Guns Versus Climate: How Militarization Amplifies the Effect of Economic Growth on Carbon Emissions." *American Sociological Review* 88, no. 3 (2023): 418–53.

Jorgenson, Andrew K., Thomas Dietz, and Orla Kelly. "Inequality, Poverty, and the Carbon Intensity of Human Well-Being in the United States: A Sex-Specific Analysis." *Sustainability Science* 13 (2018): 1167–74.

Jorgenson, Andrew, Hassan El Tinay, Jared Fitzgerald et al. "Advances in Research on Anthropogenic Drivers of Climate Change." In *Handbook of Climate Change & Society* (2024).

Jorgenson, Andrew K., Shirley Fiske, Klaus Hubacek et al. "Social Science Perspectives on Drivers of and Responses to Global Climate Change." *Wiley Interdisciplinary Reviews: Climate Change* 10, no. 1 (2019): e554.

Jorgenson, Andrew K., and Jennifer Givens. "The Changing Effect of Economic Development on the Consumption-Based Carbon Intensity of Well-Being, 1990–2008." *PloS one* 10, no. 5 (2015): e0123920.

Jorgenson, Andrew, Juliet Schor, and Xiaorui Huang. "Income Inequality and Carbon Emissions in the United States: A State-Level Analysis, 1997–2012." *Ecological Economics* 134 (2017): 40–48.

Kallis, Giorgos. "In Defence of Degrowth." *Ecological Economics* 70, no. 5 (2011): 873–80.

Kallis, Giorgos, Vasilis Kostakis, Steffen Lange, Barbara Muraca, Susan Paulson, and Matthias Schmelzer. "Research on Degrowth." *Annual Review of Environment and Resources* 43 (2018): 291–316.

Kimmerer, Robin. *Braiding Sweetgrass: Indigenous Wisdom, Scientific Knowledge and the Teachings of Plants*. Minneapolis: Milkweed Editions, 2013.

Klinenberg, Eric, Malcolm Araos, and Liz Koslov. "Sociology and the Climate Crisis." *Annual Review of Sociology* 46 (2020): 649–69.

Klinsky, Sonja, Timmons Roberts, Saleemul Huq et al. "Why Equity Is Fundamental in Climate Change Policy Research." *Global Environmental Change* (2017).

Lakhani, Nina. "Emissions from Israel's War in Gaza Have 'Immense' Effect on Climate Catastrophe." *The Guardian*, January 9, 2024. https://www.theguardian.com/world /2024/jan/09/emissions-gaza-israel-hamas-war-climate-change

Lakhani, Nina. "The Staggering Carbon Footprint of Israel's War in Gaza." *Mother Jones*, January 13, 2024. https://www.motherjones.com/politics/2024/01/gaza-israel-war -hamas-carbon-emissions-climate-damage.

Lamb, William F., Julia K. Steinberger, Alice Bows-Larkin, Glen P. Peters, J. Timmons Roberts, and F. Ruth Wood. "Transitions in Pathways of Human Development and Carbon Emissions." *Environmental Research Letters* 9, no. 1 (2014): 014011.

LeQuesne, Theo. "Petro-Hegemony and the Mmatrix of Resistance: What Can Standing Rock's Water Protectors Teach Us About Organizing for Climate Justice in the United States?" *Environmental Sociology* 5, no. 2 (2019): 188–206.

Liu, Jianguo, Thomas Dietz, Stephen R. Carpenter et al. "Complexity of Coupled Human and Natural Systems." *Science* 317, no. 5844 (2007): 1513–16.

Lockie, Stewart. "Mainstreaming Climate Change Sociology." *Environmental Sociology* 8, no. 1 (2022): 1–6.

Malin, Stephanie A., David Ciplet, and Jill Lindsey Harrison. "Sites of Resistance, Acceptance, and Quiescence Amid Environmental Injustice: An Introduction to the Special Issue on Sustainability Under Neoliberalism." *Environmental Justice* 16, no. 1 (2023): 1–9.

Malin, Stephanie A., and Stacia S. Ryder. "Developing Deeply Intersectional Environmental Justice Scholarship." *Environmental Sociology* 4, no. 1 (2018): 1–7.

Martinez-Alier, Joan. *The Environmentalism of the Poor: A Study of Ecological Conflicts and Valuation*. Northampton, Mass.: Edward Elgar Publishing, 2003.

Martínez-Alier, Joan, Leah Temper, Daniela Del Bene, and Arnim Scheidel. "Is There a Global Environmental Justice Movement?" *Journal of Peasant Studies* 43, no. 3 (2016): 731–55.

Mohai, Paul, David Pellow, and J. Timmons Roberts. "Environmental Justice." *Annual Review of Environment and Resources* 34 (2009): 405–30.

Muradian, Roldan, and Joan Martinez-Alier. "South-North Materials Flow: History and Environmental Repercussions." *Innovation: The European Journal of Social Science Research* 14, no. 2 (2001): 171–87.

Muradian, Roldan, and Joan Martinez-Alier. "Trade and the Environment: From a 'Southern' Perspective." *Ecological Economics* 36, no. 2 (2001): 281–97.

Muradian, Roldan, Martin O'Connor, and Joan Martinez-Alier. "Embodied Pollution in Trade: Estimating the 'Environmental Load Displacement' of Industrialised Countries." *Ecological Economics* 41, no. 1 (2002): 51–67.

Pellow, David Naguib. *Resisting Global Toxics: Transnational Movements for Environmental Justice*. Cambridge: MIT Press, 2007.

Pellow, David Naguib. *What Is Critical Environmental Justice?* New York: Polity, 2017.

Piketty, Thomas. *Capital in the Twenty-First Century*. Cambridge: Harvard University Press, 2014.

Purucker, David. "Critical Environmental Justice and the State: A Critique of Pellow." *Environmental Sociology* 7, no. 3 (2021): 176–86.

Rice, James. "Ecological Unequal Exchange: International Trade and Uneven Utilization of Environmental Space in the World System." *Social Forces* 85, no. 3 (2007): 1369–92.

Rice, James. "Material Consumption and Social Well-Being Within the Periphery of the World Economy: An Ecological Analysis of Maternal Mortality." *Social Science Research* 37, no. 4 (2008): 1292–1309.

Roberts, J. Timmons. "Calculating What We Owe." *Nature Sustainability* 6, no. 9 (2023): 1037–38.

Roberts, J. Timmons, Robert Brulle, and Jennifer Jacquet. "Informing Strategic Climate Action: The Climate Social Science Network." *Critical Policy Studies* 18, no. 1 (2024): 150–59.

Roberts, J. Timmons, and Bradley Parks. *A Climate of Injustice: Global Inequality, North-South Politics, and Climate Policy*. Cambridge: MIT Press, 2006.

Roberts, J. Timmons, and Bradley C. Parks. "Ecologically Unequal Exchange, Ecological Debt, and Climate Justice: The History and Implications of Three Related Ideas for a New Social Movement." *International Journal of Comparative Sociology* 50, no. 3–4 (2009): 385–409.

Roberts, J. Timmons, Julia K. Steinberger, Thomas Dietz et al. "Four Agendas for Research and Policy on Emissions Mitigation and Well-Being." *Global Sustainability* 3 (2020): e3.

Roberts, J. Timmons, Romain Weikmans, Stacy-Ann Robinson, David Ciplet, Mizan Khan, and Danielle Falzon. "Rebooting a Failed Promise of Climate Finance." *Nature Climate Change* 11, no. 3 (2021): 180–82.

Robinson, Stacy-Ann, J. Timmons Roberts, Romain Weikmans, and Danielle Falzon. "Vulnerability-Based Allocations in Loss and Damage Finance." *Nature Climate Change* 13, no. 10 (2023): 1055–62.

Schlosberg, David. "Theorising Environmental Justice: The Expanding Sphere of a Discourse." *Environmental Politics* 22, no. 1 (2013): 37–55.

Schor, Juliet. "Climate, Inequality, and the Need for Reframing Climate Policy." *Review of Radical Political Economics* 47, no. 4 (2015): 525–36.

Schor, Juliet, and Karen Elizabeth White. "Plenitude: The New Economics of True Wealth." 2010.

Taylor, Nik, Zoei Sutton, and Rhoda Wilkie. "A Sociology of Multi-Species Relations." *Journal of Sociology* 54, no. 4 (2018): 463–66.

Temper, Leah, Daniela Del Bene, and Joan Martínez-Alier. "Mapping the Frontiers and Front Lines of Global Environmental Justice: The EJAtlas." *Journal of Political Ecology* 22, no. 1 (2015): 255–78.

Temper, Leah, Federico Demaria, Arnim Scheidel, Daniela Del Bene, and Joan Martínez-Alier. "The Global Environmental Justice Atlas (EJAtlas): Ecological Distribution Conflicts as Forces for Sustainability." *Sustainability Science* 13, no. 3 (2018): 573–84.

Thunberg, Greta. *The Climate Book: The Facts and the Solutions.* New York: Penguin Press, 2023.

Tschakert, Petra, David Schlosberg, Danielle Celermajer et al.. "Multispecies Justice: Climate-Just Futures with, for and beyond Humans." *Wiley Interdisciplinary Reviews: Climate Change* 12, no. 2 (2021): e699.

Vinyeta, Kirsten, Kyle Powys Whyte and Kathy Lynn. *Climate Change Through an Intersectional Lens: Gendered Vulnerability and Resilience in Indigenous Communities in the United States.* Gen. Tech. Rep. PNW-GTR-923. Portland, OR: US Department of Agriculture, Forest Service, Pacific Northwest Research Station, 2015.

Wallerstein, Immanuel. *The Modern World-System I Capitalist Agriculture and the Origins of the European World-Economy in the Sixteenth Century.* New York: Academic Press, 1974.

Warlenius, Rikard. "Decolonizing the Atmosphere: The Climate Justice Movement on Climate Debt." *Journal of Environment & Development* 27, no. 2 (2018): 131–55.

Whyte, Kyle. "Indigenous Environmental Justice: Anti-colonial Action Through Kinship." *Environmental Justice*, 266–78. New York: Routledge, 2020.

York, Richard, Eugene A. Rosa, and Thomas Dietz. "STIRPAT, IPAT and ImPACT: Analytic Tools for Unpacking the Driving Forces of Environmental Impacts." *Ecological Economics* 46, no. 3 (2003): 351–65.

7 Climate Justice and Climate Mobility

International Migration from Central America and West Africa

ALEX DE SHERBININ, DAVID WRATHALL, SUSANA ADAMO,
SARA PAN-ALGARRA, AND ELENA GIACOMELLI

C limate mobility is an emerging field of study focusing on the contribution of climate factors to various forms of human mobility, including forced migration, displacement, and planned relocation. The field tends to have a multidisciplinary representation of geographers, economists, demographers, public health and law experts, as well as natural and physical scientists.[1] This chapter addresses international migration from a climate justice perspective. As yet, the evidence for international migration being triggered by climate factors is mixed. Studies suggest that such factors are more likely to trigger internal climate migration.[2] Yet there is also evidence of step-wise migration, whereby migrants leave rural areas for cities and then migrate to other countries after a period of time.[3] While climate or environmental factors are seldom at the top of the list of reasons that migrants cite for leaving their home countries, statistical analyses that include climate variables along with more traditional determinants of migration often find that factors like temperature or rainfall anomalies are significant.[4] Furthermore, climate extremes may be important underlying factors that affect livelihoods, food security, and poverty.[5]

In the Western dominant discourse, a politicized narrative of migration is often mixed with climate change and, as such, is portrayed as a threat itself. In other words, when climate change–induced migration gains traction in media coverage, it is usually linked to security issues rather than to efforts that reduce vulnerability to climate change effects in sending areas.[6] We adapt a climate justice approach to the specific case of international migration from the Global South to the Global North, taking as case examples the migration from Central America and West Africa to North America and Europe. Climate impacts in both regions have already been severe, with major hurricanes and flooding affecting Central America every three to four years, along with drought, and rising temperatures with periodic drought and flood in West Africa.[7] The flows from these two regions have become emblematic of the "migration crisis" more broadly, with migrants often portrayed by the media as poor, hapless, and desperate. We discuss, by contrast, how economic inequalities, historical interference in political processes by receiving countries (e.g., the United States and France), unfair terms of trade, and the ways in which Central America bears the costs of narcotics consumption in the United States while West Africa bears the costs of military efforts to contain militant Islam have created increasingly untenable situations for millions in both regions. We first review the literature on international migration from a justice perspective and then address the specificities of climate mobility justice. We then present the case studies of the two regions, followed by a conclusion.

JUSTICE PERSPECTIVES ON INTERNATIONAL MIGRATION

A justice framing of international migration was first presented through dependency and world systems theories that focus on structural explanations for migration, including colonial histories and political economies, particularly from lower-income countries to higher income countries (for more discussion, see chapter 6 this volume).[8] For example, dependency theory focuses on the extractive, colonial relationships that form between low-income nations as suppliers of raw materials, cheap labor, and markets for expensive manufactured goods from industrialized countries.[9] World systems theory adds to this the idea that the world is divided between core (industrialized) and peripheral (low income) areas, and that the latter have a low degree of autonomy and are more economically and politically disadvantaged.[10] The result, according to these theories, is that international migration is influenced by macro or structural forces and is inherently exploitative and self-perpetuating as it leads to underdevelopment in migrant source areas.[11] To sum up, global injustices and power imbalances drive the uneven distribution of wealth and opportunities that result in migration.

While these theories provide a top-down structural account of the injustices that influenced migration trends during the twentieth century, Amartya Sen's entitlement theory provides a people-centered perspective.[12] Sen's theory holds that social welfare is the result of people's success in accessing resources and enacting claims in society. In this view, the structure of entitlements within a society supports freedom, which is the expression of human agency. In entitlement theory, livelihoods become the vehicle through which people convert basic entitlements into social welfare.[13] If agricultural livelihoods fail, for example, it is not for want of resources, but due to a lack of basic guarantees in the conditions that make livelihoods successful, including land, water, and agricultural inputs.[14] The recent turn in migration studies toward aspirations and capabilities owes its lineage to these bottom-up justice critiques.[15] Human mobility, in this view, is the expression of people's "capability (freedom) to choose where to live," becoming inherently a justice proposition.[16] Where international border policies expand or circumscribe capabilities, they enter the normative domain of justice. Similarly, where displacement occurs, there is often an entitlement failure at the root, which is, in turn, ultimately an injustice.

In relation to climate mobility, Carmen Gonzalez defines the *abyssal line* as a dynamic line demarcating persons presumptively entitled to liberty, equality, and autonomy from those relegated to zones of violence and dispossession, or "sacrifice zones."[17] The abyssal line is represented in the world systems theory described earlier. From this perspective, international migration stems from the labor demands of modern industrial societies, as well as from an unbalanced world economic order that divides countries between core and periphery, being influenced by historically formed macrostructural forces that are inherently exploitative and self-perpetuating because they are linked to coloniality, extraction, and inequality in migrant source areas.[18]

Sen's justice framework rooted in equality and fairness applied to international mobility/ migration may clash with notions of national security, priorities, and sovereignty.[19] In this respect, Mimi Sheller argues that the field of international migration studies has extended to include the micro level or individual migration as an expression of agency, a continuum from immobility to mobility.[20] At the meso level, this approach to justice problematizes the border itself as the locus of differentiation and exclusion, including different flows (of people, goods, information, etc.) with governing regimes withholding access and entitlement.[21]

When people move internationally, borders are often invisible for the privileged while they are highly impermeable for others.[22] This is illustrated with the Passport Index (fig. 7.1).[23] The citizens of Central America and West Africa have far less freedom of movement than citizens of the United States and European countries. The index is inversely correlated with income and greenhouse gas emissions; those who emit the least are least entitled to move across borders, least of all the borders of the Global North.[24] As figure 1.3 in chapter 1 portrays, the map of the Passport Index scores is positively correlated with the map of damages from US-attributable greenhouse gas emissions, meaning that those who suffer the most from emissions in the United States are presumably the least entitled to travel there (see chapter 1 and fig. 7.1).

TOWARD CLIMATE MOBILITY JUSTICE

The International Organization for Migration defines climate migration as "the movement of a person or groups of persons who, predominantly for reasons of sudden or progressive change in the environment due to climate change, are obliged to leave their habitual place of residence, or choose to do so, either temporarily or permanently, within a State or across an international border."[25] Being a multidimensional phenomenon, it is possible to approach climate mobility from different angles. Approaching it from a

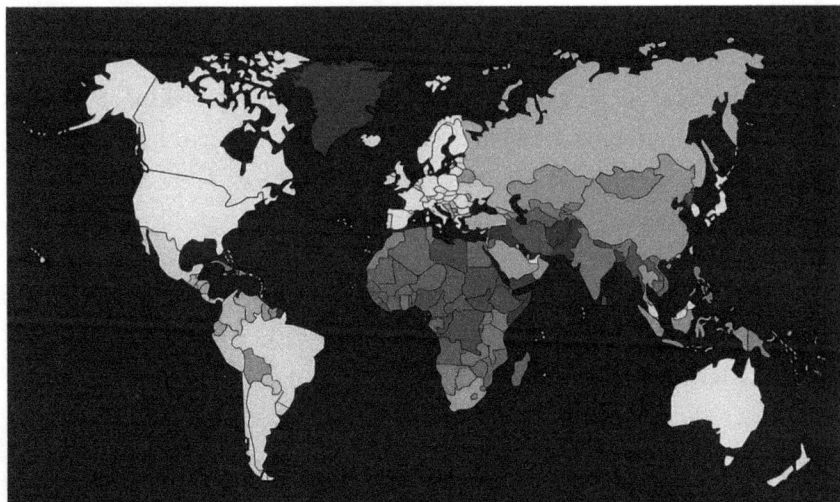

FIGURE 7.1

The Passport Power Rank as measured by the Passport Index, 2024. The darker the color, the lower the power and mobility of citizens of that country. The Passport Index ranks countries' passports based on a mobility score related to visa requirements, which includes visa-free (VF), visa on arrival (VOA), eTA, and eVisa (if issued within three days); VF portion of their score vs. VOA; and the UNDP's Human Development Index 2018 (UNDP HDI), which is used as a tiebreaker. (Source: Passport Index, https://www.passportindex.org/byRank.php. More information is available at https://www.passportindex.org/passport-power-rank -2024.php.)

climate justice angle implies focusing on the impacts of climate change on the mobility and immobility of those less responsible for the crisis, raising the principles of fairness, equity, and accountability.[26]

Whereas the previous section dealt with structural accounts of justice, this section turns to the climate justice principles surfacing in the IPCC and related literature, and their relationship to international migration. The cornerstone of these approaches is procedural justice, or "justice in the way outcomes are brought about, including who participates and is heard in the processes of decision-making."[27] In this view, the major justice conundrum concerning migrants is their limited role in the policy decisions that affect them.[28] This is particularly the case with forced migrants, such as refugees, where migrants themselves are not included in decisions concerning the basic aspects of displacement and resettlement, such as migration destinations.[29] This critique can also be extended to the general circumstances of migrants in host societies where they do not hold citizenship, and all its associated entitlements, and where they also lack influence on decision-making affecting the basic dimensions of their life.

Restorative justice, another justice frame surfacing in the IPCC AR6 and a central tenet in climate policy, is also relevant to international migration. Restorative justice is concerned with addressing historical wrongs that have produced impacts today and is satisfied whenever historical responsibility can be assigned to losses and damages occurring in the present and due compensation is provided.[30] This justice perspective finds expression in the UNFCCC's "Polluter Pays Principle" and is the basis for notions of climate reparations, hotly debated in climate policy negotiations on loss and damage.[31] Recent policy proposals on climate visa programs that would recognize countries' emissions in the allocation of visas to incoming migrants are based on notions of restorative justice.[32]

Finally, recognitional justice acknowledges the dignity, rights, and identities of people experiencing specific historical injustices.[33] Without requiring compensation for past liabilities, the simple recognition of injustices becomes a salient first step toward just futures. Communities facing environmental displacement have asked for recognition of their ties to ancestral sacred sites, and for consideration of their knowledge systems and values (e.g., trust, reciprocity, responsibility, common identity) in policies affecting them.[34] Recognitional justice is relevant to instances of forced displacement and international mobility, as will likely occur in small island developing states in coming decades.

We proceed now to an exploration of our two case studies before returning to a discussion of the similarities and differences and conclusions.

CENTRAL AMERICA CASE STUDY

The list of injustices informing Central American migration is lengthy. Structural injustices have resulted from Central America's position in the global periphery, as a locus of extraction of labor and resources. At the same time, enduring structures of coloniality have strangled Central American states' capacity to guarantee basic entitlements to their citizens.[35] These injustices create the conditions in which migrants have come to rely on income and welfare-seeking strategies by migrating outside of the region.

The creation of the Panama Canal and the development of an economy heavily based on banana plantations and other cash crops (such as coffee)—dominated by US corporations such as the United Fruit Company—in the early part of the twentieth century laid the foundation for decades of economic and political instability in the region, linked to very high levels of violence emerging from dictatorships, US interventions, civil war, and gangs.[36] Nicaragua, El Salvador, Honduras, and Guatemala were particularly affected by the political and military violence of those decades, which included the systematic violation of human rights, with Indigenous peoples being particularly targeted, especially in Guatemala. These processes led to a sharp increase in Central American migration to the United States since the 1970s, but particularly during the 1980s. While Honduras was spared the brutal civil wars and genocides of its neighbors toward the end of the twentieth century, it endured the heavy toll of intense inequality and poverty. In this context, the country was very badly affected by Hurricane Mitch in 1998, which completely destabilized it.[37] In the 2000s there was a surge in drug trafficking–related violence in several countries, particularly El Salvador, which further fueled emigration.[38] This led to further increases in migration to the United States.

Yet, despite fleeing such dangerous situations, very few migrants were admitted or granted asylum,[39] but some Central Americans irregularly in the United States were granted temporary protected status (TPS) after Hurricane Mitch.[40] At a time when more people require adaptation strategies that include migration, procedural justice problems are heaped on migrants with the hardening of the US border.[41] Central Americans seeking entry to the United States face related recognitional injustices in the raft of contemporary migration deterrent policies, including family separation, which specifically injure migrants' dignity and suspend basic rights.[42]

El Salvador, Guatemala, Honduras, and lately Nicaragua have consistently displayed the highest increases in numbers of migrants, especially unaccompanied minors. A substantial proportion of recent Central American migration to the United States is irregular, and Mexico, once the main sending country, is increasingly becoming a transit country.[43]

Contemporary root causes for international migration from the region are complex and include socioeconomic conditions (poverty, unemployment, food insecurity), environmental shocks (including climate change–related events), and security, governance, and conflict issues.[44]

Climate could be a threat and harm multiplier reinforcing and amplifying existing structural injustices. Recent changes in Central America's climate patterns (related to both climate variability and change) are affecting agricultural outcomes (of both cash and subsistence crops), which may affect emigration patterns through impacts on livelihoods (the agriculture pathway) in a context where a substantial proportion of the labor force is in the primary sector.[45] Some social groups—small/artisanal fishers, small farmers—are consequently more at risk, but gender, age, and ethnicity also play a role in differentiated and intersectoral vulnerability and inequality: women, children, and Indigenous people may be more vulnerable to the impacts, more likely to be displaced, but less likely to migrate.[46]

There is no agreement in the literature on the actual weight of climate events (be it climate variability or climate change–related events) in migration decisions vis-à-vis other drivers, with some sources indicating as little as 3 percent while others indicate about 40 percent of respondents mentioning environmental or climate factors as a reason for leaving.[47] However, it is important to keep in mind that the impacts of climatic and other environmental events could be indirect, and that income levels, food insecurity, violence, and climatic events are frequently tied together. For example, Ariel Ruiz Soto et al. found that those reporting deteriorating living standards or worsening economic conditions were more likely to have experienced a natural disaster or threat.[48]

Migrants' narratives weave these many threads highlighting the structural as well as the circumstantial reasons for leaving. Many talk about violence: for example, one said, "For a Honduran, violence is daily bread. We live with it daily."[49] Others cite lack of opportunities: "For most people in Honduras, it is very difficult to find work. San Pedro Sula and Tegucigalpa are two cities where factory employment can be found. But the rest is pure farm: bananas, beans, corn, potatoes and onions."[50] And still others cite the impacts of climate change on livelihoods: "Before, there were a lot of fish.

BOX 7.1. HONDURAS—CLIMATE MOBILITY AND GIRLS' EDUCATION

Climate disasters, in addition to violence, persecution, and poverty, drive human mobility within and across borders in Honduras.* Amid the COVID-19 pandemic, Tropical Storm Eta and Hurricane Iota hit northern Honduras's alluvial Sula Valley region, home to San Pedro Sula, the second largest Honduran city. Destructive winds and flooding affected people living in informal settlements, and many displaced by the disasters in rural settings joined their numbers. These residents confront multifaceted vulnerabilities tied to class, livelihoods, race, ethnicity, landownership, age, and gender, among others. Existing literature from other contexts discusses associations between climate disasters and gender-based violence (GBV), adolescent pregnancy, child marriage, and menstrual poverty (lack of access to menstrual products and hygiene).†

The connection between migration and climate justice is essential to understand the challenges students face to remain and complete school despite their legal right to education. To better comprehend the situation in the Sula Valley, co-author Pan-Algarra conducted a multi-stakeholder case study to examine how the complex interconnections among climate disasters, displacement, crime, violence, and dispossession can disrupt girls' education in the Sula Valley.‡ The interviews revealed that girls can face infrastructural, gendered, and social barriers to access and remain in school postdisaster. These barriers included the destruction of schools due to flooding, limited access to the internet, the use of schools as shelters, gendered violence, menstrual poverty, adolescent pregnancy, child labor within the household, psychological challenges, and trauma.§ Families' monetary loss after disasters can result in temporary or permanent displacement, parental abandonment, recruitment into gangs and prostitution, as well as insecurity and threats. Migration to "El Norte" (the United States) provides a path of escape for some.

---

* This case study research was presented under the title "Climate Displacement and the Access to Education of Communities Living in Informal Settlements in Honduras" at the 42nd Latin American Studies Association (LASA) Annual Congress celebrated in the Pontificia Universidad Javeriana, Bogotá, Colombia, June 12–15, 2024. The research supporting this presentation and case study was

conducted under Teachers College, Columbia University Institutional Review Board Protocol #23–357.

† International Refugee Assistance Project (IRAP), Enduring Change; Fry and Lei, A Greener, Fairer Future; Goulds, In Double Jeopardy; Atkinson and Bruce, "Adolescent Girls, Human Rights."

‡ In this qualitative case study, Pan-Algarra (2024) conducted semistructured interviews (n=30) with "outsiders"—individuals surrounding girls who influence distinct aspects of their capability to seek schooling. Five interviews were with parents of families hit by Eta and Iota, fourteen were with teachers and school admin, and eleven were with civil society representatives supporting people dislocated in the Sula Valley.

§ Limited internet access was particularly relevant, as Eta and Iota hit during the COVID-19 pandemic when classes were remote.

The boat used four bags, that's what we call the nets. Your hands were getting very soft from so much fish. But today that no longer happens. Boats only use two bags, at most."[51] On the other hand, those "left behind" may talk about poverty and lack of work and opportunities, the mental health impact of this (e.g., suicides), the disruptions when family members (fathers, mothers) migrate, but also the opportunities that migration provides (income, status), and the always-present gang violence.[52]

WEST AFRICA CASE STUDY

Colonial impacts in West Africa have been varied and long lasting.[53] Continuing beyond independence in the early 1960s, French and British economic hegemony remained in place through unfavorable terms of trade, debt peonage, and the extraction of agricultural goods such as cotton, peanuts, cacao, and coffee.[54] Since the early 2000s, the "war on terror" resulted in the establishment of military operations in the Sahel by Western powers; resentment of these incursions on national sovereignty gave rise to severance of diplomatic ties, the removal of French and American troops, and their replacement with Russian-backed military forces.[55] Meanwhile, high risk ratings by rating agencies continue to stifle foreign investment, even in the region's most stable countries, such as Senegal and

Ghana, which stifles economic growth and reduces opportunities for local employment.[56]

Despite longstanding trade and economic ties, up until the early 2000s the scale of international migration from West Africa to Europe and North America was small.[57] It was only in the 2000s that migrant streams moved from formal/legal to irregular types of migration, and the number of migrants began to rise steeply, peaking in 2016. The migrant "crisis" has been accompanied by a dramatic increase in the number of deaths at sea and in the Sahara, along with images of desperation as migrants scale the fences at the Spanish enclaves of Ceuta and Melilla.[58] Starting as early as the late 2000s, the European Union began to enlist the support of countries along the eastern route, starting in Libya and then, when Qadafi fell, moving south to Niger, to deter migrants before they attempted to cross the Mediterranean.[59] This became known as "Europe's new southern border," and all the while the number of migrants dying while attempting to cross the Sahara increased.[60] The recent Oscar-nominated narrative film *Io Capitano* describes the gauntlet that African migrants need to navigate—from predatory smugglers to gangs and groups out to extort migrants—to have a chance of making it to Europe.

The reasons for migration cited by West African migrants vary depending on where on the route they are interviewed.[61] Around 80 percent of those interviewed in North Africa generally cite economic reasons, while for those interviewed in Europe, the numbers drop to around 36–60 percent, while the numbers citing violence, corruption, or issues around rights and freedoms rise significantly. Family and personal reasons are also widely cited by those in Europe, at around 18 to 35 percent of respondents.[62] The percentage citing environmental or climate reasons generally hovers around 2–3 percent of respondents. Research finds that West Africans who move for environmental reasons generally stay within their countries or the region of West Africa.[63] However, there is a growing narrative in the media and reports by some NGOs that many migrants from the region are "climate refugees."[64]

While surveys can be useful, they are generally insufficient to fully understand the layered reasoning and complex emotions behind migration decision-making.[65] To understand West African migration, it is important to understand the local political economy and culture. For example, in conversations with farmers regarding the reasons young men leave the Tambacounda region of Senegal on the perilous journey to Europe, Jesse Ribot et al. find that factors behind economic and food insecurity were of critical importance, from uncertain agricultural production to opaque credit and transport conditions. According to the authors:

> When we ask today's farmers why they and their children are departing, they talk of some of the broader factors behind their food and economic security. They cite low and fluctuating agricultural production, which they attribute to non-mechanized and poor-quality agricultural equipment, expensive agricultural inputs, and an extractive system of agricultural industries and credit relations. . . . They point to usurious credit for seed and equipment (prevalent in the cotton sector), or loans from small lenders and store owners against their next crop to get through the hungry season, and government-regulated transport that allows a few businessmen to fix producer prices and control access to lucrative urban and international markets. Given these conditions, farmers see few options for long-term prosperity in farming.[66]

Thus, it is not so much climate impacts on crops that are leading many youth to give up on farming, but a system that is stacked against them. On multiple levels, farmers lack power and agency and are hobbled by debt. The researchers also found that culturally, nonmigrants die a "social death," as they are perceived as failures, poor marriage prospects, and insufficiently entrepreneurial. It is small wonder, then, that a commonly repeated phrase in Wolof is "Barça wala barsakh" (Barcelona or the afterlife)—in other words, better to die trying to get to Europe than to die a social death at home.[67]

Richard Black et al. also delve deeper into the thinking of migrants to get beyond mono-causal explanations for the "drivers of migration."[68] Based on interviews with youth in Senegal, The Gambia, and Guinea, they question the conceptualization of migration as an event or "decision," presenting evidence to suggest that most would-be migrants live in a state of contingency and uncertainty, and their orientation toward the future could best be phrased as, "if things don't work out here, I'll consider leaving for Europe." This is confirmed by research by Stephano degli Uberti et al. under the Future of Migration to Europe (FUME) project. Through in-depth interviews with migrants in Dakar, Senegal, they find that many migrants from rural areas find that they cannot meet their aspirations in the city and therefore try their hand at international migration. They conclude that "the desire to move is triggered by the great difficulties in developing stable prospects in order to succeed socially (such as building a house or getting married) and the importance of economic factors (in particular, the limited opportunities offered by the agricultural sector and the lack of vocational training or job opportunities)."[69] They, too, find that environmental factors are rarely mentioned, citing a key informant who said that migration triggered by environmental changes does not usually lead to land abandonment, but to short-term mobility patterns.

Demographic factors compound the impacts of climate change on an agricultural system that is already precarious.[70] Much of the current migration from the region to Europe can be traced to a "youth bulge" in some of the world's most rapidly growing countries, along with declining prospects in countries that have suffered from decades of resource extraction, unfair terms of trade, low rates of foreign investment, low investment in the agricultural sector, and weak governance.[71] As a result, many young people conclude that any hope of meeting their aspirations can only be achieved by leaving West Africa. Industry and services are generally underdeveloped in the region, and the job market for public sector jobs for the growing educated class is saturated.[72] Quoting degli Uberti et al., "overpopulation, the reduced prospects of finding a job in the city, the tightening of policies restricting free movement as well as a complex mix of unfulfilled expectations and feelings (along the migration path) have led to an increasing nonlinearity and fragmentation of both the internal (e.g., temporary short-term migration practices) and international migration routes (e.g., unsuccessful migration attempts and impeded mobility)."[73] It is no surprise, then, that most international migrants from the region are better educated than their peers who stay behind, and many are not unemployed but are leaving low-paying jobs to seek a better future.[74]

With the rise of militant Islam and armed groups in Mali, Burkina Faso, and Niger, the Sahel region of West Africa is becoming increasingly politically unstable. While the reasons for the breakdown in order are complex, overly simplified explanations often seek to trace instability to climate factors.[75] Most of those displaced by conflict will not have the wherewithal to travel internationally and are at risk of becoming trapped in increasingly untenable circumstances, though some may flee out of desperation. Countries in the region have sought alliances with the major powers—first France and the United States, and now Russia—to combat armed groups, a Faustian bargain that will result in further suffering on the ground.

BOX 7.2. SENEGAL—FISHING COMMUNITIES

To understand how people perceive climate change in their everyday lives, a team from the University of Bologna in Senegal researched local perceptions and the connections to migration using qualitative methodologies and drawing on visual methods (photography). *

Participants were mainly from fishing communities in Guet Ndar (Langue de Barbarie, St. Louis), which is situated between the Senegal River and the Atlantic Ocean, and the coastal communities of Rufisque, Thiaroye-sur-mer, Dalifort, and Hann Bay around Dakar. The majority of the populace in these sites relies on artisanal fishing for a living. Both interviews and images from the climate diaries show strongly how the interplay of intersecting factors, the uneven impacts of sea level rise across the globe, the extractive nature of the unequal fishing agreements, and the long histories of colonial underdevelopment become apparent. †

Paraphrasing Ian Baucom, from this "angle of inspection," the images of crumbling buildings due to sea level rise look like the result of an "act of nature" but are, in fact, the result of "a historical process of violence" rooted in colonial continuity. ‡ *In fact, in these two localities in Senegal, the urban space is experienced by many participants as a territorial and temporal abandonment, evidencing the socio-spatial inequalities. Indeed, during interviews, the lack of choice (to migrate) was a recurrent and dominant theme that emerged in the discussion of motivations, leading some people, particularly fishers, to take the risky pirogue journey to the Canary Islands. What emerged from the research is the intertwined impacts of the climate crisis on lives and livelihoods as rooted in global unequal structures, unjust migration governance, state-level fishing agreements, and global and local waste mismanagement.[§]*

---

* Giacomelli et al. For a brief video on this project, see https://www.youtube.com /watch?v=eddxMC42yuo. During a field visit in May 2021, the team conducted thirty-five semistructured interviews with people of different ethnicities, ages, genders, and levels of education who had been affected by the climate crisis. Via the visual methodology of climate diaries, participants were asked to share photos and perceptions of the climate crisis over a period of time through a WhatsApp group. In this context, the climate diary, as a diary, proved to be a tool for self-reflection, particularly about the lived environment and (im)mobility injustices.

[†] Zickgraf, "Relational (Im)mobilities."

[‡] Baucom, History 4°Celsius; Sultana, "Critical Climate Justice."

[§] Walker and Giacomelli, "Encountering Mobility (In)justice."

CLOSING THOUGHTS

This chapter has described the justice framing of international migration, and the increasing intersection of climate justice and climate im/mobilities. By focusing on two regions, we have sought to ground this in the place-specific and historic injustices that have led to differential development and societies increasingly characterized by violence and dispossession in Central America and West Africa. These factors on their own—together with rising populations, the inability to meet aspirations locally, new technologies such as social media and smartphones, and the ubiquity of human smuggling networks—have contributed to rapidly growing numbers of irregular migrants from both regions.

For Central America, we see a region with a history of migration to the United States dating back to the 1970s—itself the result of US government support for right-wing dictatorships during a period in which the United States feared the spread of communism.[76] The patterns of migration have shifted from more seasonal (associated with US agricultural labor demands) to permanent, as pathways to legal migration have been shut, and the composition of flows have shifted from almost entirely working-age male migrants to an increasing number of families and even unaccompanied minors. The rising influence of narco-traffickers and gangs has created increasing insecurity. In West Africa, by contrast, migration streams to Europe only picked up significantly in the early 2000s, peaking around 2015–16, but still at lower levels than the irregular migration across the US southern border. The pattern remains one of largely single male migrants (including minors), many of whom have grown disillusioned with the prospect of "making it" economically in the region. Failed governance and the paucity of jobs are frequently mentioned by migrants. In both regions, a system seemingly rigged against smallholder agriculturalists—whether they grow coffee or cotton—makes it hard for youth to envision a future that involves farming.[77]

Climate change and its impacts have complicated matters, but they are not generally thought of by the migrants themselves as sufficient reasons for migrating, except in those rare cases of extreme events, such as hurricanes Mitch, Eta, and Iota. Yet climate factors have been shown to negatively affect livelihoods (e.g., through crop losses), contributing to rising poverty and eroding local resilience, in ways that may escape the attention of the migrants.[78] Regardless of the causes, the migrants run a gauntlet of challenges—traversing scorching deserts and undertaking perilous sea voyages in the hands of criminal networks—in hopes of finding safety and a more secure and prosperous future.

The Global North has dealt these regions a bad hand, through brutal colonial and puppet regimes, industrial emissions, carbon capitalism, and massive consumption (including the narcotics that fuel an epidemic of crime), and the extraction of resources (including human chattel) over centuries. Furthermore, border walls notwithstanding, it has had an ambivalent approach towards migration: seeking to control it for political reasons as fear grows of the "other," while simultaneously recognizing the need for low wage and youthful laborers in countries with aging populations. Yet the elites of the countries in both regions also bear responsibility for governance failures, corruption, and the lack of basic security.

The examination of climate mobility's justice issues has been mostly restricted in this chapter to an examination of the unequal relationship between areas of origin and destination, yet other elements are equally important. These include the ill treatment migrants receive at the hands of law enforcement in the United States and Europe, and the lack of procedural justice for asylum claimants in overwhelmed immigration court systems.[79] It also includes the inadequacy of the UNFCCC Loss and Damage Fund as a mechanism for redistributive justice, as well as the growing concern over the ways in which climate impacts may strip people of the freedom to move, or climate immobility.[80] Owing to space constraints, these factors have received inadequate attention in this chapter, but readers are invited to follow the citations for further reading, or to read the excellent *Handbook of Climate Migration and Climate Mobility Justice*, by Andreas Neef et al.[81]

Notes

1. de Sherbinin et al., "Migration Theory in Climate Mobility Research"; Oakes et al. "A Future Agenda for Research."
2. Rigaud et al., *Preparing for Internal Climate Migration*.
3. Escamilla García, "When Internal Migration Fails."
4. Cattaneo et al., "Human Migration in the Era of Climate Change"; Nawrotzki and Bakhtsiyarava, "International Climate Migration."
5. Tuholske et al., "A Framework to Link Climate Change."
6. Sakellari, "Media Coverage of Climate Change."
7. Nakamura et al., "Recent Trends in Agriculturally Relevant Climate"; Biasutti, "Rainfall Trends in the African Sahel."
8. Others include dual labor market and the new international division of labor.
9. Shrum, "Science and Development."
10. Wallerstein, *The Modern World-System I*, 229–33.
11. King, "Theories and Typologies of Migration."
12. Sen, *Poverty and Famines*.
13. Leach et al., "Challenges to Community-Based Sustainable Development."
14. Adger et al., "Migration, Remittances, Livelihood Trajectories."
15. de Haas, "A Theory of Migration."
16. de Haas, 2.
17. Gonzalez, "Climate Change, Race, and Migration."
18. de Sherbinin et al., "Migration Theory in Climate Mobility Research," 4.

19. Weber and Tazreiter, "Introduction: Migration and Global Justice."
20. Sheller, "Mobility Justice."
21. Sheller, *Mobility Justice*.
22. Tschakert and Neef, "Tracking Local and Regional Climate Im/mobilities."
23. Passport Index, "Global Passport Power Rank 2024."
24. Skirbekk et al., "Religious Affiliation and Environmental Challenges."
25. International Organization for Migration, *Glossary on Migration*.
26. Cundill et al., "Toward a Climate Mobilities Research Agenda."
27. Matthews et al., "IPCC, 2021: Annex VII: Glossary," 2913.
28. Edge et al., "Procedural Environmental (In)justice."
29. Atapattu, "Climate Change and Displacement."
30. Robinson and Carlson, "A Just Alternative to Litigation."
31. Forsyth et al., "Environmental Restorative Justice."
32. Cissé et al., "Health, Wellbeing, and the Changing Structure."
33. We pick up participatory and recognitional justice as explained in chapter 2 of this volume but add restorative justice.
34. Figueroa, "Indigenous Peoples and Cultural Losses."
35. Chomsky, *Central America's Forgotten History*; Cruz, "The Root Causes of the Central American Crisis."
36. Alvarado et al., "U.S. Central American (Un)Belongings."
37. By some estimates, a quarter of the population was temporarily displaced in the aftermath, and by 2003 some 150,000 Central American migrants had been granted temporary protected status in the United States. See Reichman, "Putting Climate-Induced Migration in Context."
38. Alvarado et al., "U.S. Central American (Un)Belongings"; Masferrer et al., "Contemporary Migration Patterns"; Reichman, "Putting Climate-Induced Migration in Context"; Wrathall et al., "The Impacts of Cocaine-Trafficking"; Magliocca et al., "Shifting Landscape Suitability for Cocaine Trafficking."
39. Alvarado, "Introduction: U.S. Central American (Un)Belongings."
40. Messick and Bergeron, "Temporary Protected Status."
41. McLeman, "International Migration and Climate Adaptation."
42. The documentary *Casa en Tierra Ajena* (Home in a Foreign Land) (2017), https://www.youtube.com/watch?v=AkrZIumTRjI, collects the stories of Central American migrants leaving for the US-Mexico border and of those who did not migrate, showing the broad spectrum of injustices they are exposed to.
43. Ward and Batalova, "Central American Immigrants."
44. Meyer, *Central American Migration*; World Food Program, *Food Security and Emigration*.
45. Nakamura et al., "Recent Trends in Agriculturally Relevant Climate"; Tuholske et al., "A Framework to Link Climate Change."
46. Organización Internacional para las Migraciones (OIM), *La movilidad humana derivada de desastres*; Sanchez Peña and Adamo, "Juventudes y cambio climático." Also see chapter 8 in this volume.

47. Ruiz Soto et al., *Charting a New Regional Course of Action*; USAID, *Honduras. Climate Change, Food Security, and Migration*.

48. Ruiz Soto et al., *Charting a New Regional Course of Action*, 14.

49. "Para un hondureño, la violencia es el pan de cada día. Vivimos con eso diariamente."

50. "Para la mayoría de la gente en Honduras, es muy raro que encuentre trabajo. San Pedro Sula y Tegucigalpa son dos ciudades donde se puede encontrar empleo de fábrica. Pero el resto es puro campo: banano, frijol, maíz, papa y cebolla."

51. "Antes eran barbaridades de pescado. El barco usaba cuatro bolsos, así le decimos a las redes. Las manos se te ponían bien suavecitas de tanto pescado. Pero hoy día ya no se da eso. Sólo usan dos bolsos los barcos, a lo mucho." Wolf, *La migración forzada*.

52. Ciborowski et al., "Through Our Own Eyes and Voices."

53. For example, from the early sixteenth century to 1866, some 6.3 million West Africans were sent as slaves to plantations in Brazil, the Caribbean, and the southern United States in the most brutal forced migration in history (Statista 2024). At the Berlin West African Conference of 1884–85, the territory was divided up among the European powers—mainly the French and English, but also the Germans (Dahomey) and Portuguese (Guinea-Bissau).

54. Bond, *Looting Africa*.

55. *Al Jazeera*, "Timeline: Nine Years of French Troops in Mali," February 17, 2022, https://www.aljazeera.com/news/2022/2/17/timeline-what-led-france-to-withdraw-its-troops-from-mali; Jeff Hawkins, "Withdrawal of US Troops from Niger: A Setback for Washington?," *IRIS News*, April 26, 2024, https://www.iris-france.org/185945-withdrawal-of-us-troops-from-niger-a-setback-for-washington/.

56. Fofack, *The Ruinous Price for Africa*.

57. Starting in the 1950s, early labor migrants included former soldiers (*ancien combattants*) who traveled to France in the wake of World War II for blue-collar jobs, establishing dormitories for migrants and rarely bringing their families (Adams, *Le long voyage des gens du Fleuve*). The better educated of the region traveled to Europe or even the former Soviet Union in small numbers for advanced studies.

58. Cuttitta and Last, *Border Deaths*.

59. Bialasiewicz, "Off-shoring and Out-sourcing."

60. Williams, "African Migration Trends."

61. Data cited here are largely drawn from the Mixed Migration Centre's 2023 4mi results (https://mixedmigration.org/4mi/) and the UNDP's 2019 *Scaling Fences* report.

62. The lower percentage of migrants citing economic reasons in Europe may be influenced by asylum claims and migrants' justifiable concerns related to possible deportation.

63. Filho et al., "Where to Go?"; Kanta Kumari Rigaud et al., *Groundswell Africa*; Grace et al., "Examining Rural Sahelian Out-migration"; Romankiewicz and Doevenspeck,

"Climate and Mobility"; Henry et al., "Descriptive Analysis of the Individual Migratory Pathways."
64. Vince, "The Century of Climate Migration"; Friedman, "Out of Africa."
65. Fresnoza-Flot, "Humanising Research on Migration Decision-Making."
66. Ribot et al., "Climate of Anxiety in the Sahel."
67. Cuttitta and Last, *Border Deaths.*
68. Black et al., "Migration Drivers and Migration Choice."
69. degli Uberti et al., *Senegal: Drivers and Trajectories,* 66, 67.
70. Ribot, Faye, and Turner, "Climate of Anxiety in the Sahel."
71. Tuholske et al., "A Framework to Link Climate Change."
72. ILO, *Implementation of the Decent Work Agenda.*
73. degli Uberti et al., *Senegal,* 67.
74. UNDP, *Scaling Fences.*
75. Lamarche, *Climate-Fueled Violence*; Werz and Conley, *Climate Change, Migration, and Conflict.*
76. Blitzer, *Everyone Who Is Gone Is Here.*
77. Tuholske et al., "A Framework to Link Climate Change."
78. Tuholske et al.; Nakamura et al., "Recent Trends in Agriculturally Relevant Climate."
79. Costello and Mann, "Border Justice.
80. Uri et al., "Equity and Justice in Loss and Damage Finance"; Cundill et al., "Toward a Climate Mobilities Research Agenda"; Zickgraf, "Theorizing (Im)mobility."
81. Neef et al., eds., *De Gruyter Handbook of Climate Migration.*

REFERENCES

Adams, Adrian. *Le long voyage des gens du Fleuve*. Paris, Maspéro, 1977.
Adger, W. Neil, P. Mick Kelly, Alexandra Winkels, Luong Quang Huy, and Catherine Locke. "Migration, Remittances, Livelihood Trajectories, and Social Resilience." *AMBIO: A Journal of the Human Environment*, no. 4 (2002): 358–66, 9. https://doi.org/10.1579/0044-7447-31.4.358.
Alvarado, Karina, Alicia Estrada, and Ester Hernandez. "Introduction: U.S. Central American (Un)Belongings." In *U.S. Central Americans: Reconstructing Memories, Struggles, and Communities of Resistance*, edited by K. Alvarado, A. Estrada, and E. Hernández, 3–35. Temple: University of Arizona Press, 2017.
Atapattu, Sumudu. "Climate Change and Displacement: Protecting 'Climate Refugees' Within a Framework of Justice and Human Rights." *Journal of Human Rights and the Environment* 11, no. 1 (01 Mar. 2020 2020): 86–113. https://doi.org/10.4337/jhre.2020.01.04.
Atkinson, Holly G., and Judith Bruce. "Adolescent Girls, Human Rights, and the Expanding Climate Emergency." *Annals of Global Health* 81, no. 3 (2015): 323–30. https://doi.org/10.1016/j.aogh.2015.08.003.

Baucom, Ian. *History 4°Celsius: Search for a Method in the Age of the Anthropocene.* Durham, NC: Duke University Press Books, 2020.

Bialasiewicz, Luiza. "Off-Shoring and Out-Sourcing the Borders of Europe: Libya and EU Border Work in the Mediterranean." *Geopolitics* 17, no. 4 (2012): 843–66. https://doi.org /10.1080/14650045.2012.660579.

Biasutti, Michela. "Rainfall Trends in the African Sahel: Characteristics, Processes, and Causes." *WIREs* 10, no. 4 (2019): e591. https://doi.org/ 10.1002/wcc.591.

Black, Richard, Alice Bellagamba, Ester Botta et al. "Migration Drivers and Migration Choice: Interrogating Responses to Migration and Development Interventions in West Africa." *Comparative Migration Studies* 10, no. 1 (March 7, 2022): 10. https://doi.org/10 .1186/s40878-022-00283-3.

Blitzer, Jonathan. *Everyone Who Is Gone Is Here: The United States, Central America, and the Making of a Crisis.* New York: Penguin, 2024.

Bond, Patrick. *Looting Africa: The Economics of Exploitation.* London: Bloomsbury Publishing, 2008.

Cattaneo, Cristina, Michel Beine, Christiane J. Fröhlich et al. "Human Migration in the Era of Climate Change." *Review of Environmental Economics and Policy* 13, no. 2 (2019): 189–206. https://doi.org/10.1093/reep/rez008.

Chomsky, Aviva. *Central America's Forgotten History: Revolution, Violence, and the Roots of Migration.* Boston: Beacon Press, 2021.

Ciborowski, Haley M., Samantha Hurst, Ramona L. Perez, Kate Swanson, Eric Leas, Kimberly C. Brouwer, and Holly Baker Shakya. "Through Our Own Eyes and Voices: The Experiences of Those 'Left-Behind' in Rural, Indigenous Migrant-Sending Communities in Western Guatemala." *Journal of Migration and Health* 5 (January 1, 2022): 100096. https://doi.org/https://doi.org/10.1016/j.jmh.2022.100096.

Cissé, G., R. McLeman, H. Adams et al. "Health, Wellbeing, and the Changing Structure of Communities." In *Climate Change 2022: Impacts, Adaptation and Vulnerability. Contribution of Working Group Ii to the Sixth Assessment Report of the Intergovernmental Panel on Climate Change*, ed. D. C. Roberts et al., 1041–170. Cambridge: Cambridge University Press, 2022.

Costello, Cathryn, and Itamar Mann. "Border Justice: Migration and Accountability for Human Rights Violations." *German Law Journal* 21, no. 3 (2020): 311–34. https://doi.org /10.1017/glj.2020.27.

Cruz, José Miguel. "The Root Causes of the Central American Crisis." *Current History* 114, no. 769 (2015): 43–48. http://www.jstor.org/stable/45319276.

Cundill, Georgina, Chandni Singh, William Neil Adger et al. "Toward a Climate Mobilities Research Agenda: Intersectionality, Immobility, and Policy Responses." *Global Environmental Change* 69 (July 1, 2021): 102315. https://doi.org/https://doi.org/10.1016/j .gloenvcha.2021.102315.

Cuttitta, P., and T. Last. *Border Deaths: Causes, Dynamics and Consequences of Migration-Related Mortality.* Amsterdam: Amsterdam University Press, 2019.

de Haas, Hein. "A Theory of Migration: The Aspirations-Capabilities Framework." *Comparative Migration Studies* 9, no. 1 (2021/02/24 2021): 8. https://doi.org/10.1186/s40878 -020-00210-4.

de Sherbinin, Alex, Kathryn Grace, Sonali McDermid, Kees van der Geest, Michael J. Puma, and Andrew Bell. "Migration Theory in Climate Mobility Research." *Frontiers in Climate* 4 (May 10, 2022). https://doi.org/10.3389/fclim.2022.882343.

degli Uberti, Stephano, Frank Heins, Lancine Eric Nestor Diop, and Mohamadou Sall. *Senegal: Drivers and Trajectories of Migration to Europe: Case Study*. 2023.

Edge, Sara, Emily Lauren Brown, Sutama Ghosh, and Ann Marie Murnaghan. "Procedural Environmental (in)Justice at Multiple Scales: Examining Immigrant Advocacy for Improved Living Conditions." *Local Environment* 25, no. 9 (September 1, 2020): 666–80. https://doi.org/10.1080/13549839.2020.1812554.

Escamilla García, Angel Alfonso. "When Internal Migration Fails: A Case Study of Central American Youth Who Relocate Internally Before Leaving Their Countries." *Journal on Migration and Human Security* 9, no. 4 (2021): 297–310. https://doi.org/10.1177/23315024211042735.

Figueroa, Robert Melchior. "Indigenous Peoples and Cultural Losses." In *The Oxford Handbook of Climate Change and Society*, ed. John S. Dryzek, Richard B. Norgaard, and David Schlosberg, 232–49. Oxford: Oxford University Press, 2011.

Fofack, Hippolyte. *The Ruinous Price for Africa of Pernicious "Perception Premiums."* Report of the Africa Growth Initiative at Brookings, October 2021. https://www.brookings.edu/wp-content/uploads/2021/10/21.10.07_Perception-premiums.pdf.

Forsyth, Miranda, Brunilda Pali, and Felicity Tepper. "Environmental Restorative Justice: An Introduction and an Invitation." In *The Palgrave Handbook of Environmental Restorative Justice*, ed. Brunilda Pali, Miranda Forsyth and Felicity Tepper, 1–23. Cham: Springer International, 2022.

Fresnoza-Flot, Asuncion. "Humanising Research on Migration Decision-Making: A Situated Framework." *Open Research Europe* 3, no. 142 (2024). https://open-research-europe.ec.europa.eu/articles/3-142.

Friedman, Thomas L. "Out of Africa." *New York Times*, April 13, 2016. www.nytimes.com/2016/04/13/opinion/out-of-africa.html.

Fry, Lucia, and Philippa Lei. *A Greener, Fairer Future: Why Leaders Need to Invest in Climate and Girls' Education*. United Nations Girls Education Initiative, 2021. https://www.ungei.org/publication/greener-fairer-future.

Gonzalez, Carmen. "Climate Change, Race, and Migration." *Journal of Law and Political Economy* 1, no. 1 (2020): 109–46. https://doi.org/https://doi.org/10.5070/LP61146501.

Goulds, Sharon. *In Double Jeopardy: Adolescent Girls and Disasters*. Commonwealth Education Partnerships, 2013. https://www.cedol.org/wp-content/uploads/2013/11/2-In-double-jeopardy.pdf.

Grace, Kathryn, Véronique Hertrich, Djeneba Singare, and Greg Husak. "Examining Rural Sahelian Out-Migration in the Context of Climate Change: An Analysis of the Linkages Between Rainfall and Out-Migration in Two Malian Villages from 1981 to 2009." *World Development* 109 (September 1, 2018): 187–96. https://doi.org/https://doi.org/10.1016/j.worlddev.2018.04.009.

Henry, Sabine, Victor Piché, Dieudonné Ouédraogo, and Eric F. Lambin. "Descriptive Analysis of the Individual Migratory Pathways According to Environmental

Typologies." *Population and Environment* 25, no. 5 (May 1, 2004): 397–422. https://doi .org/10.1023/B:POEN.0000036929.19001.a4.

International Labour Organisation (ILO). *Implementation of the Decent Work Agenda in West Africa*. Geneva: ILO, 2014. https://webapps.ilo.org/wcmsp5/groups/public/-- -africa/---ro-abidjan/documents/publication/wcms_235308.pdf.

International Organization for Migration (IOM). *Glossary on Migration*. Geneva: IOM, 2019. https://publications.iom.int/books/international-migration-law-ndeg34-glossary -migration.

International Refugee Assistance Project (IRAP). *Enduring Change: A Data Review of First-hand Accounts of Climate Mobility Impacts*. New York: IRAP, 2024. https://refugeerights .org/wp-content/uploads/2024/09/Climate-Data-Report-September-2024-1.pdf.

King, Russell. "Theories and Typologies of Migration: An Overview and a Primer." *Willy Brandt Series of Working Papers in International Migration and Ethnic Relations* 3, no. 12 (2012): 1–41. https://www.researchgate.net/publication/260096281_Theories_and _Typologies_of_Migration_An_Overview_and_A_Primer.

Lamarche, A. *Climate-Fueled Violence and Displacement in the Lake Chad Basin: Focus on Chad and Cameroon*. Washington, DC: Refugees International, 2023. https://www .refugeesinternational.org/reports-briefs/climate-fueled-violence-and-displacement -in-the-lake-chad-basin-focus-on-chad-and-cameroon/.

Leach, Melissa, Robin Mearns, and Ian Scoones. "Challenges to Community-Based Sustainable Development: Dynamics, Entitlements, Institutions." *IDS Bulletin* 28, no. 4 (1997): 4–14. https://doi.org/ 10.1111/j.1759-5436.1997.mp28004002.x.

Leal Filho, Walter, Olawale Festus Olaniyan, and Gabriela Nagle Alverio. "Where to Go? Migration and Climate Change Response in West Africa." *Geoforum* 137 (December 1, 2022): 83–87. https://doi.org/https://doi.org/10.1016/j.geoforum.2022.10.011.

Magliocca, Nicholas R., Diana S. Summers, Kevin M. Curtin, Kendra McSweeney, and Ashleigh N. Price. "Shifting Landscape Suitability for Cocaine Trafficking Through Central America in Response to Counterdrug Interdiction." *Landscape and Urban Planning* 221 (May 1, 2022): 104359. https://doi.org/https://doi.org/10.1016/j.landurbplan .2022.104359.

Masferrer, Claudia, Silvia E. Giorguli-Saucedo, and Victor M. Garcia-Guerrero. "Contemporary Migration Patterns in North and Central America." In *The Sage Handbook of International Migration*, ed. Christine Inglis, Wei Li, and Binod Khadria, 342–57. London: SAGE, 2020.

Massey, Douglas S., Joaquín Arango, Graeme Hugo, Ali Kouaouci, Adela Pellegrino, and J. Edward Taylor. "Theories of International Migration: A Review and Appraisal." *Population and Development Review* 19, no. 3 (September 1993): 431–66. https://doi.org /https://doi.org/10.2307/2938462.

Matthews, J.B.R., V. Möller, R. van Diemen et al. "IPCC, 2021: Annex Vii: Glossary." In *Climate Change 2021: The Physical Science Basis. Contribution of Working Group I to the Sixth Assessment Report of the Intergovernmental Panel on Climate Change*, ed. V. Masson-Delmotte, P. Zhai, A. Pirani et al., 2215–56. Cambridge: Cambridge University Press, 2021.

McLeman, Robert. "International Migration and Climate Adaptation in an Era of Hardening Borders." *Nature Climate Change* 9, no. 12 (December 1, 2019): 911–18. https://doi.org/10.1038/s41558-019-0634-2.

Messick, Madeline, and Claire Bergeron. "Temporary Protected Status in the United States: A Grant of Humanitarian Relief That Is Less than Permanent." *Migration Information Source*, no. July 2 (2014). https://www.migrationpolicy.org/article/temporary-protected-status-united-states-grant-humanitarian-relief-less-permanent.

Meyer, Peter J. *Central American Migration: Root Causes and U.S. Policy*. Congress.gov. (2023). https://crsreports.congress.gov/product/details?prodcode=IF11151.

Nakamura, Jennifer, Richard Seager, Haibo Liu et al. "Recent Trends in Agriculturally Relevant Climate in Central America." *International Journal of Climatology* 44, no. 8 (2024): 2701–24. https://doi.org/ 10.1002/joc.8476.

Nawrotzki, Raphael J., and Maryia Bakhtsiyarava. "International Climate Migration: Evidence for the Climate Inhibitor Mechanism and the Agricultural Pathway." *Population, Space, and Place* 23, no. 4 (2016): e2033. https://doi.org/https://doi.org/10.1002/psp.2033.

Neef, A., N. Pauli, and B. Salami, eds. *De Gruyter Handbook of Climate Migration and Climate Mobility Justice*. Berlin: De Gruyter, 2024. https://www.degruyter.com/document/doi/10.1515/9783110752144/html.

Oakes, Rorbert, Kees Van der Geest, Benjamin Schraven et al. "A Future Agenda for Research on Climate Change and Human Mobility." *International Migration* 61 (2023): 116–25. https://doi.org/10.1111/imig.13169.

Organización Internacional para las Migraciones (OIM). *La movilidad humana derivada de desastres y el cambio climático en Centroamérica*. Ginebra: OIM, 2021. https://publications.iom.int/books/la-movilidad-humana-derivada-de-desastres-y-el-cambio-climatico-en-centroamerica.

Passport Index. "Global Passport Power Rank 2024." 2024. https://www.passportindex.org/byRank.php.

Reichman, Daniel R. "Putting Climate-Induced Migration in Context: The Case of Honduran Migration to the USA." *Regional Environmental Change* 22, no. 3 (July 4, 2022): 91. https://doi.org/10.1007/s10113-022-01946-8.

Ribot, Jesse, Papa Faye, and Matthew D. Turner. "Climate of Anxiety in the Sahel: Emigration in Xenophobic Times." *Public Culture* 32, no. 1 (2020): 45–75. https://doi.org/10.1215/08992363-7816293.

Rigaud, Kanta Kumari, Alex de Sherbinin, Bryan Jones et al. *Groundswell: Preparing for Internal Climate Migration*. Washington, DC: World Bank, 2018. http://hdl.handle.net/10986/29461.

Rigaud, Kanta Kumari, Alex de Sherbinin, Bryan Jones et al. *Groundswell Africa: Internal Climate Migration in West African Countries*. Washington, DC: World Bank, 2021. https://doi.org/10986/36404.

Robinson, Stacy-ann, and D'Arcy Carlson. "A Just Alternative to Litigation: Applying Restorative Justice to Climate-Related Loss and Damage." *Third World Quarterly* 42, no. 6 (May 14, 2021: 1384–95. https://doi.org/10.1080/01436597.2021.1877128.

Romankiewicz, C., and M. Doevenspeck. "Climate and Mobility in the West African Sahel: Conceptualising the Local Dimensions of the Environment and Migration Nexus." In *Grounding Global Climate Change: Contributions from the Social and Cultural Sciences*, 79–100. Dordrecht: Springer Netherlands, 2014.

Ruiz Soto, Ariel, R. Bottone, J. Waters, S. Williams, A. Louie, and Y. Wang. *Charting a New Regional Course of Action: The Complex Motivations and Costs of Central American Migration*. Rome, Washington, DC, and Cambridge, MA: World Food Program, Migration Policy Institute, and Civic Data Design Lab at Massachusetts Institute of Technology. https://reliefweb.int/report/guatemala/charting-new -regional-course-action-complex-motivations-and-costs-central-american.

Sanchez Peña, Landy, and Susana B. Adamo. "Juventudes y cambio climático: las intersecciones de género, etnia y edad en la configuración de la vulnerabilidad climática en el sector agrícola en Latinoamérica." In *Danzar en las brumas. Género y juventudes en entornos desiguales en América Latina y el Caribe*, ed. UNESCO, COLMEX and CLACSO, 89–109. 2022.

Sen, Amartya. *Poverty and Famines: An Essay on Entitlement and Deprivation*. Oxford: Oxford University Press, 1983. https://doi.org/10.1093/0198284632.001.0001.

Sheller, Mimi. "Mobility Justice." In *Handbook of Research Methods and Applications for Mobilities*, ed. Monika Büscher, Malene Freudendal-Pedersen, Sven Kesselring, and Nikolaj Grauslund Kristensen, 11–20. London: Edward Elgar Publishing, 2020.

Sheller, Mimi. *Mobility Justice: The Politics of Movement in an Age of Extremes*. New York: Verso Books, 2018.

Shrum, W. "Science and Development." In *International Encyclopedia of the Social & Behavioral Sciences*, ed. Neil J. Smelser and Paul B. Baltes, 13607–10. Oxford: Pergamon, 2001.

Skirbekk, Vegard, Alex de Sherbinin, Susana B. Adamo, Jose Navarro, and Tricia Chai-Onn. "Religious Affiliation and Environmental Challenges in the 21st Century." *Journal of Religion and Demography* 7, no. 2 (October 6, 2020): 238–71. https://doi.org/10 .1163/2589742X-12347110.

Sultana, Farhana. "Critical Climate Justice." *Geographical Journal* 188, no. 1 (2022): 118–24. https://doi.org/https://doi.org/10.1111/geoj.12417.

Tschakert, Petra, and Andreas Neef. "Tracking Local and Regional Climate Im/Mobilities Through a Multidimensional Lens." *Regional Environmental Change* 22, no. 3 (July 15, 2022). https://doi.org/10.1007/s10113-022-01948-6.

Tuholske, Cascade, Maria Agustina Di Landro, Weston Anderson, Robbin Jan van Duijne, and Alex de Sherbinin. "A Framework to Link Climate Change, Food Security, and Migration: Unpacking the Agricultural Pathway." *Population and Environment* 46, no. 1 (March 5, 2024): 8. https://doi.org/10.1007/s11111-024-00446-7.

UNDP. *Scaling Fences*. New York: UNDP, 2019. https://www.undp.org/publications/scaling -fences.

Uri, Ike, Stacy-Ann Robinson, J. Timmons Roberts, David Ciplet, Romain Weikmans, and Mizan Khan. "Equity and Justice in Loss and Damage Finance: A Narrative Review of

Catalysts and Obstacles." *Current Climate Change Reports* 10, no. 3 (September 1, 2024): 33–45. https://doi.org/10.1007/s40641-024-00196-6.

USAID. *Honduras. Climate Change, Food Security, and Migration.* USAID, 2022.

Vince, G. "The Century of Climate Migration: Why We Need to Plan for the Great Upheaval." *Guardian*, August 18, 2022. https://www.theguardian.com/news/2022/aug/18/century-climate-crisis-migration-why-we-need-plan-great-upheaval.

Walker, Sarah, and Elena Giacomelli. "Encountering Mobility (In)Justice Through the Lived Experiences of Fishing Communities in Dakar and Saint Louis, Senegal." *Mobilities* 29, no. 6 (April 2024): 1–17. https://doi.org/10.1080/17450101.2024.2334705.

Wallerstein, Immanuel. *The Modern World-System I: Capitalist Agriculture and the Origins of the European World-Economy in the Sixteenth Century.* Berkeley: University of California Press, 2011. http://www.jstor.org/stable/10.1525/j.ctt1pnrj9.

Ward, Nicole, and Jeanne Batalova. "Central American Immigrants in the United States." *Migration Information Source*, no. May 10 (2023). https://www.migrationpolicy.org/article/central-american-immigrants-united-states-2021.

Weber, Leanne, and Claudia Tazreiter. "Introduction: Migration and Global Justice." In *Handbook of Migration and Global Justice*, ed. Leanne Weber and Claudia Tazreiter, 1–12. London: Edward Elgar Publishing, 2021.

Werz, M., and L. Conley. *Climate Change, Migration, and Conflict in Northwest Africa.* Washinton, DC: Center for American Progress, 2012. https://www.americanprogress.org/article/climate-change-migration-and-conflict-in-northwest-africa/.

Williams, W. "African Migration Trends to Watch in 2024." Africa Center for Strategic Studies, 2024. https://africacenter.org/spotlight/african-migration-trends-to-watch-in-2024/.

Wolf, Sonja. *La migración forzada desde el Triángulo Norte de Centroamérica. Impulsores y experiencias.* Aguascalientes: CIDE, 2020. http://www.politicadedrogas.org/PPD/.

World Food Program. *Food Security and Emigration. Why People Flee and the Impact on Family Members Left Behind in El Salvador, Guatemala and Honduras.* Relief Web (2017). https://reliefweb.int/report/el-salvador/food-security-and-emigration-why-people-flee-and-impact-family-members-left.

Wrathall, David J., Jennifer Devine, Bernardo Aguilar-González et al. "The Impacts of Cocaine-Trafficking on Conservation Governance in Central America." *Global Environmental Change* 63 (July 1, 2020): 102098. https://doi.org/https://doi.org/10.1016/j.gloenvcha.2020.102098.

Zickgraf, Caroline. "Relational (Im)Mobilities: A Case Study of Senegalese Coastal Fishing Populations." *Journal of Ethnic and Migration Studies* 48, no. 14 (October 26, 2022): 3450–67. https://doi.org/10.1080/1369183X.2022.2066263.

Zickgraf, Caroline. "Theorizing (Im)Mobility in the Face of Environmental Change." *Regional Environmental Change* 21, no. 4 (December 2, 2021): 126. https://doi.org/10.1007/s10113-021-01839-2.

8

Climate Justice in the Field

Climate Change and the Health of
Migrant Agricultural Workers

LEWIS H. ZISKA, JEFFREY L. SHAMAN, EMILY WEAVER,
AND AMI ZOTA

P rior to the COVID-19 pandemic, if one had asked which
occupations are essential for a functional society, it is
doubtful that agricultural workers would have topped the
list. Yet, it quickly became apparent just how essential such
labor was and is. Empty shelves and food scarcity prompted
legitimate fear during the pandemic and exacerbated nutritional shortages, shortages antithetical to promoting public
health during a crisis.

It can be tempting to assume that modern agriculture is all
wheels and diesel—large space-age reapers that cut and thresh
food and whisk it away to the nearest silo. The truth, however, is
that in the United States, harvesting crops, from lettuce to
green peppers, from apples to oranges, is done by hand—hands
that are, by and large, from other countries. Hands of workers
who shuffle ladders, lift buckets, pull and pluck, run to wagons
to empty their collections, then back again, hunched and sweat-
ing, between the endless rows. Or hands that are moving along
a conveyor belt of poultry products, quickly removing cartilage,
bones, and fat; hands of individuals who can be as young as ten,

as long as parents approve and school is not missed—regulations that are often underenforced for undocumented children.[1]

From the time of the dust bowl, when desperate Midwestern families from Oklahoma migrated to California to pick oranges, to those currently fleeing poverty in Central America, US agriculture has offered steady work—and minimal wages. In 2021 it was estimated that about 75 percent of US farmworkers were Latino immigrants, and that 50 percent of hired farmworkers did not have authorization to work in the United States, although interviews with growers and labor contractors suggest the figure might be closer to 75 percent.[2]

It has been argued that such migrants are, in fact, taking jobs away from US citizens; yet, curiously, when Alabama passed the nation's strictest anti-immigration law in 2011, there was an abrupt departure of the state's farmworkers—many undocumented immigrants—and food spoiled in the fields. In one instance, growers worked to recruit unemployed US citizens for tomato harvests, but only three workers remained after the first month.[3]

There are reasons why these jobs are unappealing. The work is hard and exhausting. Workers are up before dawn, in the field stooping, picking, gathering as the first rays of light allow you to distinguish the parts of the plant that need to be harvested. Then the heat, the sweat, the contact with moving machinery that can crush a hand if you are not paying attention.

Agricultural workers face a range of physical, environmental, and chemical health exposures in their day-to-day work. The process of harvesting produce can mean long hours outdoors, resulting in extreme weather and temperature exposure, which are associated with health outcomes such as heat exhaustion.[4] Additionally, chemical pesticides applied are associated with both short-term and long-term human health implications, ranging from gastrointestinal issues and skin irritation to damage to major organs tracts such as the respiratory, renal, neurological, and reproductive systems.[5] Pregnant farmworkers face additional health impacts due to the associated risks of environmental and chemical exposures on fetal development; for example, pregnant farmworkers are more likely to miscarry than the general population, and exposure to chemical pesticides *in utero* is associated with a range of birth and childhood development outcomes including heart defects, impacts to the musculoskeletal, urinogenital, and gastrointestinal tracts, and cognitive impacts.[6] Furthermore, research has shown that many farmworkers lack access to potable water and bathroom facilities while working, resulting in dehydration and complications such as urinary tract infections among workers.[7] Housing conditions for migrant farmers are often undermaintained, with studies reporting issues such as lack of water, mold, and pests, creating additional health hazards for those living onsite.[8] Farmworkers may face unique stressors due to gender-based inequalities, such as lack of disclosure of pregnancy due to supervisors' fear, lack of access to prenatal and postnatal care, and sexual harassment and assault.[9] Such conditions reflect the highest death rates across job sectors; on average for all businesses, 3.6 deaths occur for every 100,000 individuals, but for agriculture the number is 23.[10]

When shopping at the grocery store, it is rare to think about what effort went into getting the food on the shelf—of the toil and dehumanization faced by farmworkers in our current system. Or, if acknowledged, it can be swept mentally into the old adage, "the cost of doing business."

Such a system, reflecting a series of environmental impacts, is the norm. But it will be fundamentally transformed, globally, with anthropogenic climate change.

In considering the role of climatic change—increases in temperature, unprecedented shifts in rainfall, extreme events—with regard to food security, there is a focus on production; for example, how will climate affect yields? Will warmer temperatures increase food spoilage? And there are worries about quality: Will rising levels of carbon dioxide reduce protein levels? Ultimately, the central question is whether climate change will allow production of enough safe and nutritionally relevant food to provide what is currently a global population of eight billion, soon to be nine billion.[11]

These are legitimate questions, and they generate a great deal of attention, from subsistence farmers and agrichemical corporations to grocery stores and the US Department of Agriculture. But a vital element of concern is often ignored: *the role of climate change in agricultural labor.*

There is emerging evidence that climate change will intensify the environmental health risks affecting agricultural labor. Heat stress and exhaustion are foremost concerns among several likely threats and are related to common farming conditions, such as heavy exertion, lack of shade, and inadequate potable water during the workday.[12] Interviews with farmworkers indicate that the fast work pace—exacerbated by a system where money earned is dependent on the amount of crop harvested—discourages hydration and rest.[13] Currently, the average US farmworker experiences twenty-one unsafe days due to extreme heat, and by 2050 that number is likely to almost double to thirty-nine days, or roughly 25 percent of the growing season.[14]

In Central America, an epidemic of chronic kidney disease (CKD) has been reported for sugarcane workers, consistent with rising temperatures in the region. Heat stress for these workers can be intensified by burning (done prior to harvest to remove leaves allowing easier access to stem cutting); with entering the fields following burning adding to the overall heat burden. Overall, a key risk factor in higher CKD burdens was found to be associated with heavy physical labor in heat.[15]

As the climate warms, there is evidence that agricultural pests can proliferate. It is anticipated that global warming could (a) trigger an expansion of geographic range; (b) increase fecundity and decrease generation times; and (c) promote insect-transmitted plant diseases.[16] In addition, there is increasing evidence that rising CO_2 and climate shifts may also adversely affect pesticide efficacy.[17] As pesticides are the primary means of pest control in developed countries, such changes are likely to increase pesticide applications in order to achieve pest management.[18] Agricultural workers, in turn, would be exposed to greater amounts and frequency of pesticides with subsequent health risks. In addition, there is an increased risk that agricultural workers will be directly exposed to insect vectors or waterborne pathogens.[19]

There is also strong evidence that climate change is increasing wildfire season length, frequency, and burned area.[20] Smoke emissions are sources of fine particulate matter (PM2.5) that can be dangerous to public health. Because of the outdoor and physically demanding nature of their work, farmworkers may be especially at risk for wildfire smoke inhalation. A study of agricultural workers in California reported a potential increase from six to eight million smoke exposure days for agricultural workers across the state with projected climatic changes, with the largest increases occurring in Tulare, Monterey, and Fresno counties.[21]

Because age is often ignored in field situations, children are especially at risk. For example, it is estimated that approximately 1.3 million children globally are field laborers in the tobacco industry.[22] In addition to long hours, equipment risks, and chemical exposure, tobacco also poses an immediate threat through green tobacco sickness (GTS), an occupational illness resulting from dermal absorption of nicotine from the leaves of the tobacco plant.[23] Symptoms of nicotine poisoning include headaches, nausea, difficulty breathing, dizziness, sneezing, salivating, and burning and watering of the eyes. Changes in climate change factors that increase nicotine absorption and GTS, such as temperature and rainfall, have been observed globally (Brazil, China, India, United States) and are expected to increase with projected climatic change.[24]

Beyond the intensification of health risks, there is another consideration: economic production. While the direct consequences of climate change on crop yields and humans have been thoroughly examined individually, how they interact remains unclear. Globally, such interaction is particularly relevant in locations where harvests are highly dependent on labor.[25] It is estimated that a direct effect of increased temperature (+3°C) on major crops (maize, soy, wheat, and rice) would result in a 78 billion dollar loss; however, the adverse effects of such a climate on heat and humidity exposure for agricultural workers would almost double the loss ($136 billion).[26]

There is a critical humanitarian need to better understand and address the effects of climate change on the health of agricultural workers—who are already encountering environmental vulnerability—but at the same time to recognize that these same workers are essential for maintaining production and food security. This dual need is a clarion call for climate justice for a labor group, which, as COVID 19 showed, is necessary to put food on our plates and for societal function.

In that regard, there are rights and obligations of climate justice that can begin to address circumstances that have been, for decades, underlying environmental and industrial inequality for agricultural workers. Chief among these are the unequal distribution of global change risks and the need for enhanced protection. At present, government efforts are focused on production, with recommendations and promotion of management practices such as regenerative agriculture or genetic selection. However, there is a seminal need for policy reform for agricultural laborers. Part of the challenge for the government response is that, at present, there are no US Census data that accurately describe the size, distribution, and economic and demographic characteristics of the US agricultural worker population.[27] This absence of data makes monitoring and application of routine workplace services difficult to document, and only extreme incidents are reported. For example, in Baton Rouge, Louisiana, a complaint that an employer denied temporary agricultural workers water and food, while screaming obscenities and aiming guns at the workers, finally prompted the Department of Labor to file a restraining order and preliminary injunction against Rivet and Sons LLC to prevent retaliation against the workers.[28]

The H-2A visa program provides official sanction to seasonal farmworkers, but demand by employers is rapidly increasing, and the program is in need of reform. The Biden administration had proposed more protections for migrant farmworkers under the program, making it easier for labor unions to contact and interact with H-2A workers.[29] The program also requires farmers who utilize H-2A workers to provide seat belts for vans or buses that are used to transport workers to the field. (Transportation accidents are recognized as a leading cause of death for agricultural workers.)[30] Changes in worker protection and visa status by the current Trump administration remain unclear.

The Occupational Health and Safety Administration (OSHA) is, in principle, designed to provide safe working conditions by enforcing workplace standards promoting health and well-being. OSHA oversight of farms, however, can be problematic.[31] For example, it cannot inspect or cite farms with ten or fewer employees, currently exempting 96 percent of animal agricultural workers from OSHA oversight. OSHA has no national heat standards that require farm managers to provide drinking water, shade, paid breaks, or acclimatization to heat. In addition, if a worker collapses due to heat stress, often there is no emergency plan to provide first aid quickly. At the state level, only California, Colorado, Oregon, and Washington have workplace standards requiring employers to protect outdoor workers from the heat. Conversely, high-heat states with large farmworker populations, including Texas, Florida, and North Carolina, have no such protections. OSHA is working on a federal heat standard (H.R. 2193, "The Asunción Valdivia Heat Illness, Injury and Fatality Prevention Act," introduced in 2023) which, to date, has failed to pass. Florida has even banned local heat protections for agricultural workers.[32] As of spring 2024, Health and Human Services, a federal agency, was meeting to discuss how to protect farmworkers from extreme heat and wildfire smoke (currently only California, Oregon, and Washington have protections for agricultural workers regarding wildfire smoke). Although the issue is of obvious concern, it is worth noting that OSHA has not accepted previous agency recommendations.[33]

Compensation for workers can also be affected by climatic extremes. Piece work is predominant in agricultural labor, where workers are paid by the amount harvested rather than hours worked. For workers under forty, wages increase with productivity—for example, how much product can be harvested in a given time period—however, such incentives discourage breaks. Lack of breaks can lead to greater physical stress and heat exhaustion or avoidance of personal protective equipment if such equipment slows the harvesting. A national law to ensure that piece-rate workers obtain a rest period—with pay—every four hours is currently only available in some states (for example, Washington).[34] In response to advocacy efforts and increasingly extreme weather, several policy efforts have been proposed that would set standards across the United States. In 2023 a bill was introduced in the US House of Representatives that would have mandated breaks for farmworkers in cool spaces, limited exposure time to extreme temperatures, required education on signs of heat-related illness, and required OSHA to pass a rule regulating heat exposure in work environments. In 2021 OSHA began efforts to establish a federal heat standard that would provide protections for employees working in extreme conditions; as of early 2025, the proposed rule had been drafted and was in the process of soliciting public comment.[35] Overall, there is a need to recognize these risks and to ensure compensation at the farm, state, and federal level.

The nature of the job represents unique healthcare challenges. Migrant workers frequently do not report their injuries, in part due to employment concerns, language barriers, or limited access to health providers.[36] There are migrant health centers that provide primary and preventive healthcare to migratory and seasonal farmworkers and their families.[37] However, for the health providers that are available, the itinerant nature of the work makes long-term or basic preventive care difficult.[38] A lack of sick leave, concern over losing paid work time, and if undocumented, fear of being reported to the authorities also may serve as barriers to accessing healthcare for migrant farmworkers. Despite these challenges, the Affordable Care Act (ACA) helped raise the number of seasonal farmworkers with medical insurance and increased their use of preventive medical care while reducing their use of hospitals, including emergency room visits.[39]

Overall, the health risks associated with climate change affecting these essential workers are still not being addressed within the United States and across the globe. Climate change is a threat multiplier disproportionally affecting farmworkers who are already exposed to enhanced social susceptibility factors associated with race and/or ethnicity, linguistic isolation, and immigration status as well as environmental factors including pesticide exposure and heat stress. Public health research to characterize climate change impacts at the farm level in relation to physical, chemical, and biological conditions as well as larger socioeconomic transitions, including the migration of Central American workers to the US southern border, is lacking.

Climate justice requires national standards that include climate-related risks, such as extreme heat or wildfires, access to health services, and inclusion of the migrant farmworker population in governmental planning for emergency management and climate change mitigation efforts. Yet, updates to regulations, policy, infrastructure, and operating systems to achieve environmental safety for farmworkers are limited.[40] Failure to address climate justice for these workers will result in widespread damage to their physical and behavioral health and, given their critical importance for agricultural systems, the economic and food security of the United States.

There is increasing acknowledgment of the occupational hazards faced by farmworkers that will be radically intensified with anthropogenic climate change. Currently, the public health community is doing little overall to document these interactions; instead, farmworkers and environmental justice groups are leading efforts to document climate-related stresses, from heat exhaustion to increased pesticide use to wildfires. However, at the governmental and agribusiness levels, there is a strong need to reform current laws and practices to account for the additional threats imposed by climate change. Such efforts are necessary to address climate justice for these essential workers, but also to understand and remedy the imminent production shortages that will arise in food systems if the status quo is maintained (for more discussion of the political context, see chapter 3 in this volume). Finally, it is important to stress that the effect of climate change on agricultural workers is not confined to a single region or country. *Climatic disruptions will be global in nature, and human labor is a vital and indispensable element of food security.* Yet, addressing the health and safety of this workforce at a panoptic scale remains a primary, but largely ignored, aspect of agricultural adaptation to climate change.

Notes

1. Ramos, "Child Labor in Global Tobacco Production," 235; Quandt and Arnold, "The Health of Children," 163–65.
2. Castillo et al., "Environmental Health Threats."
3. *The Guardian*, "Meet the Workers Who Put Food on America's Tables."
4. Wills and Commins, "Consequences of the American States' Legislative Action"; Fleischer et al., "Public Health Impact of Heat-Related Illness"; Holmes and Ramirez-Lopez, *Fresh Fruit, Broken Bodies*.
5. Holmes and Ramirez-Lopez, *Fresh Fruit, Broken Bodies*; Nicolopoulou-Stamati et al., "Chemical Pesticides and Human Health."
6. Rani et al., "Consequences of Chemical Pesticides"; Galarneau, "Farm Labor, Reproductive Justice," 144; and Eskenazi et al., "In Utero and Childhood Polybrominated Diphenyl Ether (PBDE) Exposures."
7. Rappazzo et al., "Maternal Residential Exposure"; Connor et al., "Providing Care for Migrant Farm Worker Families."
8. Cummins, "Tuberculosis and Poor Health"; Arcury et al., "Safety, Security, Hygiene and Privacy."
9. Vallejos et al., "Migrant Farmworkers' Housing Conditions."
10. Handal et al., "Experiences of Female Agricultural Workers."
11. Swanton et al., "Characteristics of Fatal Agricultural Injuries."
12. Wijerathna-Yapa and Pathirana, "Sustainable Agro-Food Systems," 1554.
13. Iglesias-Rios et al., "Climate Change, Heat, and Farmworker Health," 43.
14. Glaser et al., "Preventing Kidney Injury."

15. Hansson et al., "An Ecological Study of Chronic Kidney Disease," 840.
16. Skendžic et al., "Impact of Climate Change."
17. Matzrafi, "Climate Change Exacerbates Pest Damage"; Ziska and McConnell, "Climate Change, Carbon Dioxide, and Pest Biology"; and Ma et al., "Climate Warming Promotes Pesticide Resistance."
18. Delcour et al., "Impact of Climate Change on Pesticide Use."
19. Schulte and Chun, "Climate Change and Occupational Safety."
20. Jones et al., "Global and Regional Trends."
21. Marlier et al., "Exposure of Agricultural Workers in California."
22. The Guardian, "Child Labour Rampant in Tobacco Industry."
23. Arcury et al., "The Incidence of Green Tobacco Sickness."
24. Ziska and Parks, "Recent and Projected Changes."
25. De Lima et al., "Heat Stress on Agricultural Workers."
26. Hill et al., "Agricultural Labor Supply."
27. Moctezuma, "Excluded and Isolated," 183.
28. Palacios, "Reforming America's Employment-Based Immigration System," 36.
29. Darçın et al., "Accidents Involving Migrant Seasonal Agricultural Workers."
30. May and Arcury, "Occupational Injury and Illness in Farmworkers."
31. Civil Eats, "Animal Agriculture Is Dangerous Work."
32. Inside Climate News, "Florida Legislators Ban Local Heat Protections."
33. Wittenberg, "Federal Health Workers Evaluate Protections."
34. US Department of Labor, "Minimum Paid Rest Period Requirements."
35. Occupational Safety and Health Administration, "Heat and Injury Illness Protection."
36. Moyce and Schenker, "Migrant Workers."
37. See section 339(g) of the Public Health Service Act, http://www.govinfo.go.
38. Bohlke, "Lack of Portable Insurance Adds Health Care Burden."
39. Donkor and Perloff, "The Effects of the Affordable Care Act."
40. Kuehn, "Why Farmworkers Need More than New Laws."

BIBLIOGRAPHY

Arcury, T. A., M. M. Weir, P. Summers et al. "Safety, Security, Hygiene and Privacy in Migrant Farmworker Housing." *New Solutions: A Journal of Environmental and Occupational Health Policy* 22, no. 2 (2012): 153–73.

Arcury, T. A., S. A. Quandt, J. S. Preisser, and D. Norton "The Incidence of Green Tobacco Sickness Among Latino Farmworkers. *Journal of Occupational and Environmental Medicine* 43, no. 7 (2001): 601–9.

Bohlke, L. "Lack of Portable Insurance Adds Health Care Burden to Migrant Workers." *Investigate Midwest.* August 18, 2018. https://investigatemidwest.org/2018/08/13/lack-of-portable-insurance-adds-health-care-burden-to-migrant-workers/.

Castillo, F., A. M. Mora, G. L. Kayser et al. "Environmental Health Threats to Latino Migrant Farmworkers." *Annual Review of Public Health* 42 (2021): 257–76.

Civil Eats. "Animal Agriculture Is Dangerous Work, Those Who Do It Have Few Protections." November 14, 2022. https://civileats.com/2022/11/14/injured-and-invisible-1-few-protections-animal-agriculture-workers-cafos-dairy-migrants-injuries/.

Connor, A., L. Layne, and K. Thomisee, K. "Providing Care for Migrant Farm Worker Families in Their Unique Sociocultural Context and Environment." *Journal of Transcultural Nursing* 21, no. 2 (2010): 159–66.

Cummins, L., "Tuberculosis and Poor Health Among Migrant and Seasonal Farmworkers in the United States." University of Richmond, 2021.

Darçın, M., E. S. Darçın, M. Alkan, and D. Doğrul. "Accidents Involving Migrant Seasonal Agricultural Workers." *Biomedical Research* 29, no. 7 (2018): 1386–88.

De Lima, C. Z., J. R. Buzan, F. C. Moore, ULC Baldos, M. Huber, and T. W. Hertel. "Heat Stress on Agricultural Workers Exacerbates Crop Impacts of Climate Change. *Environmental Research Letters* 16, no. 4 (2021): 044020.

Delcour, I., P. Spanoghe, and M. Uyttendaele. "Literature Review: Impact of Climate Change on Pesticide Use." *Food Research International* 68 (2015): 7–15.

Donkor, K. B., and J. M. Perloff. "The Effects of the Affordable Care Act on Seasonal Agricultural Workers." *Journal of the Agricultural and Applied Economics Association* 1, no. 4 (2022): 435–45.

Eskenazi, B., J. Chevrier, S. A. Rauch et al. "In Utero and Childhood Polybrominated Diphenyl Ether (PBDE) Exposures and Neurodevelopment in the CHAMACOS Study." *Environmental Health Perspectives* 121, no. 2 (2013): 257–62.

Fleischer, N. L., H. M. Tiesman, J. Sumitani et al. "Public Health Impact of Heat-Related Illness Among Migrant Farmworkers." *American Journal of Preventive Medicine* 44, no. 3 (2013): 199–206.

Galarneau, C., Farm Labor, Reproductive Justice: Migrant Women Farmworkers in the US." *Health & Human Rights* 15 (2013): 144–66.

Glaser, J., E. Hansson, I. Weiss et al. "Preventing Kidney Injury Among Sugarcane Workers: Promising Evidence from Enhanced Workplace Interventions." *Occupational and Environmental Medicine* 77, no. 8 (2020): 527–34.

The Guardian. "Child Labour Rampant in Tobacco Industry." June 25, 2018. https://www.theguardian.com/world/2018/jun/25/revealed-child-labor-rampant-in-tobacco-industry.

The Guardian. "Meet the Workers Who Put Food on America's Tables—but Can't Afford Groceries." May 13, 2020. https://www.theguardian.com/environment/2021/may/13/meet-the-workers-who-put-food-on-americas-tables-but-cant-afford-groceries.

Handal, Alexis J., Lisbeth Iglesias Rios, and Mislael Valentin Cortés. "Experiences of Female Agricultural Workers in Michigan: Perspectives from the Michigan Farmworker Project." *ISEE Conference Abstracts*, no. 1 (2021): 43. https://doi.org/10.1289/isee.2021.O-LT-068.

Hansson, E., A. Mansourian, M. Farnaghi, M. Petzold, and K. Jakobsson, "An Ecological Study of Chronic Kidney Disease in Five Mesoamerican Countries: Associations with Crop and Heat." *BMC Public Health*, no. 1 (2021): 840.

Hill, A. E., I. Ornelas, and J. E. Taylor. Agricultural labor supply. *Annual Review of Resource Economics*, 13 (2021): 39–64.

Holmes, S. M., and J. Ramirez-Lopez. *Fresh Fruit, Broken Bodies: Migrant Farmworkers in the United States.* Berkeley: University of California Press, 2023.

Iglesias-Rios, L. M. S. O'Neill and A. J. Handal. "Climate Change, Heat, and Farmworker Health." *Workplace Health & Safety* 71, no. 1 (2023).

Inside Climate News. "Florida Legislators Ban Local Heat Protections for Millions of Outdoor Workers." March 19, 2024. https://insideclimatenews.org/news/19032024/florida-legislators-ban-heat-protections-for-outdoor-workers/.

Jones, M. W., J. T. Abatzoglou, S. Veraverbeke et al. "Global and Regional Trends and Drivers of Fire Under Climate Change." *Reviews of Geophysics* 60, no. 3 (2022): e2020RG000726.

Kuehn, B. M. "Why Farmworkers Need More than New Laws for Protection from Heat-Related Illness." *JAMA* 326, no. 12 (2021): 1135–37.

Ma, C. S., W. Zhang, Y. Peng et al. "Climate Warming Promotes Pesticide Resistance Through Expanding Overwintering Range of a Global Pest." *Nature Communications* 12, no. 1 (2021): 5351.

Marlier, M. E., K. I. Brenner, J. C. Liu et al. "Exposure of Agricultural Workers in California to Wildfire Smoke Under Past and Future Climate Conditions." *Environmental Research Letters* 17, no. 9 (2022): 094045.

Matzrafi, M. "Climate Change Exacerbates Pest Damage Through Reduced Pesticide Efficacy." *Pest Management Science* 75, no. 1 (2019): 9–13.

May, J. J., and T. A. Arcury. "Occupational Injury and Illness in Farmworkers in the Eastern United States." In *Latinx Farmworkers in the Eastern United States: Health, Safety, and Justice,* 41–81. London: Springer Nature, 2020.

Moctezuma, S. "Excluded and Isolated: Farmworker Vulnerability to Climate Change, Inadequate Regulations, and Takings Claims." *Tulane Environmental Law Journal* 36 (2023): 183.

Moyce, S. C., and M. Schenker. "Migrant Workers and Their Occupational Health and Safety." *Annual Review of Public Health* 39 (2018): 351–65.

Nicolopoulou-Stamati, P., S. Maipas, C. Kotampasi, P. Stamatis, and L. Hens. "Chemical Pesticides and Human Health: The Urgent Need for a New Concept in Agriculture." *Frontiers in Public Health* 4 (2016): 178764.

Occupational Safety and Health Administration. "Heat and Injury Illness Protection in Outdoor and Indoor Work Settings Rulemaking." 2025. https://www.osha.gov/heat-exposure/rulemaking.

Palacios, C. J. "Reforming America's Employment-Based Immigration System in a Post-Trump Era." *Notre Dame Journal of International and Comparative Law* 12 (2022): 36.

Quandt, S. A., and T. J. Arnold. "The Health of Children in the Latinx Farmworker Community in the Eastern United States." In *Latinx Farmworkers in the Eastern United States: Health, Safety, and Justice.* London: Springer Nature, 163–95.

Ramos, A. K. "Child Labor in Global Tobacco Production: A Human Rights Approach to an Enduring Dilemma." *Health and Human Rights* 20, no. 2 (2018).

Rani, L., K. Thapa, N. Kanojia et al. "An Extensive Review on the Consequences of Chemical Pesticides on Human Health and Environment." *Journal of Cleaner Production* 283 (2021): 124657.

Rappazzo, K. M., J. L. Warren, R. E, Meyer et al. "Maternal Residential Exposure to Agricultural Pesticides and Birth Defects in a 2003 to 2005 North Carolina Birth Cohort." *Birth Defects Research Part A: Clinical and Molecular Teratology* 106, no. 4 (2016): 240–49.

Schulte, P. A., and H. Chun. "Climate Change and Occupational Safety and Health: Establishing a Preliminary Framework." *Journal of Occupational and Environmental Hygiene* 6, no. 9 (2009): 542–54.

Skendžic, S., M. Zovko, I. P. Živkovic, V. Lešic, and D. Lemic "The Impact of Climate Change on Agricultural Insect Pests." *Insects* 12 (2021): 440.

Swanton, A. R., T. L. Young, and C. Peek-Asa. "Characteristics of Fatal Agricultural Injuries by Production Type." *Journal of Agricultural Safety and Health* 22, no. 1 (2016): 75–85.

Tigchelaar, M., D. S. Battisti, and J. T. Spector. "Work Adaptations Insufficient to Address Growing Heat Risk for US Agricultural Workers." *Environmental Research Letters: ERL* 15, no. 9 (2020): 094035.

United States Department of Labor. "Minimum Paid Rest Period Requirements Under State Law for Adult Employees in Private Sector." January 1, 2023. https://www.dol.gov/agencies/whd/state/rest-periods.

Vallejos, Q. M., S. A. Quandt, J. G. Grzywacz et al. "Migrant Farmworkers' Housing Conditions Across an Agricultural Season in North Carolina." *American Journal of Industrial Medicine* 54, no. 7 (2011): 533–44.

Wijerathna-Yapa, A., and R. Pathirana. "Sustainable Agro-Food Systems for Addressing Climate Change and Food Security." *Agriculture*, no. 10 (2022): 1554.

Wills, J. B., and M. M. Commins. "Consequences of the American States' Legislative Action on Immigration." *Journal of International Migration and Integration* 19, no. 4 (2018): 1137–52.

Wittenberg, A. "As Extreme Heat and Smoke Threaten U.S. Farmworkers, Federal Health Workers Evaluate Protections." *Scientific American*, 2024. https://www.scientificamerican.com/article/as-extreme-heat-and-smoke-threaten-u-s-farmworkers-federal-health-leaders/.

Ziska, L. H., and L. L. McConnell. "Climate Change, Carbon Dioxide, and Pest Biology: Monitor, Mitigate, Manage." *Journal of Agricultural and Food Chemistry* 64, no. 1 (2016): 6–12.

Ziska, L. H., and R. Parks. "Recent and Projected Changes in Global Climate May Increase Nicotine Absorption and the Risk of Green Tobacco Sickness." *Communications Medicine* 4, no. 158 (2024).

IV

Climate Justice, Capitalism, and Colonialism

From Literature to History to Science

9

Justice, the Incommensurable, and the Scale(s) of Business as Usual

A Literary Studies Approach

JENNIFER WENZEL

I often tell my students that to read literarily is to pay just as much attention to *how* something is said as to *what* is said. On the page, literary scholars consider the relationship between form and content; beyond the page, we examine how literature shapes readers' imaginations and expectations about how the world works. The plot logics of literary narratives often bleed into broader, unwritten cultural narratives. The quiet yet profound effect of literary representation means that it is a form of artifice—a human construct—that naturalizes ideas about things like progress, the good life, justice, and even nature itself. A basic tenet of literary education is that narratives consist of three elements: plot, character, and setting. Setting—that is, time and place—is generally understood as the stable ground upon which the dynamic action of plot and character unfolds. Such bedrock assumptions, however, are now challenged by climate change. As I write in *The Disposition of Nature*, "What happens to narrative when setting becomes character, plot becomes setting, and . . . agency (the capacity to be a protagonist) is distributed across human and nonhuman entities? When the relationship between cause and effect (the foundation of plot) is dilated across vast spans of space and time (the dimensions of setting)?"[1]

Literary studies thus has much to say about *energy justice*, particularly its relationship to environmental or climate justice. These terms are often used interchangeably, as synonyms, or in close relationship to each other, as imperatives that are distinct yet compatible, even allied and interwoven. Yet if these terms tend to be invoked together, as overlapping imperatives or *pillars* of justice, to what extent are they actually contradictory, incompatible, or incommensurable? Answers to this question are bound up with issues of *scale*.

The first time I wrote the words *energy* and *justice* next to each other and wondered, *is that a thing?*, what I was trying to articulate was the issue of *access* to energy, or *energy poverty* as a lack of access (see also chapter 6 in this volume). Some people use too much energy; others arguably consume far too little, and the prospect of a more equitable arrangement seems dispiritingly unlikely. How does this unevenness of access to energy reveal a form of social difference and inequality that is related to, yet distinct from, those of race, class, nation, gender, and sexuality?

In addition to unequal access to energy, energy justice might also entail exposure and vulnerability to the immediate harms entailed in the extraction, production, distribution, and consumption of energy. The Niger Delta is a quintessential site of such harms, suffered for decades by communities living amid chronic oil spills that despoil land and waterways; such communities have also been subject to political repression, without much benefit from the oil extracted there. If we consider the access issue alongside the harm issue, then energy justice might be defined in a way analogous to environmental justice: as a reckoning with how the benefits and burdens of energy are distributed. The imperative of energy justice would involve forging more just relations around the distribution of those benefits and burdens. To the extent that energy justice entails access to more energy, however, it may be difficult to align with those other pillars of environmental and climate justice, particularly in cases when fossil-fueled development is invoked as the best path to alleviating energy poverty (see also chapter 3). Redressing the problem of inadequate access to energy for some might entail harm for others, which is why energy justice is a *bind, an imperative, an aspiration*, rather than a commonplace fact to be easily found in the world. In other words, addressing the imperative of *energy justice*, if defined narrowly or primarily in terms of access, can entail new forms of *injustice*.

With climate justice, questions of scale become paramount. Climate justice is concerned with the staggering and increasingly well-known inverse relations between causes and effects, and between responsibility and harm. What's particularly confounding is that these dynamics operate over vast scales of time and space but are also multiscale, joining the local with the planetary, and the now with the long ago and with futures both distant and ever more proximate. Climate justice is an intergenerational problem that links human activity over the past two centuries with effects on the atmosphere and oceans for thousands of years into the future. Its ambit is vastly diachronic, often phrased in terms of "the world that we are leaving to our children." In using up the capacity of the Earth's oceans and atmosphere to serve as sinks for greenhouse gases, the industrialized North has arguably stolen the future of the Global South.

Yet how can we calibrate *immediate* harms at sites of extraction, mentioned earlier as a dimension of energy justice, with these more complex relations of cause and effect dilated across time and space? This distinction between immediate, proximate harm and broader effects is why it's necessary to insist on the tensions among these versions of justice. Consider, for example, the practice of gas flaring—burning off natural gas produced in drilling crude oil. It's a harmful practice with both local and planetary repercussions. Because it generates extreme heat, tremendous noise, acid rain, and numerous adverse effects on organisms and ecosystems, flaring is banned or tightly regulated in most places where oil is drilled, notable exceptions being Russia and North Dakota, where flaring is visible from space.[2] In the Niger Delta, "some children have never known a dark night though they have no electricity," observed environmentalist Nick Ashton-Jones after a 1993 visit.[3] This contradiction reflects multilayered infrastructural neglect: One alternative to flaring natural gas would be to capture and use it for local electrification. And at a planetary scale, flaring contributes to global warming. In 1995 flaring in Nigeria was estimated to be the largest single source of greenhouse gas emissions in the world.[4]

Although activism and militancy in the Niger Delta have tended to focus on immediate harm at sites of extraction, notice how questions of environmental, energy, and climate justice converge in a direct action campaign launched in December 1998 by the Ijaw Youth Movement. Named "Operation Climate Change," this campaign linked local struggles for "freedom, self-determination, and ecological justice" to the "the destructive effects of climate change principally from the burning of fossil fuel."[5] Although Nigeria's High Court ruled in 2005 that flaring violated constitutional rights to life and dignity, the practice has continued and was even inadvertently encouraged because subsequent law designated the financial penalty a "charge" rather than a "fine": To oil multinationals, gas flaring is not merely the cost of doing business; it's also tax-deductible, a boost to the bottom line.[6] This accounting trick epitomizes how the true costs of oil are externalized to faraway places and to the future.

These tensions among the three pillars of justice are at stake in discourse on energy transition as *just* transition. The imperative of decarbonizing the economy in a transition away from fossil fuels and their emissions also creates an opportunity to pursue both climate and energy justice. If these goals are not explicitly foregrounded, however, energy transition risks entrenching existing inequalities and creating new ones. A fundamental assumption in energy humanities is that energy transitions do not merely involve the substitution of one fuel source for another, but instead the transformation of myriad relations—political, economic, infrastructural, environmental, social, cultural, and affective—associated with particular energy regimes. Understood in this way, energy is bound up with values and ethics, as well as subjectivity, desire, and embodied experience. How are ideas of what it means to be human, or notions of the good life or democratic politics, premised on access to cheap, reliable energy? In other words, how does energy underwrite what we understand as justice?

I have suggested that the contradictions among energy justice, environmental justice, and climate justice are partly a function of scale: the geographic and/or temporal scales at which one contemplates questions of justice. I want to consider these issues of scale in relation to Andreas Malm's *Fossil Capital*, which narrates the birth of "business as usual": the capitalist fossil energy regime that shapes life as we know it. Malm locates the "incontestable birthplace" of the fossil economy in nineteenth-century British textile mills.[7]

Although I read *Fossil Capital* with the immersive pleasure one might associate with a nineteenth-century English novel, I also read with growing annoyance. Malm's gripping tale of conflicts—between flow and stock, water and steam power, workers and owners—felt increasingly inadequate as I wondered about parts of this story left untold, beyond the shores of nineteenth-century Britain. *Where did the cotton come from? Where did some of the capital that became fossil capital come from?* In other words, how can one separate the story of British capitalism from the story of global empire? How can one narrate the "roots" of contemporary planetary crisis from such a narrow geographic frame, quite literally from one small island?

These are the kinds of questions that I confront by examining the uneven, unpredictable ways that transnational forces shape local places, which means thinking between, say, the Niger Delta and Detroit, North Dakota, or the Mississippi Delta: sites profoundly but disparately shaped by (and indispensable to) oil extraction and hydrocarbon-fueled global capitalism. As I write in *The Disposition of Nature*, "This is the multiscalar work of reading for the planet, imagining from near to there."[8] This multiscale approach is also a contrapuntal approach, seeing one place always as juxtaposed and imbricated with another. I borrow the idea of the contrapuntal from Edward Said, for whom "contrapuntal reading" means reading the literature of empire from multiple sides of the colonial encounter, expanding the scale of one's analysis, seeing one place always as juxtaposed and imbricated with another. Said argues that "In reading a text, . . . one must open it out both to what went into it and to what its author excluded."[9] How can *Fossil Capital* be opened out to what's excluded from it, to expand the scope or scale of its analysis?

One approach would be to read Malm's *Fossil Capital* in counterpoint with the "History" and "Politics" chapters of *The Great Derangement*, where Amitav Ghosh connects the dots between European colonialism and empire, on the one hand, and climate justice and the politics of the present, on the other. Ghosh not only shows how modern Europe's military and economic dominance was enabled by the combustion of fossil fuels; he also highlights how this dominance required the active *suppression* of a carbon economy in the colonial periphery. He cites the example of local shipbuilding and shipping industries in Calcutta and Bombay, which enthusiastically turned to steam power in the 1830s; this nascent coal-driven commerce by enterprising Indian builders, merchants, and sailors posed a threat not only to English shipwrights but to the broader British economy, whose unprecedented, increasingly fossil-fueled industrialization "needed to be fed by large quantities of raw materials, produced by solar-based methods of agriculture" in places like India.[10]

Although he doesn't use the term, Ghosh offers a powerful explanation of *underdevelopment*: how capitalism simultaneously generates wealth for some, impoverishment for others. Ghosh writes, "The emerging fossil-fuel economies of the West *required* that people elsewhere be prevented from developing coal-based energy [and industrial] systems of their own, by compulsion if necessary. . . . Poor nations are not poor because they were indolent or unwilling; their poverty is itself an effect of the inequities created by the carbon economy . . . systems set up by brute force to ensure that poor nations remained always at a disadvantage in terms of both wealth and power."[11] Ghosh uses the word "poverty" in a broad sense, but we can also read it as *energy poverty*, or unequal access to energy. This aspect of energy injustice, Ghosh suggests, is the result of a *plan*.

This frank insistence on European colonial violence and its enduring economic and environmental effects in particular sites in the Global South is salutary at a moment when discourse on the Anthropocene and the human-as-species, as well as the heady pleasures of thinking at the vast scales of the planetary or deep time, have become one more occasion to forget or disavow the historical engines of inequality that continue to shape the present, what Malm calls "business as usual." Moreover, Ghosh's contrapuntal account refuses a too-easy, Manichean version of this history by tracing the role of fossil fuels in pre-twentieth-century China, India, and Burma, in order to undermine Eurocentric narratives of industrial modernity as a story of one-way technological diffusion from the West; this Eurocentrism also runs through some discourse on the Anthropocene itself.

Because, in Ghosh's account, the task of climate justice would require a global redistribution of wealth and power, he argues provocatively that "global inaction on climate change is by no means the result of confusion or denialism: to the contrary, the maintenance of the status quo *is* the plan." Twenty-first-century incumbents seek to retain for themselves the outsized wealth and power they have unjustly garnered under fossil capitalism. (And of course, confusion and denialism can also help those who benefit from the status quo to entrench their interests.) A similar frankness underwrites Ghosh's assessment of the Enlightenment ideals of equality, freedom, and justice, which he dubs "grotesque fictions"—in other words, ideological constructs that he says were "designed to secure exactly the opposite of those professed ends": that is, inequality, unfreedom, injustice.[12] These startlingly radical claims about the business-as-usual assumptions at work in business-as-usual offer a challenge to liberal apologetics that posit global warming and other environmental disruptions as the accidental, unintended consequence of capitalist development—collateral damage rather than strategic objective. Ghosh's analysis helps us to understand the twentieth-century revolutions of decolonization and democratization in relation to the twenty-first-century imperatives of decarbonization and energy, environmental, and climate justice. As Ghosh reminds us, underdevelopment and unevenness are the hard truths obscured by the ideals of equality and freedom, which at this point are difficult not to see as "grotesque fictions."

Another text that could be set in counterpoint to *Fossil Capital* is Julie Livingston's *Self-Devouring Growth: A Planetary Parable as Told from Southern Africa*, which redirects attention away from Britain and toward the "militarized, toxic, and predatory *reaches* of the carbon economy," which Livingston finds in Botswana.[13] Her parable leads me to ask how growth can be "self-sustaining" and "self-devouring" at the same time. Malm argues that growth becomes "self-sustaining" because the increasing use of fossil fuels enabled a circumvention of organic limits, creating an expanding feedback loop of further growth, along with ever-increasing greenhouse gas emissions.[14] By contrast, Livingston finds in the carbon economy a form of "self-devouring growth . . . predicated on uninhibited consumption" and characterized by a fundamental "imperative—*grow or die; grow or be eaten*."[15] This imperative is a false choice that obscures a hard truth: Growth is itself an insatiable appetite; growth *demands* a devouring that risks death. The carbon economy "is consuming itself . . . eating away at the ground beneath our feet."[16]

For Malm, the fossil economy is a fire that generates its own fuel; for Livingston, the carbon economy threatens the "planetary body" consumed by cancerous growth.[17] Perhaps these controlling metaphors in Malm and Livingston—the self-fueling fire, the cancerous body—are complementary, two views of the same problem. Malm states an easily missed caveat: Growth is self-sustaining *only* in the capacity of fossil energy to fuel economic expansion (and emissions) beyond previous natural constraints.[18] Livingston's attention to externalities—toxic waste, resource depletion, drought and desertification, urban sprawl, and debt burdens borne by individuals and nation-states alike—indicates how this growth *devours* itself and the lifeworlds on which it and we depend.

Both authors emphasize growth's normalization and naturalization. Malm observes that the fossil economy "appears indistinguishable from life itself: business-as-usual."[19] Livingston describes self-devouring growth (SDG) as "common sense," "so fundamental as to be unremarkable," an "intractable" mindset "so powerful that it obscures the destruction it portends."[20] This capacity to rationalize or direct attention away from destruction is fundamental to self-devouring growth, rather than an unintended by-product or something that happens "eventually." For Livingston, SDG is an economic and material phenomenon working on the world, but it's also a way of seeing (and not seeing): "Without us really noticing it, growth has become this unmarked category granted magical powers . . . as so much is done in its name, a cascade of unseen consequences, side effects, also become second nature."[21] When those consequences do emerge into visibility, they're rationalized as unavoidable: In the case of automobility, self-devouring growth is the logic whereby "the road commands a certain amount of death and damage as a necessary price for its freedoms and opportunities" (for more discussion of automobility, see chapter 4 in this volume).[22]

Complicating Malm's account of class conflict in the fossil economy, two forms of unevenness are at work in Livingston's parable. First, the uneven distributions of benefits and burdens, as we have seen with energy and environmental justice: Livingston writes, "Ever more intensive forms of capitalist consumption animate a system that will harm everyone, even those whose consumption mainly remains aspirational."[23] The rich consume a lot; the poor consume a lot less, while also confronting the destructions caused by the consumption of the rich. Nonetheless, the poor often aspire to consume like the rich: They also inhabit the common sense of self-devouring growth, even if that common sense *makes no sense* within the calculations of their household economies. This confrontation with the limits of common sense reveals another kind of unevenness—one not of distribution, but of perception: The contradictions of self-devouring growth are most legible to those at its margins. The capacity to understand out-of-control growth as anything other than a historical construct, not something necessary and natural like the air we breathe (at least in some bygone pre-anthropogenic world), is the product of a habitual disregard that is mostly a privilege of the rich and powerful.

I read Livingston's parable contrapuntally with and against Malm, as a corrective to a narrow history of fossil capital in Britain that is methodologically blind to the dynamics of fossil empire across the globe; this blindness is a corollary to the willful blindness that is SDG itself. Livingston describes the space that growth occupies in the social imaginary, shaping not only developmental policy and built environments but also dreams and desires. The conflation of out-of-control growth with necessity, nature, and common sense is, I argue, a crucial reason the fossil economy persists.

I now turn to a novel that imagines past the end of the fossil economy and offers, at least implicitly, a contrapuntal reading between Global North and Global South. Paolo Bacigalupi's young adult novel *Ship Breaker* is a speculative fiction that imagines the US Gulf Coast in a climate-changed, postpetroleum future. It features a group of adolescents who work at a makeshift coastal salvage yard, stripping beached oil tankers of copper and other valuable parts. The narrator describes the scene: "All around, the ocean was a glittering mirror. Breakers rolled up to the shore, white as a baby's teeth. The black hulks of the broken ships stood out in the sun, looming monuments to a world that had fallen apart."[24]

If this scenario and scene seem familiar, it's because Bacigalupi borrows them from photographs by Edward Burtynsky that document the shipbreaking industry in Chittagong, Bangladesh. In an interview, Bacig-alupi explained that his books aren't dystopias, but instead "accidental futures. The kind you get when there's a lack of forethought, a surfeit of poor planning, or just plain cynicism." He continued: "Frankly, most of the worst details of the worlds I describe are based heavily on our present. Chittagong, Bangladesh, isn't a dystopia, its [sic] just a really poor place where we dump our scrapped ships because they have fewer worker safety and environmental controls."[25]

Set in an accidental future, Bacigalupi's novel offers a cautionary tale: The people of the novel's Accelerated Age (i.e., our present) were too stupid, inert, or cynical to avoid climate change. The historical exposition that one character offers to the novel's protagonist is also implicitly addressed to readers, as a defamiliarizing invitation to re-see the unthinking excess of the oil-driven present. But what are the stakes and the perils of borrowing pho-tographic images of petro-waste work and exposure to oily toxic risk already happening in the Global South, in order to imagine what may lie ahead in a near future of the US Gulf Coast? In other words, how can we calibrate the relationship between Bacigalupi having borrowed the present misery of Bangladesh for his accidental American future, on the one hand, and fact that climate change itself can be understood as the industrialized world hav-ing borrowed against, even *stolen*, the future of the underdeveloped world, on the other? This is the question of climate justice in a nutshell.

I've written elsewhere about how twenty-first-century environmental cri-ses have reversed colonial-era progress narratives, with their promises of "civilization" and development, in which Europe and the West were said to offer to the rest of the world, in Karl Marx's words, "an image of its own future." Such temporal imaginings of Third World harbingers of First World futures emerge from across the political spectrum, from the neocon fever dreams of Robert Kaplan to the Comaroffs' "theory from the South" to Amitav Ghosh's description of climate change as a "revers[al of] the tem-poral order of modernity: those at the margins are the first to experience the future that awaits all of us."[26] In these transpositions of time and space, the consequences of carbon accumulation in the future are imagined to look a lot like being on the wrong end of capital accumulation in the present. The crucial question in making sense of such intergenerational, transnational imaginings, I would argue, is whether they acknowledge the shared but uneven history that joins these far-flung (and sometimes surprisingly proximate) pasts, presents, and futures.

In Bacigalupi's novel, the ruthless logic of the scrapyard has been elevated to a religion: "The Scavenge God promised a life of ease, if you could just find the right offering to burn with your body when you went to his scales."[27] The Scavenge God separates the saved from the damned by weighing the bodies of the dead in his scales—just as the novel's ship-breaking characters learn how much (or little) they'll earn each day by placing the scrap they've collected into the buyers' scales. Balance scales measure the weight of an object by bringing two sides (or pans) into equilibrium *This weighs as much as that.* An object's weight can be measured by filling the scale's other pan with counterweights or "reference weights" of known quantity. But in the novel, it's not only weight but also worth that is measured by the balance scale. Measuring the worth of one thing in terms of another thing—described in the novel as "balancing the scales"—is invoked as a form of justice modeled on divine authority, even if often seeming the opposite of "ethical" or "moral." In *Ship Breaker*, "balancing the scales" names practices of eye-for-an-eye retributive justice, but also other kinds of reckoning, where characters weigh the value or worth of their own lives against those of others; calculate whether they're worth more dead or alive; and keep track of interpersonal, nonmonetary obligation as a form of debt.

This recurrent image of weights and measures used to compare both tangible objects and intangible things suggests a paradox: The balance scale is supposed to work by finding equilibrium as a proxy for equivalence, but somehow it produces incommensurability and inequality instead: "The wealthy measure everything with the weight of their money." By using this idea of "balancing the scales" as a religious, ethical, and social principle, Bacigalupi's novel asks us to consider whether the scales of justice, which are supposed to balance truth and fairness, are any different from those of the shopkeeper weighing goods to calculate price, too often with rigged counterweights or his thumb on the scale. In other words, the novel suggests the incommensurability of the forms of value being reckoned in "balancing the scales," and it allows readers to consider the need for counterweights of something other than money to put in the pan.

This chapter addresses questions of scale in two senses—first, the relative geographic and temporal scale or scope of one's analysis, and second, the balance scale as a means of measuring equivalence and finding justice. Both senses of *scale* are about measurement, but that commonality obscures important differences. *Scale* in the first sense, meaning gradation, magnitude, and proportion, derives from the Latin word *scala*, which means *to climb*; one ascends from smaller to greater scales, from the cells of the human body to the planet as a whole. *Scale* in the second sense has a different etymology, from the Old Norse word *skál*, meaning bowl, as well as a device that measures equivalence by balancing two bowls.[28]

Dating back to ancient Egypt, the balance scale was associated with justice in classical Greece and Rome; it is in the English word *scale*, however, where these *discrepant*, perhaps incommensurable forms of measurement collide. In English, *scale* braids together these two etymologies, yet when these senses of *scale* are conflated, it can yield incommensurability and injustice. It's difficult to place into the pans of the balance scale, or into the columns of cost-benefit analysis, the myriad, nonlinear, and multi-scale effects of complex processes unfolding over vast scales of space and time. And so, here's the concluding gesture where I, as literary critic, suggest that one way of confronting this problem of incommensurable scales would be to draw on the capacities of literary imagining and careful, attentive reading as habits of mind that could help us to grasp both the tangled pasts that underwrite business as usual and the possible futures that might offer more just and livable alternatives.

Notes

1. Wenzel, *The Disposition of Nature*, 19.
2. The shale gas revolution in the United States has been enabled by technological innovations (e.g., horizontal drilling and hydraulic fracturing, "fracking") that have outpaced the regulatory and infrastructural capacity to capture natural gas.
3. Rowell, "Shell Shocked," 21.
4. World Bank, "Defining an Environmental Development Strategy," 58.
5. Quoted in Ukeje, "Oil Communities and Political Violence," 29.
6. Kazeem, "A Legal Loophole Has Enabled Years of Environmental Damage."
7. Malm, *Fossil Capital*, 13.
8. Wenzel, *Disposition of Nature*, 19.
9. Said, *Culture and Imperialism*, 32, 67.

10. Ghosh, *The Great Derangement*, 107.
11. Ghosh, 107, 110; emphasis added.
12. Ghosh, 149.
13. Livingston, *Self-Devouring Growth*, 106.
14. Malm, *Fossil Capital*, 11.
15. Livingston, *Self-Devouring Growth*, 5.
16. Livingston, 1, 5.
17. Malm, *Fossil Capital*, 5.
18. Malm, 11.
19. Malm, 13.
20. Livingston, *Self-Devouring Growth*, 4, 1, 5.
21. Livingston, 4–5.
22. Livingston, 92, 96.
23. Livingston, 6.
24. Bacigalupi, *Ship Breaker*, 74.
25. Ottinger, "Nebula Awards Interview," November 21, 2011.
26. Ghosh, *Great Derangement*, 62.
27. Bacigalupi, *Ship Breaker*, 12.
28. *Oxford English Dictionary*, "scale," https://www.oed.com/dictionary/scale_n1.

BIBLIOGRAPHY

Bacigalupi, Paolo. *Ship Breaker*. New York: Little, Brown, 2010.

Ghosh, Amitav. *The Great Derangement*. Chicago: University of Chicago Press, 2015.

Kazeem, Yomi. "A Legal Loophole Has Enabled Years of Environmental Damage by Global Oil Companies in Nigeria." *Quartz Africa*, January 30, 2018, https://qz.com/1192558/.

Livingston, Julie. *Self-Devouring Growth: A Planetary Parable as Told from Southern Africa*. Chapel Hill, NC: Duke University Press, 2019.

Malm, Andreas. *Fossil Capital: The Rise of Steam Power and the Roots of Global Warming*. London: Verso, 2016.

Ottinger, John, III. "Nebula Awards Interview: Paolo Bacigalupi." *Science Fiction & Fantasy Writers Association*, November 21, 2011. https://www.sfwa.org/2011/11/21/nebula-awards-interview-paolo-bacigalupi/.

Rowell, Andy. "Shell Shocked." *Village Voice*, November 21, 1995.

Said, Edward W. *Culture and Imperialism*. New York: Knopf, 1994.

Said, Edward W. *Orientalism*. New York: Pantheon, 2003.

Ukeje, Charles. "Oil Communities and Political Violence: The Case of Ethnic Ijaws in Nigeria's Delta Region." *Terrorism and Political Violence* 13, no. 4 (2001): 15–36.

Wenzel, Jennifer. *The Disposition of Nature: Environmental Crisis and World Literature*. New York: Fordham University Press, 2020.

World Bank. "Defining an Environmental Development Strategy for the Niger Delta," May 25, 1995.

10

Climate Justice in the Arctic

Multispecies Approaches in Anthropology and History

EMMA GILHEANY AND JULIA LAJUS

This chapter examines the potentialities and limitations of anthropological and historical approaches to climate justice for the social and environmental specificities of the Arctic. On the one hand, both disciplines—in their embrace of highlighting epistemological complexities, foregrounding of power dynamics, and experimentation with multispecies perspectives—are essential for both understanding and enacting climate justice. However, both disciplines need to improve commitments to ethical coproduction of knowledge with Indigenous groups, and to develop more meaningful frameworks for understanding the world through a multispecies lens. Anthropological and historical understandings of the Arctic can nuance how inhabitants can feel and experience environmental shifts. Specifically, through this nuancing, we are interested in how these disciplines can counteract climate reductionism.

We understand climate reductionism as the tendency to extract climate change from its social and environmental contexts.[1] This extraction means that *climate change* becomes an explain-all framework that obscures complex socioeconomic dynamics. Although the centrality of climate in shaping history is no longer questioned, climate reductionism, which "is driven by a hegemony exercised by the predictive natural sciences over contingent, imaginative, and humanistic accounts of social life and visions of the future," is also misleading.[2] Reductionism not only distorts the understanding of the past but also fuels apocalyptic imaginaries of a future already beyond saving.

There is no doubt, however, that climate change and climate injustice are a material reality for the Arctic. The Arctic is warming at almost four times the rate of other portions of the planet, a phenomenon known as "Arctic amplification."[3] Northern-specific environmental factors like melting sea ice and calving ice sheets create a feedback loop that amplifies planetary warming. These region-specific environmental dynamics mean that the Arctic is the subject of hyperfocus by climate scientists, often perceived as a sterile laboratory.[4] In the popular imaginary, the Arctic is often portrayed as the Anthropocene's canary in the coal mine, a melting spectacle portending the doomsday of our planet and a charismatic mega category for deleterious climate futures.

But the Arctic is not an environmental or social monolith—it is peopled and politically complex. The invention of the Arctic as a transnational region in the 1980s gave birth to institutions including those created by Indigenous people, for example, the Inuit Circumpolar Council, a political group that represents the interests of Alaska, Canada, Greenland, and Chukotka Inuit.[5] Around 10 percent of the Arctic's four million inhabitants identify as Indigenous, with forty different ethnic groups speaking up to ninety languages and dialects. Indigenous groups in the Arctic live in both urban and more remote communities, and many people depend on or choose to live lives with continued subsistence-oriented diets.

However, the prevailing global imaginary of the Arctic (by scientists as the perfect laboratory, by popular media as melting) erases Indigenous presence and politics as well as histories of colonization. Zoe Todd writes that "it is easier for Euro-Western people to tangle with a symbolic polar bear on a Greenpeace website or in a tweet than it is to acknowledge Arctic Indigenous peoples."[6] Framing the Arctic through the lonely polar bear drifting on a vanishing piece of ice means ignoring the historicity of environmental change and denies the place of Indigenous peoples in their past, present, or future.[7] This is an overdetermined image of the Arctic—one that is in perpetual and simple decline. We find this erasure to be problematic for understanding and enacting climate justice. We also think this erasure is something anthropology and history can help rectify. Historical and anthropological lenses can help us better understand how environmental changes do (or do not) dictate daily life for those who call the Arctic home. For Arctic inhabitants, much of the changing climate has been experienced as an increasing unpredictability in weather, ice and oceans—in turn affecting nonhuman species.[8] These changes govern the ability of people to move through the environment creating new and different mobilities for people and other species.

DISCIPLINES AND COLONIALITY

It is important to foreground that both history and anthropology have colonial origins and have contributed to problematic and racist perceptions and policies in the Arctic. Western culture, including historical narratives, has represented the Arctic as frontier rather than homeland, emphasizing the cultural specificity of the Arctic and "Arcticality."[9] More than just discursive violence, anthropology and anthropologists have contributed to the forced removal and resettlement of Inuit, taken remains of ancestors for study, and even trafficked Indigenous people from their communities or exposed them to fatal illnesses. While in this chapter we seek to highlight ways that our disciplines can advance conversations and praxis around climate justice, we acknowledge that our disciplines have shaped and advanced northern settler colonialism and imperialism.

Anthropology

Anthropology has long been interested in better understanding the relationship between people and their environment, with studies of Inuit in the circumpolar North figuring largely into the development of the discipline. Early Arctic anthropologist Franz Boas has been heavily critiqued by Audra

Simpson as contributing to the erasure of Indigenous political systems.[10] Anthropological work in the Arctic was historically undertaken to serve national or academic, rather than Indigenous, interests and needs.

The subfield of environmental anthropology has recently begun to consider power relations and nature—capitalism and extraction, more-than-human worlds, plant and animal ontologies, traditional ecological knowledge, and natural sciences. Susan Crate has proposed an approach she calls "climate ethnography"—a multisite research method that is tied to "the global phenomenon [of climate change] and communicates a sense of immediacy of an ethnography with a mission . . . an ethnography of the world."[11] This is a departure from earlier anthropologies that hailed the "local" as a fundamental anthropological departure from other social scientific approaches.

Ethnography is a methodology of "being there"—it is this method that is often seen to differentiate anthropology from other disciplines—a method of participant-observation over a long period of time to offer insights into what everyday life looks like. Ethnography is fundamentally a dialogue—one that might not only challenge dominant narratives and conceptions of the Arctic by illuminating lived experiences of climate change but that also can contribute meaningfully to climate justice analyses and actions. This potentiality includes room for the unexpected. In chapter 13 of this volume, for example, Sheng Long highlights the ways that anthropology can contribute to alternative imaginaries of climate change through an examination of how rural residents in Meixian, China, use terms like "Heavenly Year" to describe weather shifts, rather than nonsecular or scientific terms like "climate change." Her ethnographic study shows that anthropology can be a generative window into the multiplicity of theories and languages of climate change.

Recently, some anthropologists have adopted explicitly unsettling or anticolonial approaches. Ellam Yua et al. "define co-production of knowledge (CPK) as a process that brings together Indigenous Peoples' knowledge systems and science to generate new knowledge and understandings of the world that would likely not be achieved through the application of only one knowledge system." They emphasize that a true CPK approach is rare, and that approaches to research must address "past and current inequalities that start with shared understandings of the historical and present trauma experienced by Indigenous Peoples as well as using Indigenous approaches to address systematic problems."[12] Indigenous political entities like the Inuit Circumpolar Council have developed their own written Equitable and Ethical Engagement Protocols in response to the immense global interest in Arctic research.

History

Social scientists, in their efforts to reorient climate research and action toward people and ethics, appeal to historians by arguing that "social science research should look to situate the challenges of climate change within a broader historical and geopolitical context."[13] For depicting the historically rooted unevenness that causes higher vulnerability and limitations of adaptive capacities of Arctic inhabitants, approaches of critical transnational and global history are necessary because most of the unevenness was generated beyond the Arctic itself.[14]

To reveal the construction of injustice, both the process and outcome, is a task for historians whose vocation is to question the ability to take an apparently immutable existing status quo and show that the present seemingly "natural order" is "unnatural"—that it is anthropogenically constructed.[15] A key value of learning about the past is to defamiliarize the present. If we agree that historical responsibility lies at the core of environmental justice, "historical" needs to be better defined. For that kind of history, the best-suited is a long-durée approach, which means a study of a long period of time during which social processes in their connections with environment develop.[16] This approach comes close to, although is less formal than, what social scientists refer to as longitudinal analysis—the examination of change over time. However, for producing long-durée narratives, it is not easy to find proper baselines: "How far back would we go, if we wanted to find the origins of our current discontent, both to save our oceans and to protect the rights of poor people to food and water?"[17] There is also a danger of what was named as the "dirty long-durée," when nonhistorians, using an impoverished array of historical evidence, appropriate the sense of the deep past of institutions and movements for drawing broad-gauge conclusions. The problem is that professional historians were often too squeamish to be involved in the production of such simplified narratives. This decreases the application of what might be named the "usable pasts" approach for informing political decisions, including ones connected with injustice.

However, more than a decade after Mike Hulme's publication on the danger of climate reductionism, where he rightly noted that professional historians are more inclined not toward climate determinism but toward climate indeterminism, which does not consider climate influences as having explanatory power, this situation is gradually changing, especially in connection with the studies of the Anthropocene as a new geological epoch. The Anthropocene in itself is both a historical context and a result of long-term unintentional human actions. Because of that, environmental historians are involved in defining it together with natural scientists.[18] The distance between geological time and the chronology of human histories is now claimed as collapsed.[19] The search for justice that was considered earlier in the form of oppression foisted by other humans or human-made systems now inevitably includes climate justice that is closely tied to the histories of capitalism. According to Dipesh Chakrabarty, humankind paid the price for the pursuit of freedom since the Enlightenment epoch in becoming a geological agency that put enormous pressure on our planet. Most of our freedoms so far have been energy intensive.[20] Social and technological progress became possible only at the expense of geological changes of the planet and its atmosphere, first of all due to the burning of fossil fuels. Thus, questions arose on how "should we conceive of the immediate predicament, which requires plans for climate change mitigation alongside an Enlightenment inspired quest for climate justice? Does responding to climate change require that humanists abandon their ongoing explorations of disciplinary power and social difference—particularly those produced by the sharp inequalities of global capital—in favor of species history?"[21]

Debates on climate justice commonly focus on both mitigation and adaptation.[22] More work, however, has been done by historians on adaptation.[23] Historians know well that humans' encounters with climate are pronounced during history when humans learned to endure climatic influences and to adapt. By following the path of critical history, there are the possibilities of raising historical questions about institutions, institutional structures, and their mode of governance, which is important for better defining climate justice on both international and local levels. Incorporating Traditional Ecological Knowledge into knowledge production or coproduction of such a knowledge in a dialogue that leads to decision-making is central to providing climate justice. Historians might help in this by looking at the unevenness of knowledge production related to power asymmetry that emerged deep in the past.[24]

CASE STUDIES IN COLONIALISM, MOBILITY, AND MULTISPECIES JUSTICE

The most conventional understanding of climate justice is bounded by an appreciation of the highest value of strong connections to the place—solastalgia. Forced displacement—be that relocation or undermining of the quality of place caused by climate disruption when "home is shaking away" because the environment is rapidly changing—is traumatic.[25] Further, settler colonialism has more recently been examined as a policy of containment. Settler colonialism works to curtail the mobility of Indigenous peoples through "rather unsustainable means: deforestation, extraction, water and air pollution, commodity agriculture, urban sprawl, widespread automobile adoption, and so on."[26] If we consider these processes as refracting in Arctic environmental change—leading to sea ice melt, for example—it is clear that these global processes of colonialism, capitalism, and industry are at work in the circumpolar North in unique ways that proscribe mobilities.

Climate justice frameworks have been critiqued as a Western schema that do not pay enough attention to the nonhuman in their approach.[27] Zoe Todd has made this argument—that there is an underconsideration of the nonhuman and more-than-human.[28] The advantage of reconceptualizing climate justice via a multispecies perspective is that it becomes more inclusive, decentering the human and recognizing the everyday interactions that bind individuals and societies to networks of close and distant others. It acknowledges the differential histories and practices of social, environmental, and ecological harm while opening just pathways into uncertain futures. Such a relational lens provides a vital scientific, practical, material, and ethical road map for navigating the complex responsibilities and politics in the climate crisis. Multispecies justice is also central for interstate justice because of the high mobility of nonhuman species. In this section, we discuss three examples of intersecting mobility and multispecies entanglements for Inuit in the Arctic to highlight the relationship between climate change, colonialism, and conceptualizations of climate justice.

Over the past several centuries, missionaries, and later governmental officials in the United States and Canada, have worked to relocate and "settle" Inuit groups in centralized communities, discouraging seasonal mobility and subsistence practices. Attempts to settle Inuit were based on logics of assimilation as well as staking claims to land for purposes of bolstering national sovereignty during the Cold War. Geopolitical tensions were partly the rationale for violent forced relocations of Inuit to the High Arctic—to ensure Canadian claims over territory in the far North. In 1953 and 1955 the Royal Canadian Mounted Police, operating as an arm of the Canadian Department of Resources and Development, moved ninety-two people from Inukjuak and Mittimatalik to two High Arctic communities, promising abundant wildlife and services.[29] There was little familiar wildlife (Inukjuak is about two thousand kilometers away from Grise Ford and Resolute) and almost zero infrastructure when they arrived. One relocatee recalls his mother crying when she saw an endemic fish— the Arctic char—in 1961, her first time in nine years.[30] Here, injustice in forced relocations is linked to a denial of accumulated environmental knowledge and ecological familiarity. These kinds of forced relocations and sedentarization both alter human-environmental relations and in some ways have made Inuit more vulnerable to climate change.

Having an anthropological and historical perspective on present community concerns allows us to understand the ways that recent and historical settler colonialism contributes to climate vulnerability and to specify the nature of this vulnerability, working against climate reductionism. Moreover, the study of how animal species and human societies have adapted to historical and contemporary climate changes challenges the notion of reductionism: Significant changes in both human societies and animal populations—including population crashes—do not simply follow climatic fluctuations in straightforward, domino-like cascades.[31]

Anthropologist Elizabeth Marino has conducted long-term research with the Iñupiat community of Shishmaref, Alaska, which is increasingly flood-prone because of the legacy of colonial decision-making that has led to infra-structural dependency. Mobility and rotating seasonal settlements were pre-viously used as adaptive strategies in the face of extreme northern climates. Marino points out that "colonization and sedentarization ended high mobil-ity as an adaptation strategy to climate variability and extreme weather without replacing it with other readily identifiable adaptation strategies for rural communities. . . . In the past, movement was always an option."[32] One way forward for Shishmaref has been identified by outside entities: a community-wide relocation. However, Marino's ethnographic work shows that there is interest within the community in staying at Shishmaref, and that this has inherent value for residents who want continued access to his-torical subsistence and familial places despite flooding. Ethnographic research can reveal what communities, rather than outside policymakers, find to be an appropriate framework or application of climate justice.

Ethnographic engagement can highlight how climate change articulates with the local, how multispecies and mobility politics are related. During the Cold War, policies of relocation were not the only ones that affected Inuit communities. Dozens of military radar sites were built within such communities by the Royal Canadian Air Force and the US Air Force to detect incoming Soviet ballistic missiles. Missionaries who had descended on the North in the last three hundred years attempted to settle Inuit-constructed (with the expertise and labor of Inuit) fixed infrastructures, such as churches and mission houses. The fixed infrastructures made it easier for surveyors to enter communities in order to plan and then build radar complexes in places like Hopedale, Nunatsiavut.[33] When the air force abruptly departed Hopedale in 1968, they left decaying infrastructures in their wake. Heavy metals and polychlorinated biphenyls have since leaked into the ground near the community. This has made flora and fauna within Hopedale unsafe for consumption. As a result, residents have had to leave the community to obtain wild food. However, climate change makes practices of leaving the community more difficult, as sea ice and weather have both become increasingly unpredictable. There is higher variability year to year, and months when the sea ice would dependably freeze have in more recent years yielded open water. Further, leaving the community is difficult for the many residents who cannot afford gas—often community members who were forcibly relocated from more northern Inuit communities who are economically disadvantaged. Toxicity causing multispecies harm means that exposure risk is higher for residents who cannot afford to be mobile. This is another example of violation of freedom of mobility on a local level in comparison to the international level, discussed in chapter 7. Thus climate injustice is better understood through combined anthropological and historical perspectives.

OCEANIC NONHUMAN MOBILITY AND CLIMATE JUSTICE

Understanding mobility in the Arctic means foregrounding not only human mobility but the mobility of other species. It is largely connected with the Arctic Ocean and its resources, which have a past far beyond that of human history.[34] In the narrative on the history of the Bering Strait, Bathsheba Demuth demonstrated that states are ecological processes as much as ideological ones, shaped by energy flows. Because of that they experience strong influence from the more-than-human world.[35] This approach gives us hope for coproduction of spaces of justice through the development of more-than-human climate cosmopolitics.[36]

Understandings of contemporary colonialism and climate justice must be discussed through the recognition of embeddedness of humans and animals in shared social, cultural, political, and economic relationships.[37] Further, conceptualizations of climate justice in the Arctic can serve as a way to consider the relationships between Indigenous self-determination and interspecies justice, as it requires attention to the ways that climate change affects nonhuman animals in communities that rely on subsistence in a region with unique ecosystems and relationality.[38]

Migratory nature is a fundamental characteristic of many commercially important aquatic animals, and climate is among the main drivers changing their migration patterns as well as their population dynamics. Studies of multispecies mobility for historians are possible by considering places like fisheries and examining documents and archives from these practices and institutions. Through such analysis we can delve into the history of other species in their relations with such a unifying historical agency as climate. Archaeologists hypothesized that changes in human-marine animal interactions in combination with climatic, global economic, regional political, and cultural changes might be an influential factor for the ill fate of Norsemen settlements in Greenland.[39] Historians use the term "animal culture" for the description of changing behavior of marine animals as a result of interactions with humans in the seventeenth century.[40] Another period of significant migrations of marine species took place a hundred years ago during regional Arctic warming that had predominantly natural causes.[41] This warming and then subsequent cooling of the Arctic in the 1960s–1970s became one of the causes of shifts in the economy from one species to another as it happened in Greenland with the rising of cod fisheries in the warm period and cod-to-shrimp transition in the cold one.[42] During the same period, experiments on moving marine species into the European Arctic as well as the increasing appearance of invasive species took place.[43] This situation could also be considered from the framework of justice when invasive species like the Kamchatka crab are producing more harm or becoming more valuable for the local coastal communities in relations with climate.

The migratory nature of nonhuman species in the Arctic and sub-Arctic waters complicates the possibilities of equity for all parties involved in the 'blue economy," which is defined as sustainable use of ocean resources for economic growth, improved livelihoods, and jobs while preserving the health of ocean ecosystem. The patterns may change so that particular countries and Indigenous or local coastal communities gain an advantage or disadvantage of access to resources. For instance, the snow crab population in the Bering Sea has collapsed due to warming in its eastern US waters, while the western waters that are warming more slowly still provide good catches.[44] Similar—but not as dramatic—changes are also observed in the world's largest fishery targeting Alaska pollock.[45] One of the mechanisms that could facilitate equity and justice in the access to marine resources in the future is international agreements, for instance, joint distributions of quotas, as between Norway and Russia in the Barents Sea, where the Joint Fisheries Commission began to work during the Cold War in 1974.[46] Such a management scheme is crucial for equity because the Russian fleet catches nonmature migratory fish that, when overfished, will cause the collapse of the whole stock. A more sophisticated agreement on conservation and equity of access in the future is the recently signed Agreement to Prevent Unregulated High Sea Fisheries in the Arctic Ocean, which will prevent commercial fishing by the signatory states for the next sixteen years.[47] This is a critical period in furthering our understanding of an ecosystem that is emerging below a retreating sea ice cover. Could we consider such an agreement as an act of justice? Probably, because it protects the resource that could be used by many, and ensuring it for future generations. This agreement also is a step toward multispecies justice because more fish remain in the oceans for the consumption of marine mammals and birds. And it can be considered good for fish themselves. However, the question remains: Would we be able to hear the 'voices" of nonhuman species when scientists collect more information on changes in their distribution pattern? The problem of who could speak for other species, and how, remains central to multispecies justice.[48]

• • •

Reflecting on climate justice in the Arctic is a reminder that "the Arctic is now placed into a larger global context that had not been part of the earlier exceptionalism discourse."[49] Recent scholarship that highlights Indigenous politics and coproduction and acknowledges multispecies worlds provides hope for a better articulation of the climate justice agenda in the Arctic. In this chapter, we have sought to highlight the ways that anthropological and historical attention to the Indigenous and localized Arctic might illuminate both colonial and climate-related experiences often glossed over by popular or scientific understandings of the circumpolar North. The realities of mobility politics and multispecies worlds in the Arctic mean that interspecies climate justice is necessary.

Anthropological and historical approaches and methodologies can illuminate the ways in which climate change articulates with Arctic communities, drawing out the complexities of colonial histories and presents to squash climate reductionism. This chapter is a decisive argument against climate reductionism. An interspecies climate justice that advances Indigenous self-determination—which fundamentally requires listening, relationality, and obligation and might not look like what Western governments or policymakers initially imagine—can be better understood through community-based and long-term research relations. This sort of scholarship reckons with how and by whom the narratives that could be empowered by climate justice actions are created.

Notes

1. Hulme, "Reducing the Future to Climate," 245; Rigg and Mason, "Five Dimensions of Climate Science Reductionism."
2. Hulme, 245.
3. Ford et al., "The Rapidly Changing Arctic"; Rantanen et al., "The Arctic Has Warmed."
4. Stuhl, *Unfreezing the Arctic.*
5. Keskitalo, "International Region-Building."
6. Todd, "An Indigenous Feminist's Take," 6.
7. Huntington et al., "Climate Change in Context."
8. Kubelka et al., "Animal Migration to Northern Latitudes."
9. Doel et al., "Science, Environment, and the New Arctic," 2–14; Palsson, "Arcticality: Gender, Race and Geography," 275–310.
10. Simpson, "Why White People Love Franz Boas."
11. Crate, "Climate and Culture," 185.
12. Yua et al., "A Framework for Co-production of Knowledge."
13. Mattar et al., "Climate Justice in the Arctic," 2.
14. Roberts and Howkins, "Introduction: The Problems of Polar History"; Sutter, "Climate Change and the Uses of History," xi–xii.
15. Boone and Buckley, "Historical Approaches to Environmental Justice," 222–30; Swart, "Feral Historians?"
16. Guldi and Armitage, *The History Manifesto*, 28.
17. Guldi and Armitage, 32.
18. Steffen et al., "Planetary Boundaries"; Head et al., "The Great Acceleration Is Real."
19. Chakrabarty, "The Climate of History," 208.
20. Chakrabarty.
21. Emmett and Lekan, eds., "Whose Anthropocene?," 10.

22. Schlosberg and Collins. "From Environmental to Climate Justice"; see also chapter 1 in this volume.
23. Desjardins et al., "Looking Back While Moving Forward."
24. Sörlin, "Rituals and Resources of Natural History"; see also chapter 2 in this volume.
25. Cunsolo and Ellis, "Ecological Grief as a Mental Health Response."
26. Whyte et al., "Indigenous Mobility Traditions, Colonialism, and the Anthropocene," 326.
27. Mattar et al., "Climate Justice in the Arctic."
28. Todd, "An Indigenous Feminist's Take."
29. Wright and Paradis, "High Arctic Relocation in the '50s."
30. Sponagle, "We Called It 'Prison Island.'"
31. Krupnik, "'Arctic Crashes.'"
32. Marino, "The Long History of Environmental Migration," 378.
33. Gilheany, "Circumpolar Materialities."
34. Sörlin, "The Arctic Ocean," 271.
35. Demuth, *Floating Coast*, 509.
36. Tschakert et al., "Multispecies Justice," 6–7.
37. Todd, "Fish Pluralities," 232.
38. Palmer, "Does Nature Matter?"; Mattar and McCauley, "Climate Justice in the Arctic"; Whyte, "Indigenous Environmental Justice."
39. Dugmorea et al., "Cultural Adaptation, Compounding Vulnerabilities."
40. Degroot, "Blood and Bone, Tears and Oil."
41. Hegerl et al., "The Early 20th Century Warming."
42. Hamilton et al., "West Greenland's Cod-to-Shrimp Transition."
43. Höhler et al., "Troubling the Northern Seas."
44. Szuwalski et al., "The Collapse of Eastern Bering Sea Snow Crab."
45. NOAH Fisheries, "Study Shows Pollock Stocks Are Mixing."
46. Hønneland, *Arctic Politics, the Law of the Sea and Russian Identity.*
47. Arctic Council, "Introduction to International Agreement."
48. Appadurai, "Introduction: Place and Voice in Anthropological Theory"; Chao et al., eds., *The Promise of Multispecies Justice.*
49. Wormbs et al., "Bellwether, Exceptionalism and Other Tropes."

BIBLIOGRAPHY

Appadurai, Arjun. "Introduction: Place and Voice in Anthropological Theory." *Cultural Anthropology* 3, no. 1 (1988): 16–20.
Arctic Council. "Introduction to International Agreement." 2021. https://arctic-council.org/news/introduction-to-international-agreement-to-prevent-unregulated-fishing-in-the-high-seas-of-the-central-arctic-ocean/.
Boone, Christopher G., and Geoffrey L. Buckley. "Historical Approaches to Environmental Justice." In *The Routledge Handbook of Environmental Justice*, ed. Ryan Holifield, Jayajit Chakraborty, and Gordon Walker, 222–30. New York: Routledge, 2018.

Cassotta, Sandra. "Ocean Acidification in the Arctic in a Multi-regulatory, Climate Justice Perspective." *Policy and Practice Reviews* 3 (2021). https://doi.org/ 10.3389/ fcilm.2021.713644.

Chakrabarty, Dipesh. "The Climate of History: Four Theses." *Critical Inquiry* 35, no. 2 (2009): 197–222.

Chao, Sophie, Karin Bolender, and Eben Kirksey, eds. *The Promise of Multispecies Justice*. Durham: Duke University Press, 2022.

Crate, Susan A. "Climate and Culture: Anthropology in the Era of Contemporary Climate Change." *Annual Review of Anthropology* 40 (2011): 185. https://doi.org/10.1146/annurev .anthro.012809.104925.

Cunsolo, Ashlee, and Neville R. Ellis. "Ecological Grief as a Mental Health Response to Climate Change-Related Loss." *Nature Climate Change* 8, no. 4 (2018): 275–81. https:// doi.org/10.1038/s41558-018-0092-2.

Degroot, Dagomar. "Blood and Bone, Tears and Oil: Climate Change, Whaling, and Conflict in the Seventeenth-Century Arctic." *American Historical Review* 127, no.1 (2020): 62–99. https://doi.org/10.1093/ahr/rhac009.

Demuth, Bathsheba. *Floating Coast: An Environmental History of the Bering Strait*. New York: Norton, 2019.

Desjardins, S.P.A., T. M. Friesen, and P. D. Jordan. "Looking Back While Moving Forward: How Past Responses to Climate Change Can Inform Future Adaptation and Mitigation Strategies in the Arctic." *Quaternary International* 549 (2020): 239–48. https://doi .org/10.1016/j.quaint.2020.05.043.

Doel, Ronald E., Urban Wråkberg, and Suzanne Zeller, "Science, Environment, and the New Arctic." *Journal of Historical Geography* 44 (2014): 2–14. https://doi.org/10.1016/j .jhg.2013.12.003.

Dugmorea Andrew J., Thomas H. McGovern, Orri Vésteinssonc, Jette Arneborgd, Richard Streetera, and Christian Kellere. "Cultural Adaptation, Compounding Vulnerabilities and Conjunctures in Norse Greenland." *PNAS* 109, no. 10 (2012): 3658–63. https:// doi.org/10.1073/pnas.1115292109.

Emmett, Robert, and Thomas Lekan, eds. "Whose Anthropocene? Revisiting Dipesh Chakrabarty's 'Four Theses.'" *RCC Perspectives: Transformations in Environment and Society* no. 2 (2016). https://doi.org/10.5282/rcc/7421.

Ford, J. D., T. Pearce, I. V. Canosa, and S. Harper. "The Rapidly Changing Arctic and Its Societal Implications." *WIRE Climate Change*, 12, no. 6, e735 (2021). https://doi.org/10 .1002/wcc.735.

Gilheany, Emma. "Circumpolar Materialities: Sea Ice, Cold War Radar Stations and Inuit Environmental Politics." Doctoral dissertation, University of Chicago, 2024.

Guldi, Joe, and David Armitage. *The History Manifesto*. Cambridge: Cambridge University Press: 2014.

Hamilton, Lawrence C., Benjamin C. Brown, Rasmus Ole Rasmussen. "West Greenland's Cod-to-Shrimp Transition: Local Dimension of Climate Change." *Arctic* 56, no. 3 (2003): 271–82.

Head, Martin J., Will Steffen, David Fagerlind et al. "The Great Acceleration Is Real and Provides a Quantitative Basis for the Proposed Anthropocene Series/Epoch." *Episodes* 45, no. 4 (2022): 359–76. https://doi.org/10.18814/epiiugs/2021/021031.

Hegerl, Gabriele C., Stefan Brönnimann, Andrew Schurer, and Tim Cowan. "The Early 20th Century Warming: Anomalies, Causes, and Consequences." *WIREs Climate Change* 9, e522 (2018). https://doi.org/10.1002/wcc.522.

Höhler, Sabine, Julia Lajus, Peder Roberts, and Urban Wråkberg. "Troubling the Northern Seas: The Turbulent History of Norwegian and Russian Fisheries." In *The Politics of Arctic Resources: Change and Continuity in the "Old North" of Northern Europe*, ed. Carina Keskitalo, 181–99. New York: Routledge, 2019.

Hønneland, Geir. *Arctic Politics, the Law of the Sea and Russian Identity: The Barents Sea Delimitation Agreement in Russian Public Debate*. London: Palgrave Macmillan, 2014.

Hulme, Mike. "Reducing the Future to Climate: A Story of Climate Determinism and Reductionism." *OSIRIS* 26 (2011): 245–66. https://doi.org/10.1086/661274.

Huntington, Henry P., Mark Carey, Charlene Apok et al. "Climate Change in Context: Putting People First in the Arctic." *Regional Environmental Change* 19 (2019): 1217–33. https://link.springer.com/article/10.1007/s10113-019-01478-8.

Keskitalo, Carina. "International Region-Building: Development of the Arctic as an International Region." *Cooperation and Conflict*, 42 (2007): 187–205. https://doi.org/10.1177/0010836707076689.

Krupnik, Igor. "'Arctic Crashes': Revisiting the Human-Animal Disequilibrium Model in a Time of Rapid Change." *Human Ecology* 46 (2018): 685–700. https://link.springer.com/article/10.1007/s10745-018-9990-1.

Kubelka, Vojtěch, Brett K. Sandercock, Tamás Székely, and Robert P. Freckleton. "Animal Migration to Northern Latitudes: Environmental Changes and Increasing Threats." *Trends in Ecology & Evolution* 37, no. 1 (2022): 30–41. https://doi.org/ 10.1016/j.tree.2021.08.010.

Mattar, Sennan D., Michael Mikulewicz, and Darren McCauley. "Climate Justice in the Arctic: A Critical and Interdisciplinary Climate Research Agenda." *Arctic Yearbook* (2020).

Marino, Elizabeth. "The Long History of Environmental Migration: Assessing Vulnerability Construction and Obstacles to Successful Relocation in Shishmaref, Alaska." *Global Environmental Change* 22, no. 2 (2012): 374–81.

NOAH Fisheries. "Study Shows Pollock Stocks Are Mixing More Due to Changing Ocean Conditions and Weather." October 26, 2020. https://fisheries.noaa.gov/feature-story/study-shows-pollock-stocks-are-mixing-more-due-changing-ocean-conditions-and-weather.

Palmer, Clare. "Does Nature Matter? The Place of the Nonhuman in the Ethics of Climate Change." In *The Ethics of Global Climate Change*, ed. Denis G. Arnold, 272–91. Cambridge: Cambridge University Press, 2011. https://doi.org/10.1017/CBO9780511732294.014.

Palsson, Gisli. "Arcticality: Gender, Race and Geography in the Writings of Vilhjalmur Stefansson." In *Narrating the Arctic. A Cultural History of Nordic Scientific Practice*, ed. Sverker Sorlin and Michael Bravo, 275–310. Canton, MA: Science History Publications, 2002.

Rantanen, M., Karpechko, A. Y., A. Lipponen et al. "The Arctic Has Warmed Nearly Four Times Faster than the Globe Since 1979." *Communications Earth & Environment* 3, no. 168 (2022). https://doi.org/10.1038/s43247-022-00498-3.

Rigg, Jonathan, and Lisa Reyers Mason, "Five Dimensions of Climate Science Reductionism," *Nature Climate Change* 8 (2018): 1030–32. https://doi.org/10.1038/s41558-018-0352-1.

Schlosberg, David, and Lisette B. Collins. "From Environmental to Climate Justice: Climate Change and the Discourse of Environmental Justice." *WEREs Climate Change* (2014). https://doi.org/10.1002/wcc.275.

Roberts, Peder, and Adrian Howkins. "Introduction: The Problems of Polar History." In *Cambridge History of the Polar Regions*, ed. Peder Roberts and Adrian Howkins. Cambridge: Cambridge University Press, 2022.

Simpson, Audra. "Why White People Love Franz Boas; or, the Grammar of Indigenous Dispossession." In *Indigenous Visions: Rediscovering the World of Franz Boas*, ed. Ned Blackhawk and Lorado Wilner. New Haven: Yale University Press, 2018.

Sörlin, Sverker. "The Arctic Ocean." In *Oceanic Histories*, ed. David Armitage, Alison Bashford, and Sujit Sivasundaram, 269–95. Cambridge: Cambridge University Press, 2018.

Sörlin, Sverker. "Rituals and Resources of Natural History: The North and the Arctic in Swedish Scientific Nationalism." In *Narrating the Arctic. A Cultural History of Nordic Scientific Practices*, ed. Sverker Sorlin and Michael Bravo, 73–124. Canton, MA: Science History Publications, 2002.

Sponagle, J. "We Called It 'Prison Island': Inuk Man Remembers Forced Relocation to Grise Fiord." *CBC News*, June 30, 2017. https://www.cbc.ca/news/canada/north/larry-audlaluk -inuit-relocation-1.4181689.

Steffen, Will, Katherine Richardson, Johan Rockström et al. "Planetary Boundaries: Guiding Human Development on a Changing Planet." *Science* 347, no. 6223 (2015). https:// doi.org/10.1126/science.1259855.

Stuhl, Andrew. *Unfreezing the Arctic: Science, Colonialism, and the Transformation of Inuit Lands*. Chicago: University of Chicago Press, 2016.

Sutter, Paul S. "Foreword. Climate Change and the Uses of History." *In Making Climate Change History. Documents from Global Warming's Past*, ed. Joshua P. Howe, xi—xiv. Seattle: University of Washington Press, 2017.

Swart, Sandra. "Feral Historians?" In *Seeing the Woods. A Blog by the Rachel Carson Center*. 2017. https://seeingthewoods.org/2017/09/06/uses-of-environmental-history-sandra -swart/.

Szuwalski, Cody S., Kerim Aydin, Erin J. Fedewa, Brian Garber-Yonts, and Michael A. Litzow. "The Collapse of Eastern Bering Sea Snow Crab." *Science* 382, no. 6668 (2023): 306–10. https://doi.org/10.1126/science.adf6035.

Todd, Zoe. "Fish Pluralities: Human-Animal Relations and Sites of Engagement in Paulatuuq, Arctic Canada." *Etudes/Inuit/Studies* 38, no. 1/2 (2014): 217–38.

Todd, Zoe. "An Indigenous Feminist's Take on the Ontological Turn: 'Ontology' Is Just Another Word for Colonialism." *Journal of Historical Sociology* 29 (2016): 4–22. https://doi.org/10.1111/johs.12124.

Tschakert, Petra, David Schlosberg, Danielle Celermajer et al. 2021. "Multispecies Justice: Climate-Just Futures with, for and Beyond Humans." *WIREs Climate Change* 12:e699 (2021). https://doi.org/10.1002/wcc.699.

Whyte, Kyle. "Indigenous Environmental Justice." In *Environmental Justice*, ed. Brendan Coolsaet, 266–78. New York: Routledge, 2020. https://doi.org/10.4324/9780429029585-23.

Whyte, Kyle, Jared L. Talley, and Julia D. Gibson. "Indigenous Mobility Traditions, Colonialism, and the Anthropocene." *Mobilities* 14, no. 3 (2019): 319–35. https://doi.org/10.1080/17450101.2019.1611015.

Wormbs, Nina, Döscher, Ralf, Nilsson, Annika E., and Sörlin, Sverker. "Bellwether, Exceptionalism and Other Tropes: Political Coproduction of Arctic Climate Modelling." In *Cultures of Prediction in Atmospheric and Climate Science*, ed. Matthias Heymann, Gabriele Grammelsberger, and Martin Mahony, 133–55. London: Routledge, 2017.

Wright, T., and D. Paradis. "High Arctic Relocation in the '50s Still Lingers with Inuit Elders," *APTN News*, November 21, 2023, 145. https://www.aptnnews.ca/national-news/high-arctic-relocation-in-the-50s-still-lingers-with-inuit- elders/.

Yua, E., J. Raymond-Yakoubian, R. A. Daniel, and C. Behe. "A Framework for Coproduction of Knowledge in the Context of Arctic Research. *Ecology and Society* 27, no. 1 (2022), art. 34. https://doi.org/10.5751/ES-12960-270134.

11

Bridging the Gulf

Intersections of Geology, Biology, and Climate Justice

KAILANI ACOSTA AND GISELA WINCKLER

S tories of demographic and environmental change have been repeated many times throughout history and around the world. Legacies of change, due to climate and anthropogenic pressures, are tied to and reflected in landscapes. For example, the 100-million-year-old southern coastline of North America is connected to present-day Black communities in the United States through slavery and the proliferation of environmental racism through time. This chapter portrays the succession and interconnectivity of geology, biology, race, agriculture, and climate justice tying contemporary injustice to history through a three-part story:

1. How the Late Cretaceous coastline preconditioned agricultural use and led to the creation of cotton fields;
2. The Black Belt—the interconnectivity of race, agriculture, and environmental and climate justice in the southern United States;
3. The proliferation of exclusion—patterns of burden, justice, and science.

Understanding the linkages between racial and environmental change are fundamental to the ways in which we do science and create scientific communities. We invite readers to deeply reflect on the ways in which race, environment, and climate are fundamentally intertwined. This is an appeal to scientists and the public to think holistically about the legacies, communities, and histories of the environments in which they study and work, and what the past means for charting a more inclusive future in our changing world.

Climate change issues are climate justice issues. Geoscientists approach climate justice theoretically, by considering how climate change affects different regions disproportionately based on their geography, geology, biology, natural resources, and vulnerability to natural hazards. However, studies of how climate change is affecting the oceans, air, soil, and land do not often include a climate justice lens. Though the evidence of human-caused climate change is solidly backed up by geoscience research, few geoscientists go beyond anthropogenic causes to investigate the ongoing disproportionate harms that people with lower socioeconomic standing, people of color, and people living on island nations must face. As a discipline, the geosciences do not often include themes of climate justice in their work and research. Recently, there have been attempts to change the way we conduct geoscience research to move away from the parachute science norm, with more concerted efforts to co-create science with local Indigenous peoples. "Parachute science" refers to the common extractive practice of scientists from highly resourced countries and institutions conducting fieldwork in lower-income countries (often in the Global South), where they extract data, samples, and knowledge without properly including or acknowledging local expertise and communities. However, these efforts have not resulted in longstanding or widespread changes in the discipline as a whole or in the incorporation of climate justice lenses into geoscience. In this chapter, we hope to elucidate the inherent, deep connections between geoscience, culture, and people, to show just one example of the histories and futures of climate injustice.

HOW THE LATE CRETACEOUS COASTLINE
LED TO THE CREATION OF COTTON FIELDS

During the Cretaceous period (145 to 66 million years ago), the climate was warm, 5 to 10 degrees Celsius warmer than today's climate, and high sea levels created shallow inland seas filled with life, especially phytoplankton.[1] Over millions of years, these phytoplankton—especially calcium carbonate–shelled plankton called coccolithophores—prospered, died, and accumulated as their calcium carbonate shells sunk to the seafloor along the Late Cretaceous coastline, creating large swaths of carbonate rocks (for example, chalk or limestone) throughout much of the modern-day southern United States (see fig. 11.1).

The changing climate created ideal conditions for the explosion of life in the shallow seas of the region for millions of years. As climate got colder, ice caps formed and seas retreated, moving coastlines further offshore, uncovering the thick chalk beds left behind by the productive seas. These chalks weathered into clays over time, which led to the growth of diverse prairie grasses. These grasses grew and died, with organic matter decaying and forming rich black soils that were ideal for farming, thus called the "Black Belt."[2]

Climatic changes over millions of years led to successions of geological and biological events. In recent history, climate changes have had widespread justice impacts for communities that are based in the Black Belt region.

THE BLACK BELT—INTERCONNECTIVITY OF RACE,
AGRICULTURE, ENVIRONMENTAL, AND CLIMATE JUSTICE
IN THE SOUTHERN UNITED STATES

For hundreds of years before European colonization, the fertile lands of the Black Belt were home to Native peoples that depended on the land to grow crops. In the early 1800s, Native peoples were removed from their ancestral homelands and forced to relocate by Europeans, with thousands dying on the Trail of Tears and in other forced removal practices.[3]

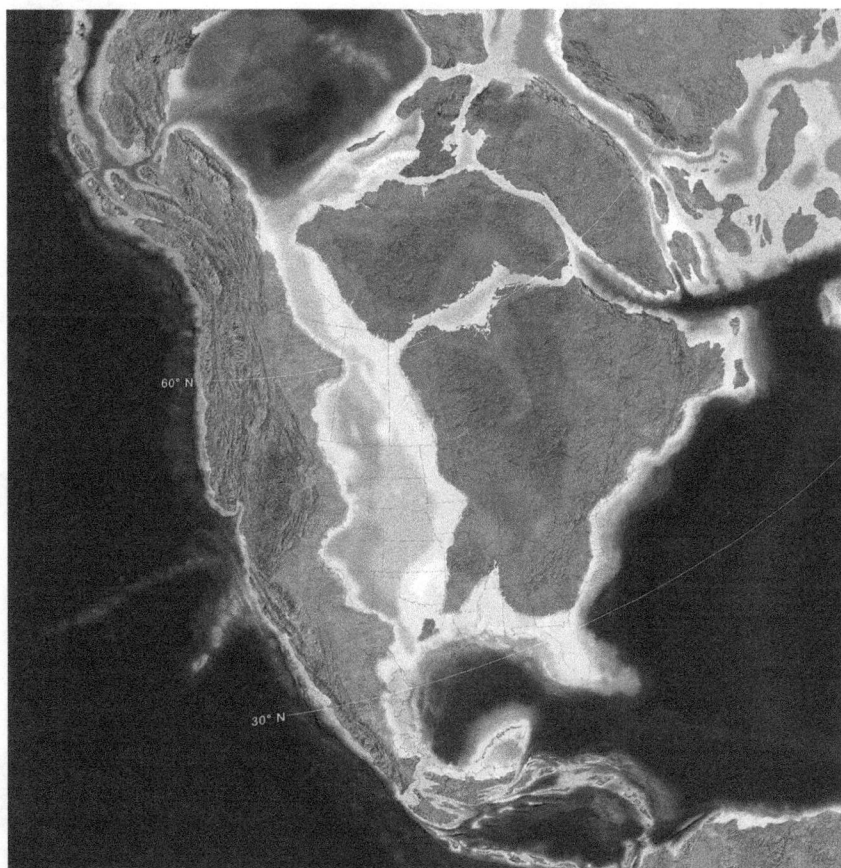

FIGURE 11.1

Cretaceous Map of Paleo North America, 100 million years ago. (Source: Colorado Plateau Geosystems, Inc., Late Cretaceous North America, in *North America in Deep Time*, 2023.)

The fertile Black Belt region became the center of agriculture in the South, and the Transatlantic Slave Trade greatly increased the numbers of enslaved Africans brought to the Black Belt to farm cotton, sugar, and oats on plantations.[4] This trade drove the industrialization of plantation economies of sugar and cotton and solidified the region as a hub of resource extraction and agriculture.[5] By the early twentieth century, the term "Black Belt" was also used to describe the region's sizable Black population. The Black Belt of Alabama and Mississippi is well known for its

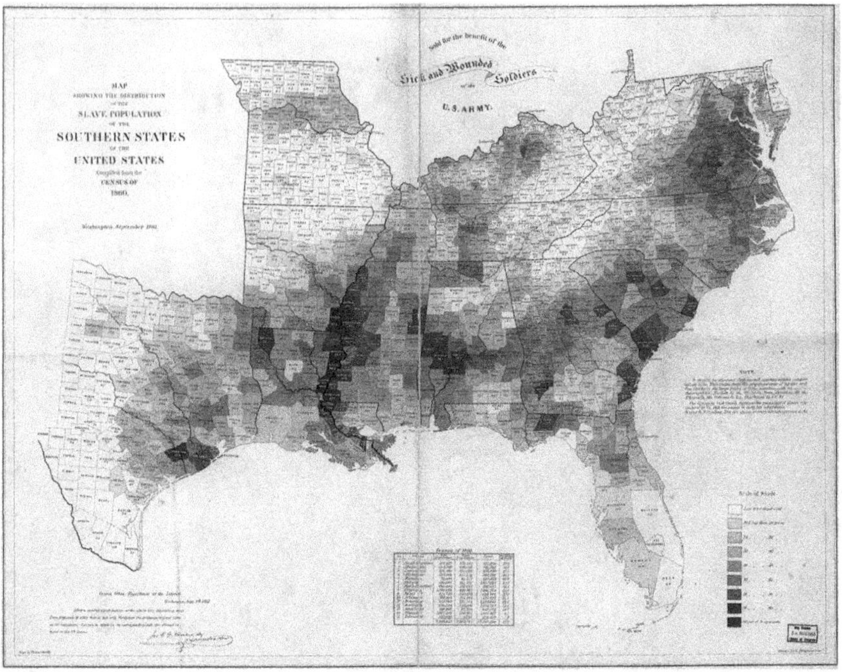

FIGURE 11.2

Distribution of the slave population of the southern states of the United States in 1860. Source: E. Hergesheimer, *Map Showing the Distribution of the Slave Population of the Southern States of the United States*, compiled from the census of Washington Henry S. Graham, 1861, https://www.loc.gov/item /99447026/.

cotton plantations and history of slavery, with even larger populations of slaves and farms arising from the ability to transport goods downriver to New Orleans (see fig. 11.2).[6] The lower Mississippi region rivaled the Black Belt in the extent of anthropogenic environmental and climate change that occurred starting in the 1600s with industrialization of farming and large-scale industry. Today, these geographic patterns of race and disenfranchisement and disproportionate climate impacts in the Black Belt follow the scars of the history of colonization, and large-scale agriculture. Though these patterns vary and exist throughout the South, the Black Belt is a stark and extreme example of how geology, biology (phytoplankton, plants, soil, and succession), and environmental justice are fundamentally linked.

FIRST AND WORST—IMPACT OF CLIMATE CHANGE ON BLACK COMMUNITIES IN THE SOUTHERN UNITED STATES

Climate justice issues are also civil rights matters, especially in the Black Belt and more broadly in the southern United States, where Black communities often feel the impacts of climate change first and worst. Climate change is exacerbating public health, economic, and environmental injustices that are already centered in communities of color, and specifically in Black communities of the Black Belt region. This correlation is directly related to the history of slavery based on agriculture in the Black Belt. In Lowndes County, Alabama, a rural county located within the Black Belt, its predominantly low-income Black residents have been fighting for their rights to water and sanitation for generations.[7] Natural soil conditions, climate change-driven increased rain and temperatures, and inadequate sanitation infrastructure combine to create a powerful example of climate and environmental injustices that are continually intensifying. The roots of this injustice lie in history and geological patterns, as well as in the racial, economic, and cultural divides that continually limit access and connection to municipal systems.

"Climate justice" can be defined as envrionmental justice issues that intensified by changing climate patterns. The impacts of the lack of adequate sanitation in Lowndes County are extensive. Public health issues stemming from water contamination are rampant, and though these issues have been known for years, no actions have been taken.[8] Because all these issues will be further exacerbated by climate change, each year that nothing changes, the situation becomes worse for those living in such conditions—predominantly low-income Black people living in the Black Belt. Though these dire health and well-being threats to people living in this region have existed for centuries, attention has only recently been called to Lowndes County. In 2022 the US Justice Department rolled out a Comprehensive Environmental Justice Enforcement Strategy and reached an interim resolution agreement with the Alabama Department of Health in the county to attempt to make conditions better for residents. Though progress has been made, there is no future action planned to fully address these climate and environmental injustices.

Past patterns of geology, biology, and climate of the Black Belt are reflected in all facets of culture in the United States. To address the proliferation of exclusion and burden in the Black Belt region, we must acknowledge and confront these historical patterns. The history of the formation and change of the land itself has myriad impacts on the demographics, economics, and lives of the people in the Black Belt today. Over time, land use has drastically changed, as have the perceptions of the value and import of the region and the people who live there. Humans manipulated this one example of a "fertile crescent" to become the agricultural and economic foundation of the nascent nation of the United States, with those in power directly profiting from the labor of enslaved peoples on stolen lands. The systems of oppression in the Black Belt have been deeply entrenched for hundreds of years. In 2023 Today, the scars of continuous oppression and governmental neglect redefine the Black Belt in its current state.

THE PROLIFERATION OF EXCLUSION: PATTERNS OF BURDEN, JUSTICE, AND SCIENCE

Exclusionary patterns extend far beyond the geological, biological, and climatic domains that permeate the cultural tapestry of the United States. The legacy of systemic racism has deeply impacted education, with unequal resource allocation, limited access to quality education, and high dropout rates. Addressing climate justice in the geosciences requires systemic changes in education, workplaces, and research institutions. Understanding the linkages between racial and climate justice is fundamental to the ways in which we conduct science and create scientific communities. Underrepresented minority communities (URM; Black, Hispanic, Asian, and Indigenous) are the most environmentally burdened and the least involved in the science that impacts their communities and environments (see fig. 11.3).[9]

To create inclusive robust scientific communities, we must upend the institutional norms that proliferate exclusion and inaccessibility of science writ large. Climate injustice affects us all, from increasing temperatures and rainfall in the eastern part of the United States to multibillion-dollar climate disasters from 2018 to 2022.[10] The compounding impacts and

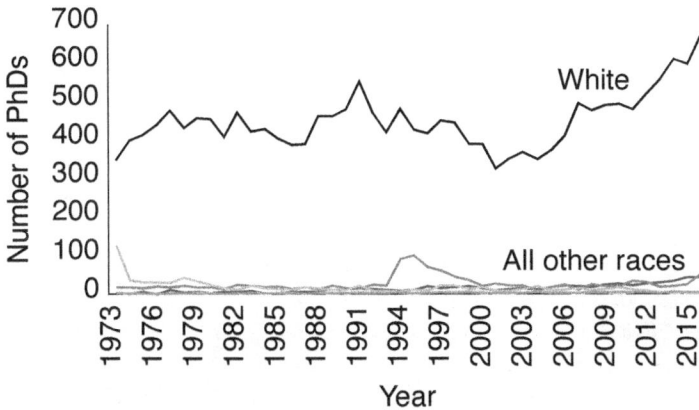

FIGURE 11.3

Demographics of geoscience PhDs. (Source: R. E. Bernard and E. H. G. Cooperdock, "No Progress on Diversity in 40 Years," *Nature Geoscience* 11, no. 5 (2018): 292–95.)

causalities of climate change are always in communities of color worldwide. Science must change to focus on these impacts and include climate justice lenses into the science we do every day. We must recognize that climate justice is the foundation for science moving forward, and this integration is crucial to our success and survival as scientists and as humans. We cannot advance science and fix climate change issues when the ways in which we do science are currently exclusionary and unjust.

EPILOGUE

The linkages of geology, race, agriculture and climate justice have been showcased as an art exhibition under the name "Bridging the Gulf" by Kailani Acosta. Through combining maps, rocks, artifacts, fossils, and live phytoplankton, the exhibit bridges the gulf of our understanding and provides viewers with a creative space to deeply reflect on the ways in which race, the environment, and justice are fundamentally intertwined. The exhibit, designed as a traveling exhibition, was first shown on February 9, 2024, in Brooklyn, New York, and will continue its journey through the United States and worldwide.

Notes

1. Campbell and Seymour, "A Review of Native Vegetation Types"; O'Brien et al., "Cretaceous Sea-Surface Temperature Evolution"; Tierney et al., "Past Climates Inform Our Future."
2. Barone, "Historical Presence and Distribution of Prairies."
3. Garrison, *The Legal Ideology of Removal.*
4. Webster and Samson, "On Defining the Alabama Black Belt."
5. Rosol et al., "Introduction: The Mississippi River Basin."
6. Moore, *The Emergence of the Cotton Kingdom.*
7. Winkler and Flowers, "America's Dirty Secret."
8. Clarke, "Assistant Attorney General Kristen Clarke."
9. American Geosciences Institute, "Diversity in the Geosciences"; Centers for Disease Control, "The Environmental Justice Index"; Acosta et al., "Past as Prologue."
10. Jay et al., "Understanding Risks Impacts, and Responses."

BIBLIOGRAPHY

Acosta, K., B. Keisling, and G. Winckler. "Past as Prologue: Lessons from the Lamont-Doherty Earth Observatory Diversity, Equity, and Inclusion Task Force." *Journal of Geoscience Education* 71, no. 3 (2022): 307–19. https://doi.org/10.1080/10899995.2022.2106090

American Geosciences Institute (AGI). "Diversity in the Geosciences." 2020. https://www.americangeosciences.org/geoscience-currents/diversity-geosciences.

Barone, J. "Historical Presence and Distribution of Prairies in the Black Belt of Mississippi and Alabama." *BioOne* 70, no. 3 (2005): 170–83.

Campbell, J.J.N., and W. R. Seymour, Jr. "A Review of Native Vegetation Types in the Black Belt of Mississippi and Alabama." *Journal of the Mississippi Academy of Sciences* 56, no. 2 (2011): 166–84.

Centers for Disease Control (CDC). "The Environmental Justice Index." 2020. Environmental Justice Index (EJI) (cdc.gov).

Clarke, K. "Assistant Attorney General Kristen Clarke Delivers Remarks to Announce Agreement in Civil Rights and Environmental Justice Investigation of Alabama Department of Public Health." Office of Public Affairs, U.S. Department of Justice, 2023. https://www.justice.gov/opa/speech/assistant-attorney-general-kristen-clarke-delivers-remarks-announce-agreement-civil.

Garrison, T. A. *The Legal Ideology of Removal: The Southern Judiciary and the Sovereignty of Native American Nations.* Athens: University of Georgia Press, 2002.

Jay, A. K., A. R. Crimmins, C. W. Avery et al. "Chapter 1. Overview: Understanding Risks Impacts, and Responses." In *Fifth National Climate Assessment*, ed. A. R. Crimmins, C. W. Avery, D. R. Easterling et al. Washington, DC: U.S. Global Change Research Program, 2023. https://doi.org/10.7930/NCA5.2023.CH1.

Moore, J. H. *The Emergence of the Cotton Kingdom in the Old Southwest: Mississippi, 1770–1860.* Baton Rouge: Louisiana State University Press, 1988.

O'Brien, C. L., S. A. Robinson, R. D. Pancost et al. "Cretaceous Sea-Surface Temperature Evolution: Constraints from TEX86 and Planktonic Foraminiferal Oxygen Isotopes." *Earth Science Review* (172): 2017, 224–47, https://doi.org/10.1016/j.earscirev.2017.07.012.

Rosol, C., T. Turnbull, and J. Renn. "Introduction: The Mississippi River Basin—a Model for Studying the Anthropocene in Situ." *Anthropocene Review* 8, no. 2 (2012): 99–114.

Tierney, J. E., C. J. Poulsen, I. P. Montanez et al. "Past Climates Inform Our Future," *Science* 370 (2020). https://doi.org/10.1126/science.aay3701.

Webster, G. R., and S. A. Samson. "On Defining the Alabama Black Belt: Historical Changes and Variations." *Southeastern Geographer* 32, no. 2 (1992): 163–72.

Winkler, I. T., and C. C. Flowers. "America's Dirty Secret: The Human Right to Sanitation in Alabama's Black Belt." *Columbia Human Rights Law Review*, 2017.

12

How Much Positive Influence on the Climate Problem Can One Have as an Academic or Industry Climate Scientist?

ADAM SOBEL AND MELANIE BIELI

DISCIPLINARY ORIENTATION AND CLIMATE JUSTICE PERSPECTIVE

Our discipline is climate science; our specific subfield of it is sometimes called atmospheric science or meteorology. Our own research aims to understand and predict the behavior of the atmosphere on time scales ranging from days to decades. Our primary tools are numerical simulation and statistical analysis. We are arguably as similar to physicists and applied mathematicians as we are to those working in other areas of the earth sciences.

In our view, our field does not have a sophisticated or closely examined view of what climate justice means. Our science is about how the earth system works; human beings enter it mainly as sources of carbon emissions, or sometimes as the recipients of "impacts." But considerations of justice, per se, have been construed as outside the field, almost by definition. At least until quite recently, to the extent that scientists in our field may have read, thought, or talked explicitly about climate justice, they did so outside the normal boundaries and expectations of their workplaces. The discipline is evolving rapidly now, and some climate science curricula are beginning to include classes that address climate justice. None were part of our educations.

It is nonetheless safe to say that most climate scientists consider human-induced global warming to be a bad thing and further understand that those who suffer the most from it are generally the least responsible for it.[1] It follows that any action that either reduces greenhouse gas emissions or lessens the impacts of climate change, particularly on relatively vulnerable or marginalized people, serves the interests of climate justice (but see chapter 5 for cases where environmental justice, such as improving air quality in marginalized communities, may come into conflict with carbon-reducing policies). This very simple view will be (almost) enough for the purposes of this essay.

PROBLEM STATEMENT

We consider the problem faced by a young person who has studied climate science and acquired some expertise in it, who would like their work to address the climate problem in some material way, and who now has to decide what to do in their career.

Such a person has to balance their passion for intellectually satisfying work, their desire to make a difference in the climate problem, and their need to make a living. Of these, the extent to which a given job "makes a difference" or "has an impact" is the hardest to assess. Here, we attempt to think through some of the issues that inform that assessment, and the choice of a career path in climate more generally. We focus on the potential for positive impact through two possible career choices: academic or government climate science, and careers assessing and managing climate risk in the private sector. We do not consider government policy work, as critically important as it is, due to the limitations of space and our own experience.

Our private sector focus should not in any way be construed as an argument that "the free market" (scare quotes intentional) can solve the climate problem on its own. We agree with Naomi Oreskes and Erik Conway that this is a myth, that unfettered capitalism has so far been much more a source of climate injustice than justice, and that government action is essential to any just solution of the climate problem, as it is to the provision and maintenance of most public goods.[2] (Consider the voluntary carbon market: its recent struggles have multiple causes,[3] but clearly a fundamental one is the absence of any true underlying price on greenhouse gas emissions, something only governments can set.) Rather, it reflects the practical fact that the private sector has seen huge growth in the number of climate-related jobs in the last few years. We have seen many of our students and peers take these jobs—many more than have gone into either government policy or advocacy, and perhaps a number comparable to those who have gone into academic careers—for whatever combination of pragmatic and idealistic reasons. This alone is in our view sufficient justification to try to think through the possibilities that the private sector offers for working toward climate justice.

This is not quantitative or rigorous scholarship, but a reflection based on our own experiences, and those of our peers, friends, colleagues, students, and teachers. We are also informed by about forty long-form interviews, conducted for a podcast we have produced over the last four years, with climate scientists and a few others from adjacent fields.[4] To foreground the personal nature of our approach to the problem, we start with our own stories.

PERSONAL STATEMENT: MELANIE

I was in graduate school when I first thought about my career options in systematic and concrete terms. The result of that process was a Venn diagram of four intersecting sets, labeled "weather and climate," "finance," "quantitative," and "sustainability," that symbolized my professional aspirations, interests, and desire to contribute positively to the world. It was meant to guide my transition from academia to the private sector, because at that point I had decided to leave university life behind (though it would yet be a couple more years, as a postdoc, before I actually did that), joining the approximately 80 percent of STEM PhD graduates in the United States who transition out of academic careers.[5]

Given my background in climate science, I seemed well qualified to work on one of humanity's biggest challenges. Yet I thought that the kind of climate-related work I was most interested in, namely, climate science research, was not an effective way of addressing the climate problem. Climate advocacy, which I considered to be a more promising way to create impact, simply did not interest me. I had to admit that impact alone would not do it for me; I would have to enjoy the work itself. While that realization did not come without guilt, it led to my husband and me taking an idea more seriously that has most prominently been propagated by the Effective Altruism (EA) movement.

In its extreme form, the idea is often termed "earning to give," and it posits that one can make a significant contribution to society by pursuing a lucrative career and donating a substantial portion of the earnings to charities. The "effective" in "Effective Altruism" refers to the use of reason and evidence to find out how we can best use our resources to help others.[6] The EA movement, despite facing much (and, at least in part, valid) criticism,[7] has influenced our approach to doing good in two ways: First, it has affected the priorities in our bucket of donations aimed at global development. (We have other buckets, for example, for charities that lobby for climate change action or foster healthy democracies, which, due to their less quantifiable work, might not survive an EA analysis.) Second, and more important, it dramatically increased the amount we donate by opening our eyes to the idea that you can have impact in many different ways, and that the ways that require the most direct personal engagement are not necessarily the most effective ones.

I took a job modeling natural catastrophe risk in the reinsurance industry and then more recently moved to a quantitative finance firm. I have not viewed either job primarily as a form of climate action or social change; my contribution on that front is mainly through my donations. Nonetheless, the insurance industry does have an important role to play in climate adaptation. We will discuss this further shortly.

PERSONAL STATEMENT: ADAM

I double majored in music and physics as an undergraduate and tried to be a musician in my early twenties. By a couple of years after graduation, my music career was not going the way I'd hoped, and I thought about doing a PhD degree in physics. My girlfriend (now wife), an ardent environmentalist, suggested a PhD in meteorology and work on global warming. This would not have occurred to me, but I became interested in it. Mostly, I was drawn to fluid dynamics, chaos, numerical simulation, and other aspects of the science. But also, I liked that the subject had the potential to contribute to societal welfare.

My doctoral degree was on aspects of stratospheric dynamics motivated by the ozone problem; as a postdoc I switched to tropical meteorology. From then through my first five years or so as a junior faculty member, in the early 2000s, I was not motivated directly by any societal application. I wanted to understand how the atmosphere works. As time went on, though, I, like much of the rest of the world, began to feel the urgency of the climate problem more strongly. My research began to have a climate change dimension, though still no direct application.

Hurricane Sandy (2012) changed my life and broadened my view of what my job entails. The storm and its aftermath drew me into a period of speaking to the media frequently for several months; teaching an interdisciplinary class about the multiple dimensions of the event and its consequences; writing the book *Storm Surge*, aimed at a nonscientist audience; and beginning to write op-ed pieces for the mainstream media. At the same time, several colleagues and I began interacting with the insurance industry.

Four years later, the election of Donald Trump to the US presidency in 2016 shocked and demoralized me. It drew my life choices into stark relief, making me ask how a future me looking back on this time would judge them. I thought a lot about how I might have an impact through my work.

My attempts at societal relevance have taken two forms. First, our group's work with the insurance industry has continued to the present and has led us to work with nonprofit organizations as well. This work has involved developing models and data sets that quantify risk from hurricanes and other extreme weather events, and that considers how human-induced climate change is altering that risk. The latter is a question with which the industry did not seriously engage until relatively recently but now does much more so; broadly, I would like to think our work has played a small part in that. Beyond insurance itself, I hope to apply what we have learned from this work to the development of models, data, and understanding for climate risk assessment in the public sector, to support climate adaptation more broadly.

Second, I have put increasing effort into public communication. Most of my effort has taken the form of op-ed style pieces and articles in mainstream media outlets, talks to relatively diverse audiences, and the podcast. The content in these pieces has evolved from being mostly about scientific questions to include broader social, human, and political concerns as well. While I am under no illusions that I can change the world on any large scale by doing this, it is something I can do that takes advantage of the status and privilege of my position as a professor at an elite university, and it might have a greater (or at least different) impact than my research does. This public communication has also led to some new relationships with scholars outside my field, and to new ideas that have deeply informed my thinking over the last decade, particularly on climate justice.[8]

THE USABLENESS (OR NOT) OF ACADEMIC CLIMATE SCIENCE

The primary job of a climate scientist in a research university or government laboratory is research, and one's primary output consists of peer-reviewed articles in professional journals. To what extent does this work serve to mitigate the climate problem and thus contribute to the goal of climate justice?

Historically, it is almost entirely academic and government research that has described, defined and explained what human-induced global warming is, how it works, and what it will take to stop it. From this perspective, it is reasonable to think that this research contributes to solving the problem.

But it is no longer 1950, or 1980, or even 2000. The basic workings of the climate system are understood well enough to know what the problem is and, at least on the largest scales, what it will take to address it. Climate mitigation—the reduction or cessation of greenhouse gas emissions—depends on many kinds of science and engineering, as well as politics and policy. But it does not, at this point in history, depend very much on advances in climate science per se. One of us has thus made the argument that academic climate science research can have a greater positive impact in the near term if it is oriented toward climate adaptation rather than mitigation[9].

Adaptation encompasses a very broad set of activities. Some of them could benefit from climate information that is more local, detailed, and impacts-oriented than is necessary for mitigation. Climate scientists can choose to orient our research toward producing such information, and see a shorter route between the production of research and its use in the external world.

There are, of course, still routes for climate science to influence mitigation as well. Henri Drake and Geoffrey Henderson argue that some advances in climate science support and motivate collective action on mitigation.[10] One of their examples is extreme event attribution; a recent study provides evidence that the media has been the primary consumer of these studies, and in our view it is plausible to claim that they have had an influence on public opinion.[11] Synthesis or assessment reports, like those from the Intergovernmental Panel on Climate Change (IPCC), to which many climate scientists contribute, can also have such an influence—if perhaps less so than in the past, since the essential messages have not changed much in the decades since they began.

The time and effort it takes to do impact-oriented work can sometimes come at the expense of the type of productivity that is most straightforwardly rewarded by academic institutions (peer-reviewed articles, grant funding, and number of students taught or mentored). Effective adaptation science, for example, must be based on engagement with the science's "users" or "stakeholders" in the external world, a potentially time-consuming activity.[12] While academic institutions value external impact in principle, they do not have well-defined mechanisms for incentivizing it. One way academics can contribute to climate justice may be to advocate for institutional changes to incentivize external engagement, or what our own institution calls the "Fourth Purpose."[13]

Our point is not to criticize climate scientists, but simply to point out that times have changed, and so has the relationship of science to action. The basic messages—"it's real, it's us, it's bad, we can fix it"—have now been transmitted to the global public about as well as can be expected. It is certainly still possible for climate science research to influence climate action; it is just clear now that it does not happen automatically, but generally through conscious effort, by scientists who have an explicit theory of change that involves more than simply publishing research in academic journals.

LEVERS OF CHANGE IN THE PRIVATE SECTOR

There are some industries whose business models are, or at least can in principle be, inherently aligned with climate goals. Renewable energy and carbon capture and storage are obvious examples. More broadly, multiple industries offer opportunities to decarbonize: Utilities and transportation are examples. Others might include environmental consulting, architecture firms focused on green buildings or sustainable urban planning, agriculture and food systems, and so on. In fact, because nearly every industry produces emissions in one way or another, it has been said that "all jobs are climate jobs." This should not be taken literally, but nearly every industry does now contain workers with some mandate to reduce the climate harm being done by their firms.

Here we focus on the financial sector. This sector has created many new companies and jobs that have hired many climate scientists and use their expertise directly. Also, companies in nearly every other sector participate in the financial markets and thus are affected by what the financial sector does around climate.

Many companies have made pledges to reduce their emissions to "net zero" by some target date, or to work toward the goal of the Paris Agreement to limit global mean warming to 1.5°C, or similar. In Europe there are rules requiring disclosure of climate risks (and under the Biden administration, similar rules were proposed in the United States by the Securities and Exchange Commission, though as we write, under the second Trump administration, those are not being enacted). These developments and others have created new jobs for climate scientists, both at the companies making the pledges or complying with the rules, and at consulting or data-providing firms offering their services to help them do it. These jobs offer decent salaries, freedom from the constant publish-or-perish and grant-writing pressures of academia, and the promise of doing something about the climate problem that is, in principle, more direct than writing scientific papers is.

Carbon budgets and net-zero pledges address companies' "transition risk," i.e., the risks to their businesses posed by decarbonization. The other component of climate risk is "physical risk," meaning the risk posed by actual physical hazards associated with climate. The real estate, financial, and insurance industries also—together with government, as the regulator and insurer of last resort—can participate in the creation of physical risk by incentivizing people to live in hazardous areas and structure and thus can reduce risk by changing those incentives.[14] Many new companies and jobs have been created to quantify physical climate risk, and these offer the hope of contributing to climate adaptation.

Skepticism about voluntary private sector action is, however, warranted. A for-profit corporation's primary explicit purpose is to make money, and the corporation is likely to do the right thing—on climate or any other societal issue—only if and when doing so can be justified as consistent with that goal. Net zero pledges are much easier to make than to keep, for example, and the incentive for greenwashing is strong. A recent report by Columbia's Center on Sustainable Investment found that although "there has been a proliferation of bottom-up models, tools, metrics, methodologies, and initiatives designed to measure, evaluate, and coordinate the climate performance of financial institutions" nonetheless "meaningful progress in realigning global finance to support climate goals has been limited." It also concludes that while the private sector is important, its voluntary commitments cannot be a substitute for government action. We have not seen similar studies about companies' efforts to address physical climate risk through adaptation and resilience, but we suspect similar critiques could be raised there.

While skepticism is healthy, it need not become cynicism or erase ambition. The private sector comprises by far the lion's share of the economy and employment in the United States and many other nations, and it has to be part of any meaningful climate solutions. To be sure, government leadership would make private sector efforts much more powerful and effective. For example, a real price on carbon would motivate capitalist ingenuity toward mitigation in a way that volunteerism cannot. This should motivate us as voters and political actors. But as workers, as long as voluntary efforts are mostly what we have, strengthening them and making them more effective are worthy goals. We address this further in the section on personal agency.

ETHICAL, SOCIAL, AND POLITICAL QUESTIONS IN THE USE OF SCIENCE

Besides the difficulty of doing work that has an impact, there are additional challenges associated with being certain that the impact serves climate justice. Nearly every positive climate action has trade-offs of some kind. Solar and wind energy generation facilities occupy land that could be used for other purposes; increasing high-density urban development to reduce emissions associated with transportation and housing causes local opposition and raises complex political questions about the roles of local vs. state or federal government. Research that quantifies increasing physical risk in high-hazard areas, for example, near sea level along hurricane-prone coastlines, could potentially lead insurers to charge higher rates or to pull out of markets entirely, making it more difficult to get mortgages. People will be displaced from their homes, threatening local tax bases and the broader housing market. Overall climate risk may decrease, but at substantial cost to many individuals and families, and probably the greatest suffering will fall on those least able to absorb it.

Complex ethical, social, and political challenges are by no means the sole province of the private sector. To the extent that academics orient their work explicitly toward influence in the external world, the same issues must arise. Once scientific information has any impact on anything, there are likely to be consequences beyond those intended, regardless of who produced the information.

This topic is much more profound than we have either space here or expertise to address properly, so we simply counsel recognition of it, vigilance, and humility. In general, our training as scientists does not prepare us adequately to think through ethical issues, or the complex societal dynamics through which science can raise them. Openness to evidence is important, but perhaps even more so is openness to diverse ways of thinking. Education in the social sciences and humanities, including the history and philosophy of science, can be helpful, and we think it would generally be good if scientists got more of it and saw it as part of their education as scientists per se.

PERSONAL AGENCY AND THE STRUCTURES OF INSTITUTIONS AND CAREER PATHS

When trying to assess the potential to have a positive societal impact through one's job, one question is whether the organization's mission and goals align with the societal goals one wishes to work toward. To the extent that the alignment is not perfect, however, another question is the extent to which the job allows one the agency to shape one's own work to increase it.

Academic careers have a distinct advantage on the latter. A typical academic has a degree of agency that is uncommon elsewhere in the professional world, at least in the earlier stages of one's career. An academic scientist whose research is externally funded can work on whatever they want, if they can raise grants to support it. Also, many academic scientists—particularly faculty members at research universities—do not have a boss in the sense that a typical private sector professional does, for example, someone to whom one is accountable in detail regarding how one's time is spent day-to-day. They may be busy and stressed out with many responsibilities, but if an academic can manage their time well enough, they can spend some of it on things that contribute to societal aims. These can include writing or speaking for the public, working with community or other external "stakeholder" groups, even activism or political organizing. In addition to flexibility in time management, academics also enjoy academic freedom, which in principle allows them to express their views without oversight or fear of punishment. While academic freedom is not absolute and has recently been under attack in the United States, nothing comparable is available in most private sector jobs.

Compared to an academic, a private sector employee's ability to effect positive change through their work depends a bit more strongly on the priorities, goals, strategies, and concrete practices of their particular employer, and also of their direct superiors within the organization. On the other hand, it is generally easier and more common to change jobs in the private sector compared to academia. In principle, this can give the individual worker a kind of agency that is harder to come by in academia, where slow, hidebound, and standardized hiring processes, the tenure system, and the geographic dispersion of institutions (such that changing jobs more often means moving one's home) tend to keep academics in one place to a greater extent than private sector workers.

Work outside of academia also has the advantage of being less removed from the rest of the world. One standard meaning of the word "academic" is a synonym for "irrelevant." The societal value of academic research is often expressed as occurring through its potential to influence "decision-makers," "stakeholders," or some other words describing actors external to academia. If one is outside academia, such an influence will involve action at a lesser distance. And while a typical junior employee in the private sector (or, for that matter, perhaps in government or nonprofits) will have less individual agency than one in academia will, this may change as one becomes more senior. One's control over one's own time will increase with seniority, and one may gain some ability to shape the organization's overall direction, and perhaps even that of one's industry as a whole.

CLOSING THOUGHTS

In general, it may be hard to assess the climate impact, or any other metric of broad societal impact, of most jobs. The sort of data-driven assessment that effective altruism asks for (for example) risks looking for the key under the lamppost (that is, seeking answers where the evidence is of highest quality, even if they are unlikely to be there), whether applied to charitable donations or to career choices. So while we view critical, fact-based examination of oneself, one's employer, and one's job as worthwhile and important exercises, at the same time one cannot make all life decisions purely based on those. Some aspirational thinking, and going with one's gut, may be in order.

Reasons for skepticism about any organization's claims of societal benefit can also be viewed as opportunities to make a difference. If a given job does not guarantee meaningful climate impact, does that not mean that the worker who finds a way to achieve that impact has done something even more important? While adjusting investment portfolios to account for climate risk (for example) could be an exercise in greenwashing, perhaps a thoughtful innovator can find a way to do it so that it materially reduces emissions or supports meaningful adaptation to a greater degree than would have happened otherwise. While a typical climate science research project may not have any impact outside academia, maybe that means the one that does is all the more meaningful. And in general, there will be increasing opportunities to influence the overall direction of many organizations—to affect what the organization does at a larger scale than one's own output—as one becomes more senior.

The difficulty of assessing impact in the broader world may also be an argument for doing work whose direct, visible, local impacts are positive, whether they address climate or other values. In academia, we educate people and create new knowledge (including about climate). Those are good things! Insurance is a necessity for individuals, families, and businesses in the modern world, so the insurance industry provides a useful service. And so on. If one sees value in one's daily work, that can provide the steady motivation that one needs to keep going while one chips away at the bigger and harder problems.

It can also be helpful to practice some realism, compassion for oneself and others, and what the stoics call "negative visualization"—that is, imagining specific ways in which the situation could be worse. There used to be very few jobs that had anything to do with climate. Now there are a lot. Maybe some of them do not live up to the most naive version of their promise, but what are the realistic alternatives? The capitalist economy was not designed to solve collective societal problems like climate change (at least not without government regulation and oversight), but would we not rather it take climate into account, in whatever ways it can, than not? Are over-promising, greenwashing, and hype not inevitable precursors and accompaniments to any real action, and so should we not accept that and do what we can to move toward that action? Did academic climate science not need to define the problem first, before it could contribute to solving it, and so is it not natural that we are still figuring out how to do the solving part?

As promised, in the end we have no simple solutions or recipes for success. The world is changing quickly, and we do not really know what we are doing either. We offer our support and encouragement to those who are thinking critically about the real or potential societal benefit of their work, and choosing (or changing) jobs with that thinking as at least part of their motivation. And also to those who push against the boundaries of their jobs, using whatever freedom they have to seek ways to increase the positive impact of their work. And (while it is not the subject of this essay) to those who work for the benefit of the climate and other societal goals outside their day jobs, through volunteering, donation, or activism. Keep it up![15]

Notes

1. For more discussion of distributive justice concerns, see chapter 3 in this volume.
2. Oreskes and Conway, *The Big Myth*.
3. Battocletti et al., "The Voluntary Carbon Market."
4. See, for example, deep-convection.org.
5. Edwards et al., "Mapping Scientists' Career Trajectories."
6. E.g., Singer, *The Most Good You Can Do*.
7. For some critiques of effective altruism, as well as some responses to them, see Gabriel, "Effective Altruism and Its Critics"; Mounk, "The Problem with Effective Altruism"; or Alexander, "In Continued Defense of Effective Altruism."
8. The book connected me to historian Deborah Coen, with whom I've since collaborated to develop ideas of "usable science" (e.g., Coen and Sobel, "Critical and Historical Perspectives") and to writer Amitav Ghosh (as documented in his books *The Great Derangement* and *The Nutmeg's Curse*), which in turn led to a study of cyclone risk to Mumbai (Sobel et al., "Tropical Cyclone Hazard to Mumbai") and much else.
9. Sobel, "Usable Climate Science Is Adaptation Science," 8.
10. Drake and Henderson, "A Defense of Usable Climate Mitigation Science."
11. Jézéquel et al., "Singular Extreme Events and Their Attribution."
12. Coen and Sobel, "Introduction: Critical and Historical Perspectives."

13. See Fourth Purpose, fourthpurpose.columbia.edu: "In addition to research, edu-cation, and public service, Columbia University identified a Fourth Purpose: to leverage scholarly knowledge to create societal and global impact, in close partner-ship with organizations outside academia."
14. E.g., Boomhower et al., "Wildfire Insurance, Information, and Self-Protection"; Sas-try et al., "When Insurers Exit."
15. We thank the editors, Paul Gallay, and Mary Claire Morris for helpful comments on the first draft of this chapter, and the anonymous reviewers for theirs on the first submission.

BIBLIOGRAPHY

Alexander, Scott. "In Continued Defense of Effective Altruism." *Astral Codex Ten*, Novem-ber 28, 2023. https://www.astralcodexten.com/p/in-continued-defense-of-effective.

Battocletti, Vittoria, Luca Enriques, and Alessandro Romano, "The Voluntary Carbon Market: Market Failures and Policy Implications" *University of Colorado Law Review* 95, no. 32 (2024). https://scholar.law.colorado.edu/lawreview/vol95/iss3/2.

Boomhower, J., M. Fowlie, and A. J. Plantinga. "Wildfire Insurance, Information, and Self-Protection." *AEA Papers and Proceedings* 113 (2024): 310–15.

Coen, D. "A Brief History of Usable Climate Science. *Climatic Change* 167, no. 51 (2021). https://doi.org/10.1007/s10584-021-03181-2.

Coen, D., and A. H. Sobel. "Introduction: Critical and Historical Perspectives on Usable Climate Science. *Climatic Change* 172, no. 15 (2022). https://doi.org/10.1007/s10584-022 -03369-0.

Columbia Center for Sustainable Investment (CCSI). "FINANCE FOR ZERO: Redefining Financial-Sector Action to Achieve Global Climate Goals." https://ccsi.columbia.edu /sites/default/files/content/docs/Finance_for_Zero_CCSI_June_2023.pdf.

Drake, Henri F., and Geoffrey Henderson. "A Defense of Usable Climate Mitigation Sci-ence: How Science Can Contribute to Social Movements." *Climatic Change* 172, no. 10 (2022). https://doi.org/10.1007/s10584-022-03347-6.

Economic Report of the President (ERP), with the Annual Report of the Council of Eco-nomic Advisors. "Opportunities for Better Managing Weather Risk in the Changing Climate" (2023), chap. 9.

Edwards, Kathryn, Hanna Acheson-Field, Stephanie Rennane, and Melanie Zaber. "Map-ping Scientists' Career Trajectories in the Survey of Doctorate Recipients Using Three Statistical Methods." *Scientific Reports* 13, no. 8119 (2023). https://doi.org/10.1038/s41598 -023-34809-1.

Gabriel, Iason. "Effective Altruism and Its Critics," *Journal of Applied Philosophy* 34 (2017): 457–73. https://www.jstor.org/stable/10.2307/26813081.

Ghosh, A. *The Great Derangement*. Chicago: University of Chicago Press, 2016.

Ghosh, A. *The Nutmeg's Curse*. Chicago: University of Chicago Press, 2021.

Jézéquel, A. V., H. Dépoues, A. Guillemot et al. "Singular Extreme Events and Their Attribution to Climate Change: A Climate Service–Centered Analysis." *Weather, Climate, and Society* 12 (2020): 89–101.

MacAskill, William. *What We Owe the Future.* New York: Basic Books, 2022.

McMahan, Jeff. "Philosophical Critiques of Effective Altruism." *The Philosophers' Magazine* 73 (2016): 92–99.

Mounk, Yascha. "The Problem with Effective Altruism." *Persuasion*, September 19, 2024. https://www.persuasion.community/p/the-problem-with-effective-altruism.

Oreskes, Naomi, and Erik M. Conway, *The Big Myth: How American Business Taught Us to Loathe Government and Love the Free Market.* London: Bloomsbury, 2023.

President's Council of Advisors on Science and Technology (PCAST). *Extreme Weather Risk in a Changing Climate: Enhancing Prediction and Protecting Communities.* April 2023. https://www.whitehouse.gov/wp-content/uploads/2023/04/PCAST_Extreme-Weather-Report_April2023.pdf.

Sastry, Parinitha, Ishita Sen, and Ana-Maria Tenekedjieva. "When Insurers Exit: Climate Losses, Fragile Insurers, and Mortgage Markets." December 23, 2023. http://dx.doi.org/10.2139/ssrn.4674279.

Singer, P. *The Most Good You Can Do.* New Haven: Yale University Press, 2016.

Sobel, A. H. *Storm Surge: Hurricane Sandy, Our Changing Climate, and Extreme Weather of the Past and Future.* New York: Harper-Collins, 2014.

Sobel, A. H. "Usable Climate Science Is Adaptation Science." *Climatic Change* 166 (2021): 8.

Sobel, A. H., C.-Y. Lee, S. J. Camargo et al. "Tropical Cyclone Hazard to Mumbai in the Recent Historical Climate." *Monthly Weather Review* 147 (2019): 2355–66.

V

Rethinking Knowledge at a Time of Crisis

Climate Justice, Expertise, and Community

13

Climate (In)Justice for Whom?

Alternative Theories and the Absence of Scientific Language

SHENG LONG

I do not know who told him what, but later at the hotel he explained to me: Now I know that these people call this that the earth is heating up; that is how I will explain it to them [the World Bank] next time. And half seriously, half-jokingly we talked about how, after all, in Spanish to heat up, calentarse, *can also mean to be mad.*

—Marisol de la Cadena, *Earth Beings*

In his influential review article, "The Case for Letting Anthropology Burn," Ryan Jobson starts by discussing the smoke released from wildfires in California, shrouding the entire American Anthropological Association conference.[1] He reflects on the need for anthropology's examination of environmental crises to move away from the assumption of the universal liberal subject. During the same period when North American anthropologists were facing a crisis within their discipline in the wake of environmental issues, droughts and wildfires were also troubling pomelo farmers in Meixian, a Hakka-speaking area in the southeastern mountains of China. At that time, I was conducting fieldwork in this area. In the evening, villagers gathered in the courtyard outside a mahjong store. The crimson sunset filled part of the sky, serving as signs of weather for the villagers. They speculated that there would be no rain the next day, which caused them great anxiety: "The pomelo trees are dying!" Months of drought severely affected their income, but they felt helpless. The mainstream media in Meixian described this situation as "extreme weather," while the villagers complained that it was a "bad heavenly year (*tian nian*)." The word *tian* refers to the sky, heaven, or nature.

In this chapter I explore the relationship between language discrepancies and (in)justice. Rural residents in Meixian, who were profoundly affected by climate change, did not employ the widely accepted scientific terminology. This raises questions: How do they interpret severe weather phenomena? More radical inquiries arise: Is global warming equally "real" for those who have never contemplated it? When do scientifically established facts become inaccessible to those directly experiencing them? Anthropological studies examine the movements of marginalized groups, such as Indigenous people, regarding climate change and analyze their understandings of climates.[2] Attention to climates has shifted ethnography from a community focus to an increasingly cross-scale global context.[3] Some anthropologists focus on climate justice, which involves equality and human rights under climate change, and actively engage in policy discussions and collaborations.[4] This chapter considers the audience of climate justice as both an academic and a mainstream concept. The underlying question raised by this ethnographic story is: To whom are the academic discourses on climate justice speaking? As the authors of this book have demonstrated, academic discourses should not be confined to a single discipline but should instead be interdisciplinary. Moreover, academic research is inseparable from public voices.

When discourses such as "global warming" and "climate change" entered the international political sphere, becoming focal points of scientific and political debate, some of the vulnerable populations remained unaware. Hence, climate justice also concerns the (un)fairness of "knowing/not knowing." This also raises the questions: Can scientific terminology better incorporate and appreciate alternative knowledge, and can these alternative theories wield influence in communication? Sara de Wit's research on climate change revealed that global scientific discourse, facilitated by various "translation mechanisms," became integrated into the local narratives of Cameroonian grassland farmers, empowering them with avenues for political actions.[5] China, after years of socialist movements, has embraced globalization and shaped the environmentalist movements that led to the intertwining of scientific discourses with history.[6]

While the term "climate change" may have suggested climate as a singular entity, it is crucial to recognize that climates should also be plural, and they are not always reducible in various theories. These theories should not be simplistically categorized as either scientific or unscientific. I examine the situation where alternative narratives of environmental crisis are expressed in the absence of Western/modern scientific terminology. I explore how pomelo farmers in southeastern China relied on their existing knowledge to understand climate anomalies, even without terms such as "climate change" or "global warming." This does not suggest that the knowledge of pomelo growers remained stagnant; rather, they actively updated their vocabulary to interact with governmental bodies for discussions on agricultural insurance compensation, negotiations with forestry departments regarding penalties, and management of kinship/neighborhood ownership. That is, the vocabulary they already mastered, such as "precipitation," served as a common language for them to communicate with the government. While they were adept at incorporating new vocabulary, their shielding from some other scientific discourses may be due to limited dissemination of these terms, and, more significantly, a lack of avenues for them to use them to seek justice.

THE LANGUAGES OF CLIMATES

The scientific discourse of global warming has now become general vocabulary for certain groups, establishing authority on factual interpretation and political accountability. However, numerous other groups remain unexposed to or do not engage with this discourse. Does this imply they inhabit a different reality from others? Could their alternative theories about climate potentially facilitate their involvement in broader conversations?

During my years of fieldwork in Meixian, residents encountered droughts, floods, and frosts. This Hakka-speaking mountainous area is located in the eastern part of Guangdong Province, along the southeastern coast of China. Since the 2000s, rural residents in this region have shifted from primarily cultivating rice to growing economical crops like pomelos. The local government sought to establish a regional brand, positioning Meixian as famous for golden pomelos. Before the widespread use of smartphones, people would diligently gather in front of their televisions, waiting for the 19:35 weather forecast on China Central Television (CCTV), considered the most authoritative forecast. Later, as many rural residents acquired smartphones, they would often discuss which weather forecast was more accurate during their leisure time. Of course, weather was often challenging to predict accurately, except in cases of prolonged rainfall or drought. While reading these scientific forecasts issued by governments or institutions, fruit growers also had their own theories to predict and explain the weather. They were particularly adept at judging cultivation timing based on the "twenty-four solar terms," which were the markers of weather changes in the lunar calendar.

The "Heavenly Year" was not a term used to forecast weather. Its usage typically emerged when people struggled to explain unusual weather conditions. Individuals also employed various methods to influence the Heavenly Year. In this rural region, people observed the "Patch Up the Sky" festival on the twentieth day of the lunar new year. This festival was rooted in a well-known Chinese myth, recounting a massive hole in the sky causing floods on earth, later mended by the goddess Nüwa. On this occasion, middle-aged and older individuals visited the earth shrines to offer prayers for favorable weather. While some villagers may not entirely believe that such rituals alter the year's fortune, they would refrain from farmwork on this day and caution non-Hakka people that "working today might bring no profit this year."

What the Heavenly Year refers to seems to be an enigmatic and resilient presence. This term serves to encapsulate all weather phenomena, including both favorable and unfavorable conditions for agricultural production. However, more often than not, the Heavenly Year was used to lament extended periods of adverse weather. Following droughts, floods, or frosts, fruit growers frequently voiced complaints, such as "It's a bad Heavenly Year, we can't make any money," or "It's a bad Heavenly Year, what can we do?" Unlike the moral implications associated with terms such as "global warming," complaints about the Heavenly Year did not trace to industrialization, pollution, urbanization, or specific political and economic entities. The Heavenly Year is elusive, making it difficult to trace back to its origins or understand directly; it manifests itself to people without clear explanations.

Languages of climates are generated from different discourses, which shape and influence the actions people may take. Terms such as "global warming" and "heavenly years" not only describe climate phenomena but also provide possibilities for accountability. These scientific terms might be understood as "historical objects," as described by Webb Keane, objects that contain affordances for moral judgment and behavior.[7] So, under what circumstances do people ask who is responsible for the frequent occurrence of extreme weather? In fact, on the issue of "climate change," the Chinese government has sought to maintain an open attitude and has strongly supported climate change adaptation in national policies.[8] Such policies have also been disseminated to provincial and municipal governments.[9] However, in greater rural areas, residents affected by climate anomalies have not used these terms.

In rural Meixian, the mainstream media accessible to residents used the term "extreme weather" without referencing "global warming" or "climate change." While in academic discourses, extreme weather and climate change are correlated, neither these government announcements nor mass media offered explanations for extreme weather occurrences. Extreme weather was viewed as a phenomenon that was challenging to forecast and manage. In this regard, extreme weather bore some resemblance to what fruit growers called the Heavenly Year. While villagers may not use terms like "extreme weather" in their conversations, it did not impede their comprehension of it. Villagers did not express surprise or confusion when discussing news related to extreme weather.

Discourses have specific audiences for dissemination, and their communicative competence varies among different groups. Linguistic anthropologist Charles Briggs analyzed the communicative competence of fieldworkers and research subjects in a classic article, noting that anthropologists often demonstrate incompetence in both knowledge and linguistic ability.[10] When bureaucratic or academic discourses are used to communicate with nonoriginal speaker groups, they create barriers, hierarchies, and authority. However, this does not mean that nonoriginal speakers cannot learn and utilize these discourses. On the contrary, these discourses can also become "shared discourses," establishing channels of connection and communication between authorities and nonauthorities. That is to say, the obscurity and professionalism of these discourses should not be the sole reason for their lack of circulation. In fact, many of the terms farmers used regarding climate and natural disasters, such as drought, floods, and frost, are consistent with official reports and mainstream media descriptions.

Thus, what I am suggesting here is not that terms such as "global warming" or "climate change" are not simple enough to understand; on the contrary, these words can be readily grasped by those who have never heard of these theories before, as mentioned in the opening quote about the Peruvian Indigenous rights activist who immediately learned the phrase "earth is heating up." I want to reflect, through the example of Chinese pomelo farmers, on the circumstances where the right to learn and use these scientific terms is denied or rendered ineffective.

Anthropologists have encountered many such instances in their research where Indigenous peoples learned and employed scientific or official language while advocating for their rights. This suggests that it's not solely anthropologists striving to grasp field languages; individuals in the field also acquired foreign languages to establish linguistic competence, facilitating communication with officials or institutions.

In her book *Earth-beings*, Marisol de la Cadena recounts the journey of her friend Nazario Turpo, a shaman from the mountainous regions of Peru, who gained recognition in the tourism industry and public sphere.[11] As illustrated in the epigraph of this chapter, Turpo acquired new climate-related expressions through his interactions with World Bank personnel. Initially, he attributed the drying up of irrigation canals in reports on World Bank loan projects to the anger of the mountains caused by excessive airplane traffic overhead. Later, he learned phrases like "the earth is heating up" and opted for more impactful language to advance his interests with bureaucratic officials. De la Cadena posits that Turpo possesses intricate and intersecting forms of knowledge and "had the ability to visit many worlds."[12] Turpo excelled in translating between different ecological theories. However, often Indigenous theories are difficult to absorb or comprehend within official discourses. In legal documents in Peru, the familiar concept of earth-beings, the combination of humans and nonhumans, became merely geographic features.[13]

The sharp division between nature and culture is a prominent characteristic of modern Western world theory. This dichotomy also delineates the boundary between scientific and nonscientific disciplines, a point criticized by Bruno Latour.[14] It is precisely the modern politics and academia that rigidly separate the concepts of life and nonlife, obstructing mutual communication between more diverse theories.[15]

Is climate change theory referring to the same thing as the Heavenly Year? Are people with different theories experiencing the same reality? Some anthropologists emphasize that when researchers cannot comprehend the theories of those they study, they should recognize that they have their own reality rather than dismissing it as a distorted theory.[16] However, theories that excessively emphasize differences and incommensurability may strengthen social barriers and overlook the potential for dialogue.[17] Paul Nadasdy introduces the concept of indeterminacy, describing it as a quality or status of things. In his research with Indigenous peoples in the Yukon region of Canada, he examines how they perceive the landforms they inhabit as sentient beings, legal property, homeland, and national territory through various sets of material-semiotic practices.[18] Following this perspective, climates can also be regarded as "indeterminate." That is, climates can be seen as part of the environment influenced by humans, or can be autonomous, irresistible, or possessing agency. The next question is, who can recognize this uncertainty, and what possibilities can the inclusiveness of these diverse theories bring about?

The dissemination and use of languages are often intertwined with political (in)justice.[19] As Jennifer Hadden demonstrated in her study of political activism in Copenhagen, slogans like "climate justice" can serve as a framework for civil groups.[20] However, this does not mean that such terms necessarily lead to justice. Who uses this language matters. Some slogans, when used by governments or authoritative institutions, might legitimize their extraction or restriction of resources.

WHERE (IN)JUSTICE IS UNSPOKEN

Returning to Meixian, the Hakka-speaking region, in that year, several months of consecutive droughts left fruit growers concerned about their harvest. However, a few large orchard owners received notices from the town agricultural station instructing them to negotiate compensation with insurance companies. In the previous year, Meixian had introduced government-backed agricultural insurance to compensate for losses caused by extreme weather and other disasters. This marked the first implementation of agricultural insurance in Meixian, providing participating fruit growers with substantial discounts. As Meixian experienced months of rainfall shortage severely affecting yields, these large orchard owners were pleased to learn they would receive compensation. A large orchard usually covered at least 200 *mu* (approximately 33 acres), receiving an average compensation of 200 RMB per *mu*.

However, the vast majority of farmers were unaware of such agricultural insurance. Although mentions of Meixian promoting agricultural insurance had appeared in county news, it seemed more like a secret agreement between town agricultural station officials and the large orchard owners. In the subsequent year, officials from the agricultural station continued contacting some large orchard owners but did not renew contracts with others who had participated in the insurance. Those not selected started privately inquiring about the situation of other orchard owners. They discovered that the town government had only selected a few of the largest orchards. One orchard owner went to the town agricultural station to ask why they could not continue purchasing agricultural insurance, and the response he received was, "The pomelo yield is not significant." His orchard partner suspected it was because they "hadn't treated the officials to a dinner." These large orchard owners were skilled at navigating the bureaucratic system and thus gained access to more government resources. Most of them were village cadres or children of cadres who had acquired the use rights to the newly developed large orchards in the 1990s after the decollectivization reforms. Political connections were closely intertwined with the economic compensation for climate disasters.

The majority of pomelo growers in the 2000s converted their rice fields into pomelo orchards. These orchards were small in size, with scattered trees and insignificant yields. In everyday conversations of pomelo farmers, government agricultural insurance was never mentioned, nor did they expect the government to be responsible for their economic losses caused by weather. During severe droughts, the Meixian government conducted several rounds of cloud seeding. Villagers shared short videos of these operations on their social media. However, the effectiveness of cloud seeding was consistently limited, disappointing almost every time. On one occasion, heavy rain fell in the county seat, but "only a few drops fell in the village." During a challenging Heavenly Year, people placed their hopes on the following year. Pomelo farmers often analyzed Big Years and Small Years, referring to years with favorable or unfavorable weather conditions or large and small harvests. Many held the belief that Big Years and Small Years alternated, resulting in a fluctuation between despair and hope.

The unequal allocation of resources in addressing climate anomalies extends beyond urban-rural divides or industrial-agricultural sectors; it also encompasses disparities among pomelo farmers due to variations in social and economic capital. The majority of pomelo farmers lacked direct communication channels with the government, state-owned enterprises, or academic institutions to mitigate their losses and navigate financial challenges. Many had never even expected such a prospect. Government ecological initiatives are frequently associated with gentrification, disproportionately benefiting a select few while perpetuating poverty and marginalization among other vulnerable groups.[21] As argued in the following chapter by Raffaella Taylor-Seymour and Courtney Bender, interpretations of the weather are often nonsecular. During the Chinese imperial era, it was the duty of rulers, ranging from the emperor to local officials, to offer prayers to the gods for favorable weather on behalf of the people.[22] These rituals served ceremonial as well as practical purposes, aligning with popular expectations and assessments. The imperial bureaucratic system responded to disasters such as famine by constructing large granaries and extensive hydraulic projects across regions.[23] When the bureaucratic system failed to provide adequate assistance to disaster-affected people, regional elites and grassroots organizations also played pivotal roles in relief efforts.[24] After the modernization and socialist movements of the twentieth century in China, the link between weather, deities, and the bureaucratic system largely waned following criticisms of superstition. In Meixian, worshiping the Earth God and observing festivals such as "Patch Up the Sky" have become more individualized acts. The notion of the Heavenly Year that people refer to has a tenuous connection with deities and is no longer tied to government involvement.

Just as there is diversity in climate theories, the concept "justice" is also a hybrid of various meanings. The Western conception of justice is considered similar yet different from the concept of *yi* in early Chinese moral thought, which refers to the moral standards beyond individuals.[25] These public discussions on justice have often focused on legal and moral righteousness.[26] The contemporary Chinese legal justice system is seen as an interaction of government, civil society, third parties, and imported ideas.[27]

With the emergence of the concept of the "Anthropocene" and the reevaluation of ontology to encompass human and nonhuman relationships within anthropology, the field of multispecies ethnography has gained prominence.[28] This has sparked discussions about the notion of multispecies justice, which focuses on addressing unequal vulnerabilities among humans and other species, with political, legal, and ethical concerns.[29] In chapter 10 authors Emma Gilheany and Julia Lajus explore multispecies justice in the Arctic amid climate change, using historical and anthropological approaches. As climate change continues to induce ecological disruptions, the dynamics of interactions between humans and animals undergo transformation.[30] Eduardo Kohn, whose extensive research is situated in the Amazon rainforest, advocates for attention to multispecies relationships, stressing the importance of cross-species communication. Despite many animals lacking systematic symbolic signs, Kohn suggests they can communicate with other species in a pidgin-like manner.[31] Multispecies studies suggest that achieving justice among unequal parties relies on sign processes that can be translated and engaged with. Put differently, injustice is more likely to occur when a group is neither heard, seen, nor understood.

Just as anthropologists have observed within Indigenous rights movements, the public also analyzes and comprehends the discourse utilized by governments and corporations. Stuart Kirsch introduced the term "reverse anthropology" to depict how Indigenous peoples in South America, over years of grappling with mining companies, scrutinized and assessed their counterparts—much like quintessential anthropologists.[32] These residents, in their political engagements, absorbed and applied diverse theories while also contemplating the repercussions of these external discourses on them.

When several large orchard owners were informed by town officials that they could participate in agricultural insurance, they expressed great surprise. The Chinese government has introduced agricultural insurance in recent years to guarantee "food production security." Although agricultural insurance was often associated with global warming and climate change in news and academic works, such connections were not mentioned in the local implementation, especially in the communication between grassroots cadres and orchard owners. However, this does not imply that fruit growers cannot grasp the content of agricultural insurance; on the contrary, they were very familiar with the units used for analyzing insurance compensation: rainfall, temperature, humidity, and so on. These were also indicators they relied on in their daily agricultural activities. Fruit growers often felt they understood the relationship between weather and pomelo cultivation better than government officials and insurance company personnel did. For instance, they paid close attention to the sequence of weather patterns, whether it was drought before flood or flood before drought, as these sequences and timing points can have different effects on pomelo production. These complex situations were not accounted for in the more rigid measurement methods of insurance companies.

It is precisely this collision of languages from different climates that, to some extent, fosters new prospects for cooperation. In her research on water in the Andes, Astrid Stensrud analyzes how Indigenous communities use various practices, including rituals, water infrastructure, and protests, to address the drying up and erratic weather patterns brought about by global warming.[33] Terms such as "global warming" make people aware of how phenomena previously difficult to understand are interconnected with the broader world. These diverse discourses—historical, bureaucratic, scientific—shape the multiple ontologies of water.[34] Sarah Vaughn, in her research on sea defense and mangroves in Guyana, reflects on what constitutes "expert knowledge" and analyzes how different forms of expertise allow certain facts to be seen by governments, colonizers, scientists, and grassroots organizations while obscuring others.[35]

The inequality of vulnerability is nested and intersecting in this way. Where there are silenced groups, there are more silenced ones. In Meixian, even the scant government agricultural insurance compensations were allocated to the wealthier large orchard owners within the communities most severely impacted by climate change, who were better equipped to manage such crises. Rural women there constituted the primary labor in pomelo cultivation, yet they were not encouraged to directly communicate with the government or the merchants who purchased pomelos. During poor Heavenly Year, they must exert additional labor to sustain agricultural production and support their families. They even needed to expend their remaining energy on temporary jobs during the agricultural off-season to earn cash for their households. Climate issues transcend climates alone; they encompass various intersecting factors— gender, ethnicity, class. Therefore, studies on climate justice should draw from feminism, postcolonialism, and critical race studies.[36]

This is a two-way or multiway interaction. Not only are scientific discussions on "climate change" and "global warming" absent for pomelo farmers, but also the theories of farmers regarding the weather have not been acknowledged and adopted by governments and companies. In their examination of generative terms such as the "Anthropocene," Heather Davis and Zoe Todd suggest that while these new terminologies can encompass various environmental crisis phenomena, they also grapple with escaping the pervasive logic of colonialism and imperialism, thus severing the entanglement of body, mind, and land. Consequently, they call for mutual listening from all sides, especially urging power entities like governments to listen to the voices of those with rich traditions who understand how to coexist with the Earth.[37]

• • •

This could be a story about climate change, yet "climate change" is conspicuously absent. What's absent is not just these scientific terms, which have become part of the general vocabulary elsewhere, but also the endorsement and avenue these scientific terms provide for seeking justice. This chapter unveils an unspoken narrative of climate injustice. After the drought significantly diminished the pomelo yield in this mountainous regions of southeastern China, one pomelo farmer, skilled in arithmetic, revealed that he earned a mere 0.8 RMB, approximately 12 cents, for every 1 kilogram, or 2.2 pounds, of pomelos harvested, even assuming his and his family's labor was free. During bad Heavenly Years, pomelo farmers struggled with meager incomes, sometimes resorting to dipping into their savings. Villagers often said that the Heavenly Year hinged on luck, and they likened the market to gambling. It is these very individuals who never mentioned "climate change" who bear the brunt of the erratic weather.

This untold story of "climate change" may prompt alternative imaginings. There are already instances where these scientific discourses, upon reaching vulnerable communities, may engender different understandings, thereby generating more government or societal resources. The language of climate is intricately intertwined with politics, from the national to the individual level.[38] These authoritative discourses on environmental rhetoric strategies may also lead to extractivism and new inequalities.[39] This chapter has explored language, and the theories that make these languages possible, in influencing how people make sense of their realities. The aim here is not to argue that theories of climates should be converging into the only reality, but to acknowledge the plurality of climates and inquire into the possibility of engaging in dialogues among various climate theories.

In recent years, there has been a growing recognition of the significance of traditional environmental knowledge. Concepts such as "Heavenly Years" and alternative climate theories may hold the potential to be integrated into scientific discourses. However, in this specific context, Heavenly Years did not provide pomelo farmers with avenues to seek assistance from governmental or institutions. Even the seemingly neutral term "extreme weather," as used in media, lacks implications of accountability. I wanted to emphasize that these residents were not incapable of comprehending climate theories beyond traditional frameworks. On the contrary, they demonstrated an ability to rapidly acquire and utilize new terminologies in their interactions with government and insurance companies. This case invites a critical examination of how scientific discourses permeate agricultural communities and the extent to which they facilitate individuals to assert their rights and seek redress.

Notes

1. Jobbson, "The Case for Letting Anthropology Burn."
2. Morris, "Managing, Now Becoming, Refugees"; Verma, "The Role of Anthropology in Climate Change Research."
3. Crate, "Climate and Culture"; Whitington, "What Does Climate Change Demand?"
4. Crate and Nuttall, eds., *Anthropology and Climate Change.*
5. De Wit, *Global Warning.*
6. Hathaway, *Making the Global in Southwest China.*
7. Keane, *Ethical Life: Its Natural and Social Histories.*
8. Gippner, *Creating China's Climate Change Policy*; Kopra, *China and Great Power Responsibility.*
9. Qi et al., "Translating a Global Issue Into Local Priority."
10. Briggs, "Learning How to Ask."
11. Marisol de la Cadena, *Earth Beings.*
12. de la Cadena.
13. de la Cadena.
14. Latour, *Pandora's Hope.*
15. Povinelli, *Geontologies.*
16. Henare et al., eds., *Thinking Through Things.*
17. Bessire and Bond, "Ontological Anthropology and the Deferral of Critique"; Graeber, "Radical Alterity Is Just Another Way."
18. Nadasdy, "How Many Worlds Are There?"
19. Liu et al., eds., *Global Language Justice.*
20. Hadden, *Networks in Contention.*
21. Checker, *The Sustainability Myth.*
22. Keliher, *The Board of Rites.*

23. Lee, *Gourmets in the Land of Famine.*
24. Li, *Fighting Famine in North China.*
25. Chen, 公義觀念與中國文化.
26. Lee, *A Certain Justice.*
27. Huang, 中国的新型正义体系; Zhao, *Power and Justice.*
28. Haraway, *Staying with the Trouble*; Tsing, *The Mushroom at the End of the World.*
29. Chao et al., *The Promise of Multispecies Justice.*
30. Caine, "Herding at the Edges."
31. Kohn, *How Forests Think.*
32. Kirsch, *Reverse Anthropology.*
33. Stensrud, "Climate Change, Water Practices."
34. Stensrud.
35. Vaughn, "Disappearing Mangroves."
36. Sultana, "Climate Change, COVID-19."
37. Davis and Todd, "On the Importance of a Date."
38. Zee, *Continent in Dust.*
39. West, *Dispossession and the Environment.*

BIBLIOGRAPHY

Bessire, Lucas, and David Bond. "Ontological Anthropology and the Deferral of Critique." *American Ethnologist* 41, no. 3 (2014): 440–56.

Briggs, Charles L. "Learning How to Ask: Native Metacommunicative Competence and the Incompetence of Fieldworkers." *Language in Society* 13, no. 1 (1984): 1–28.

Caine, Allison. "Herding at the Edges: Climate Change and Animal Restlessness in the Peruvian Andes." *Ethnos* (2022): 1–21.

Chao, Sophie, Karin Bolender, and Eben Kirksey, eds. *The Promise of Multispecies Justice,* Durham: Duke University Press, 2022.

Checker, Melissa. *The Sustainability Myth: Environmental Gentrification and the Politics of Justice.* New York: New York University Press, 2020.

Chen, Ruoshui 陳弱水. 公義觀念與中國文化 [The Ideas of Gong and Yi in Chinese Culture]. 聯經 Lianjing Press, 2020.

Crate, Susan. "Climate and Culture: Anthropology in the Era of Contemporary Climate Change." *Annual Review of Anthropology* 40, no. 2011 (October 21, 2011): 175–94.

Crate Susan, and Mark Nuttall, eds. *Anthropology and Climate Change: From Encounters to Actions.* Walnut Creek, CA: Left Coast Press, 2009.

Davis, Heather, and Zoe Todd. "On the Importance of a Date, or, Decolonizing the Anthropocene." *ACME: An International Journal for Critical Geographies* 16, no. 4 (2017): 761–80.

de la Cadena, Marisol. *Earth Beings: Ecologies of Practice Across Andean Worlds.* Durham: Duke University Press Books, 2015.

De Wit, Sara. *Global Warning. An Ethnography of the Encounter Between Global and Local Climate-Change Discourses in the Bamenda Grassfields, Cameroon.* Langaa Research and Publishing Common Initiative Group, 2015.

Gippner, Olivia. *Creating China's Climate Change Policy: Internal Competition and External Diplomacy.* Cheltenham, UK: Edward Elgar Publishing, 2020.

Graeber, David. "Radical Alterity Is Just Another Way of Saying 'Reality': A Reply to Eduardo Viveiros de Castro." *HAU: Journal of Ethnographic Theory* 5, no. 2 (2015): 1–41.

Hadden, Jennifer. *Networks in Contention: The Divisive Politics of Climate Change.* New York: Cambridge University Press, 2015.

Haraway, Donna J. *Staying with the Trouble: Making Kin in the Chthulucene.* Durham: Duke University Press Books, 2016.

Hathaway, Michael J. *Environmental Winds: Making the Global in Southwest China.* Berkeley: University of California Press, 2013.

Henare, Amiria, Martin Holbraad, and Sari Wastell, eds. *Thinking Through Things: Theorising Artefacts Ethnographically.* London: Routledge, 2006.

Huang, Philip 黄宗智. 中国的新型正义体系:实践与理论 [China's New Justice System: Practice and Theory]. Guangxi Normal University Press, 2019.

Jobson, Ryan Cecil. "The Case for Letting Anthropology Burn: Sociocultural Anthropology in 2019." *American Anthropologist* 122, no. 2 (2020): 259–71.

Keane, Webb. *Ethical Life: Its Natural and Social Histories.* Princeton: Princeton University Press, 2015.

Keliher, Macabe. *The Board of Rites and the Making of Qing China.* Oakland: University of California Press, 2019.

Kirsch, Stuart. *Reverse Anthropology: Indigenous Analysis of Social and Environmental Relations in New Guinea.* Stanford: Stanford University Press, 2006.

Kohn, Eduardo. *How Forests Think: Toward an Anthropology Beyond the Human.* Berkeley: University of California Press, 2013.

Kopra, Sanna. *China and Great Power Responsibility for Climate Change.* London: Routledge, 2018.

Latour, Bruno. *Pandora's Hope: Essays on the Reality of Science Studies.* Cambridge: Harvard University Press, 1999.

Lee, Haiyan. *A Certain Justice: Toward an Ecology of the Chinese Legal Imagination.* Chicago: University of Chicago Press, 2023.

Lee, Seung-Joon. *Gourmets in the Land of Famine: The Culture and Politics of Rice in Modern Canton.* Stanford: Stanford University Press, 2011.

Li, Lillian M. *Fighting Famine in North China: State, Market, and Environmental Decline, 1690s–1990s.* Stanford: Stanford University Press, 2010.

Liu, Lydia H., Anupama Rao, and Charlotte A. Silverman, eds. *Global Language Justice.* New York: Columbia University Press, 2023.

Morris, Julia. "Managing, Now Becoming, Refugees: Climate Change and Extractivism in the Republic of Nauru." *American Anthropologist* 124, no. 3 (September 2022): 560–74.

Nadasdy, Paul. "How Many Worlds Are There?" *American Ethnologist* 48, no. 4 (2021): 357–69.

Qi, Ye, Li Ma, Huanbo Zhang, and Huimin Li. "Translating a Global Issue Into Local Priority: China's Local Government Response to Climate Change." *Journal of Environment & Development* 17, no. 4 (December 1, 2008): 379–400.

Povinelli, Elizabeth A. *Geontologies: A Requiem to Late Liberalism*. Durham: Duke University Press, 2016.

Stensrud, Astrid B. "Climate Change, Water Practices and Relational Worlds in the Andes." *Ethnos* 81, no. 1 (January 1, 2016): 75–98.

Sultana, Farhana. "Climate Change, COVID-19, and the Co-Production of Injustices: A Feminist Reading of Overlapping Crises: Social & Cultural Geography." *Social & Cultural Geography* 22, no. 4 (May 2021): 447–60.

Tsing, Anna Lowenhaupt. *The Mushroom at the End of the World: On the Possibility of Life in Capitalist Ruins*. Princeton: Princeton University Press, 2015.

Vaughn, Sarah E. "Disappearing Mangroves: The Epistemic Politics of Climate Adaptation in Guyana." *Cultural Anthropology* 32, no. 2 (2017): 242–68.

Verma, Ritu. "The Role of Anthropology in Climate Change Research and Policy in Bhutan: Reflections from a Carbon-Negative Country." *American Anthropologist* 122, no. 4 (2020): 947–52.

West, Paige. *Dispossession and the Environment: Rhetoric and Inequality in Papua New Guinea*. New York: Columbia University Press, 2016.

Whitington, Jerome. "What Does Climate Change Demand of Anthropology?" *PoLAR: Political and Legal Anthropology Review* 39, no. 1 (May 2016): 7–15.

Zee, Jerry C. *Continent in Dust: Experiments in a Chinese Weather System*. Oakland: University of California Press, 2022.

Zhao, Xudong. Power and Justice: Disputes Resolution in a North China Village. New York: Springer, 2019.

14 | Climate Justice and Religion

RAFFAELLA TAYLOR-SEYMOUR AND
COURTNEY BENDER

T wo images of "religion" circulate in conversations about climate justice. On the one hand, an image of the white conservative Christian has come to stand in for ideologies of climate skepticism and denial. In its most notorious manifestations, this stock figure uses religious doctrine as ballast for asserting human dominion over the natural world and to buttress ideologies of imperialism, colonialism, and unfettered resource extraction.[1] On the other hand, images of Indigenous people are often invoked as romanticized representative of non-Western philosophies and spiritualities. This stock figure is the inversion of the Christian imperialist, one that represents cultures that offer ecological paradigms with more estimable resources for the pursuit of climate justice. As these figures manifest in the myth-making projects of the Global North, Indigenous people are reinscribed as "spiritual" and more proximate to nature, while the specificities of Indigenous philosophies and the lived expressions thereof are frequently collapsed.[2]

While we do not deny the rhetorical or political significance of these two images of religion in current climate justice debates, we argue that neither serves climate justice scholarship well. We observe these stock ideas of religion being deployed frequently, to "spiritual-wash" otherwise extractive practices and obscure neo-liberal agendas, or to delegitimate the activities of climate activists and campaigners.[3] Neither image offers much to scholars and activists who are invested in thinking about the myriad ways that religious people, concepts, movements, and claims are at work in climate justice projects today. We thus argue for approaches to religion that are built on, and can engender, greater critical awareness of the ways that religious ideas and institutions might intersect with climate justice.

As scholars working within the interdisciplinary field of religious studies, we approach climate justice as a growing area of scholarship and activism that engages with the reality that "climate change is having the most severe effects on those with the least responsibility for causing it, and who, at the same time, are often excluded from decision-making processes regarding responses."[4] This wide field of scholarship, activism, and policymaking is naturally filled with diversity of approaches and internal debates, to which religious studies scholars have already contributed in a variety of ways.[5] Here, we follow the lead of work that has argued that concepts of "justice" deployed in climate justice movements need to engage critically with the limits of universalist philosophies rooted in Western traditions. Universalist framings of justice are premised on notions of equal rights, entitlements, and humanity yet serve to obscure current and past material inequalities that render access to these rights—and the forms of humanity on which they are premised—highly unequal.[6] At the same time, scholars have argued that we must acknowledge the value and, indeed, ethical imperative to expand understandings of climate justice and recognize the "diversity of people's views on planetary well-being and their skills in protecting it."[7] Articulations of justice in this vein embrace "pluriversality" and seek to advance the ethical, social, and political world-building projects that this moment demands in a variety of directions (for more discussion of justice concepts, see chapter 2). Given that climate justice engages with "multiple cultures, subjective representations and practices of well-being, justice, and sustainability across the globe,"[8] we suggest that religious "stuff in the world" is necessarily part of this multiplicity.

The contemporary study of religion takes its object to be a highly complex and historically varied set of human practices and structures.[9] Critical studies of religion investigate the practices, social communities, and epistemological claims that shape the everyday lives of religious practitioners. Equally, they call attention to the embedded political theologies of secular states, institutions, and individuals, seeking to understand the impact of religious ideas in ostensibly secular contexts and political debates. Religious studies scholarship on climate justice begins with the paradox that religion simultaneously underpins universalist accounts of climate justice— given that Western legal and philosophical ideals of justice emerge in part out of certain traditions of Christian theology—*and* promises to provide greater substance to more expansive approaches to climate justice that many activists and scholars are calling for. As scholars of religion, we view religion and religious traditions as holding a wide range of ways of living and thinking justice at numerous scales and in more pluriversal ways. Religious traditions promulgate practices of well-being and understandings of relationality and responsibility in a variety of iterations, ranging from the local to the planetary. These working definitions and practices of justice are not all suited to the pursuit of climate justice, and some risk doing harm to it, but they do important work in the world and represent long histories of human investment in thinking about justice that do not exclusively make recourse to the modern liberal subject or nation-state.

One way to give greater substance to "pluriversal" approaches to climate justice, we suggest, is for scholars and activists to attend to a wider range of religious stuff in the world. Doing so will attune thinkers to a plurality of modes of envisaging and acting for climate justice and, through further conversation and argument, find new resources to inform present and future challenges of building just planetary futures. We propose two different and distinct paths toward thinking about and including religion in the pursuit of climate justice. The first approach focuses on the practical activities of self-identified religious actors, including those who explicitly foreground religious identity or who derive authority or motivation for climate justice via religious traditions. This approach focuses on how such actors contend with climate change and advocate for justice in a variety of circumstances. This approach, which we term "religion, situation normal," identifies how we might acknowledge religion acting within the conventional framework of universalist justice initiatives, which are currently the most widely represented. The second path focuses on religion in a more expansive register, gesturing to the many ways that religious thought has established, worked with, and sought to address core ethical and justice questions of scale, environmental community, and existential futurity, through theological debate, ritual, spiritual experience, and engagements with the unseen. We argue that such efforts can establish and in fact are establishing more transformative forms of climate justice from the ground up. Such an account, we suggest, assists us in thinking across scales and envisages a broader range of actors with stakes in the climate emergency.

RELIGION, SITUATION NORMAL

Contemporary climate justice actors and most scholars in the fields of political science, economics, and policy see religion as operating "below" or "outside" the level of secular policymaking.[10] In secular ideologies, religion is the "other" of both scientific research and bureaucratic decision-making on climate change, which for the most part envisage technocratic and market-driven solutions to the climate crisis. When invited into the conversation, religious actors and institutions are understood to represent discrete sets of interests, primarily those of their members, and occasionally as avenues for the pursuit of certain types of rights, especially for those with limited political opportunity. Religion thus enters debates about climate justice either to present the perspectives of religious actors or to operate as a voice for morality in an otherwise technocratic debate. Scholarly attention to religion in climate justice has often followed this model, focusing on particular cases where religious social movements have taken action to raise awareness or advocate for persons and communities facing environmental catastrophe.

The activities of religious people in the struggle for climate justice in this vein are substantial. As many have observed, the US environmental justice movement—a precursor to the climate justice movement—grew out of a collaboration between religious and nonreligious organizers in North

Carolina in the early 1980s, in the context of Black communities that sought to redress racialized patterns of exposure to pollution and contamination.[11] In this campaign, religious groups and leaders drew on established organizing skills as well as their powerful repertoire of ritual, symbolic, and rhetorical resources to argue for environmental justice and forge bonds across communities. At an institutional scale, while bodies like the Roman Catholic and Anglican Churches have historically been closely entangled with the political, economic, and cultural drive toward resource extraction and climate destruction, they have been increasingly vocal about climate change.[12] These shifts are evident in initiatives like Pope Francis's treatise *Laudato Si'*, which argued for "the inseparable bond between concern for nature, justice for the poor, commitment to society, and interior peace," and the Anglican Church's massive tree-planting project under the banner of "Communion Forests."[13] The Dalai Lama has been an important figure in climate activism since the 1990s and made his most recent intervention in his book *Our Only Home: A Climate Appeal to the World*. Beyond these prominent initiatives, numerous declarations have been authored by faith groups, including the Islamic Declaration on Climate Change, the Rabbinic Letter on Climate Change, the Hindu Declaration on Climate Change, and the Statement of the Indigenous Peoples' Caucus.[14] Not all these responses to climate change extend to social or political action, and much religious activity remains resistant to ready translation.[15] Nonetheless, they often arise out of traditions of theology that actively engage questions of material inequality alongside the environment, and we would be remiss in assuming that these are not important efforts to respond to climate emergencies and to build ethical and just lives.

While this "situation normal" approach is important, however, we do not believe that this represents the most interesting or potentially significant way to approach religion and climate justice. As "situation normal" suggests, such an approach plots religious actors in a secular political landscape that figures certain spheres—including science, justice, and policy—as inherently universal, while religion and religious subjects come to be configured as "particular," and the relevance of religious ideas remains confined to specific communities.[16] Many religious actors invested in climate change find themselves required to shape and comport within these secular frames of what religious actors can and should do, in order to gain a place at the table.[17] From the vantage point of religious studies, a critical analysis of the currently hegemonic framing of justice allows us to see that specificity of religious and particularly modernist Christian concepts of universality that inform these conversations.[18] These same concepts were woven into colonial projects and thus have played a role in the construction of the contemporary world order while sanitizing its many injustices.[19] Yet as Mayra Rivera has argued, religion is "not only a tool of colonial ideology, but . . . also a resource for its subversion."[20] We suggest that better awareness of the religious genealogies of our notions of justice can help de-universalize and decolonize these concepts, and at the same time create opportunities for an array of contemporary religious actors and practices in a different light.

We must be open, then, to continuing to interrogate the presumptions that feed into our conceptions of justice and consider who and what is legible within them. "Religion normal" renders the landscapes, communities, and evaluations that religious actors live within, contend with, and seek to foster illegible. If climate justice from the perspective of "religion, situation normal" involves occasional consultations of religious actors without a more thoroughgoing deconstruction and rebuilding of the grounds of justice, in the next section we shift our focus from religious leaders to religious studies approaches that query the roles that religious activities, practices, and traditions can play in undertaking some of the world-making labor necessary for this moment.

RELIGION, FROM IMPRACTICALITIES TO POSSIBILITIES

Religions remain the biggest carriers of cosmological distinctions in the world and as such provide a grounding for a human experience that has been and continues to be an exemplum of radical cosmological plurality. Thinking about religion and looking at the ways religions are actually practiced, we suggest, illuminates potential avenues toward the next stage of the transformative climate justice that many seek. These traditions are without question not always ready made for recovery or extraction. Here, we focus attention on the ways that apparently "impractical" ideas and activities, which do not fit universalist frames of justice, are not fantasies or made-up strategies, but rather things that people—including modern secular people—are actually doing now.

One of the central complexities of the climate crisis, in Dipesh Chakrabarty's view, is the predicament of needing to find ways to live "in an order that presently seems *unimaginable*: an order that is not necessarily human dominant."[21] For Chakrabarty, climate change gives rise to two different kinds of temporality: the "now" we subjectively experience and the extended "now" of geological or epochal time. This collision of scales requires new ways of understanding the human and the political—and by extension, new conceptions of justice—a demand that combines practical efforts with imaginative risks. Many scholars have pushed back against claims about the novelty of this situation, demonstrating that for Indigenous peoples the crushing of temporalities and experiences of apocalypse are far from historically unprecedented.[22] Yet Chakrabarty is among numerous ostensibly secular scholars who gesture to the intellectual and imaginative limitations of secular thinking and strive to find alternative sources of inspiration, and whose work often implicitly invokes religious motifs and thinking.[23]

Notwithstanding our agreement on the current condition of the planet from an ecological perspective, we take exception with claims by secular scholars that humans have never lived or thought on multiscale or planetary scales (see also chapter 13). The numerous histories of thinking and practice that have been encoded and embedded in religious traditions, which long predate the nation-state and the emergence of modern liberal imaginings of the human—even as they have contributed to and been transformed by them—give lie to such premises. We recall a long history of human endeavors to explicate just places for themselves in the world, often in ways that challenge or slip beyond an exclusively human-centric view.[24] Our discipline observes practices and rituals that frequently make alternative ways of being-in-the-world an object of consideration, and (for good or ill) make such potentials politically salient. In the remaining pages of this essay, we briefly outline two dimensions of religious studies approaches that add depth and nuance to "pluriversal" understandings of climate justice. First, we consider how religious groups organize their understandings of community, responsibility, and relationality in ways that interact with yet are distinct from the nation-state. Second, we examine how religious groups and organizations regularly extend personhood, agency, and thus a range of rights and responsibilities to more-than-human actors.

Border-Crossings

A significant complication of universalist approaches to climate justice is that they embody something of a paradox. At once attached to the institutions of the state and asserting claims to universality, conventional climate justice frameworks quickly run into the issue of borders. Given that climate justice must reckon with the deeply unequal burden of climate harms—everything from the sources of the most polluting forms of resource extraction to the burden of waste to the erosion of coastlines and soils overwhelmingly affects the communities with the least resources—conventional frameworks often struggle to imagine justice beyond the borders of the nation-state.[25] At the same time, while borders frequently serve to delimit spheres of political responsibility, the realities of climate change force a recognition of the relations that proliferate far beyond the contemporary lines on a map.

Religions and their gods may have a surprising role to play here as itinerant border- and boundary-crossers. As climate change contributes to the displacement and relocation of humans (among many other species), "religion" arises as a pathway for establishing new networks of care and maintaining circuits of relationship and solidarity. For peoples excluded from or demoted within the structures of nation-state politics and who have had unequal access to the rights of citizenship, religions have long played an important role not only in sustaining life but also in shaping political conscience and concepts around the possibilities of other human futures and connections. Religions work to forge relays between landscapes separated by colonial borders and to recapitulate or refuse the incursions of imperial power through missionary and kinship-making projects.[26] Religious traditions are practiced today in ways that are trans- or cross-national, boundary and species crossing, and multiscale, allowing them to shift and act at different scales when confronted with climate change.[27] Scholars have sought to reflect these expansive images of religion in their theory building, eschewing older ideas of religions as unproblematic wholes, and moved toward viewing them as "confluences" that seek to grapple with the crossing of boundaries as well as promote ideas of dwelling and making home.[28] As such, religions come into view as vast flows that comprise many currents and are characterized by endless movement and flux, reflecting their historic and contemporary character as itinerant boundary-crossers.

These boundary-crossing flows speak to efforts among scholars of climate justice that have sought to identify forms of action that do not make the nation or state their central object. At the same time, these streams are often integral to making life in ostensibly marginal spaces. Such communities—marginalized along lines of race, wealth, nationality, and caste—frequently become "sacrifice zones" where environmental and economic harms are disproportionately concentrated.[29] Though such zones are figured as being outside the mainstream economy and society—and as such render the damages of elite patterns of extraction and consumption invisible to their perpetrators—they are in fact deeply connected to the wider economy through interlocking chains of extraction, production, and consumption.[30] Recently, scholars have sought not only to advocate on behalf of such regions but to examine the practices that emerge within them as extensions of much longer histories of resistance and fugitivity, especially among Black and Indigenous communities.[31] These modes of political and social resistance often have spiritual dimensions, as these communities transform border zones like rivers and swamps into sites of escape, refuge, and religious significance.[32]

Climate justice actors must find ways of reckoning with these relations and the environmental harms that exist on the margins of or extend beyond the borders of the nation-state. Religion appears as precisely the kind of force and metaphor that scholars advocating for climate justice seek—projects of solidarity, meaning-making, and world-building that exceed and transcend colonial borders. These realities offer a counterpoint to the concerns of scholars like Chakrabarty, illustrating that the world today is filled with people who are already at work imagining and living just connections and relations to each other and to the worlds they live in beyond the nation-state. In doing so, they challenge and trouble the centrality of nation-states in hegemonic approaches to justice, rights, and human thriving.

Nonhuman Actors

Contemporary scholars of climate justice argue for a radical decentering of the human and a reattunement to the vast scale of interconnectedness within which humans are enmeshed. Many suggest that this demands conceptions of justice that extend beyond the human, which might give rise to new political visions and possibilities (see also chapter 10). As Sophie Chao and Danielle Celermajer contend, this moment "requires new political imaginaries that take into account the ontological diversity, relational complexity, and incommensurable forms of communication and desire, within which just arrangements and outcomes can be co-crafted."[33] The study of religion has a crucial role to play here, we suggest. As we have noted, scholars seeking to consider realms of ontological difference and connectedness often find resources in Indigenous and "animist" traditions as antidotes to or alternatives to colonial, modernist, and human-centric paradigms of justice, despite the regularly voiced concern that many such uses reintroduce a false and damaging dichotomy between "modern" and "indigenous." Here, the study of religion (alongside Indigenous Studies and related fields) provides important correctives to projects that, as Mayanthi Fernando has argued, risk recolonizing Indigeneity as a pristine "other."[34]

Recent work in religious studies takes people seriously as they engage in worlds where they have obligations—and understandings of justice—that extend beyond liberal accounts of personhood to encompass ancestors, saints, spirits, other species, and nonhumans. In religions with rich legal traditions, there exist histories of juridical argument about the legal status of animals, including those that argue for animals' existential rights to life that exceed or are orthogonal to any relation they have to humans or human society.[35] In the context of the United States, J. T. Roane has examined how Black communities in the Tidewater region of Virginia and Maryland have resisted evolving legal and commercial regimes that have curtailed Black access to coastal waters. In places configured as "sacrificial" or abandoned,

Roane demonstrates how "intimate relationships between the dead, the living, and possible futures" are cultivated and endure on the edges of settler corporate interests.[36] In the context of West Papua, Sophie Chao has shown how Indigenous people grapple with the destruction of their forests and the encroachment of palm oil plantations.[37] This work demonstrates that ecological change creates existential crises that span human and more-than-human worlds, and that nonhumans are intrinsic to the labor of imagining new worlds. Such examples tell us that there are a host of ways that people have sought to forge just relations with other species, even within conditions of ecological devastation where human agency is complexly figured and not always ascendant.

In the history and present practice of religions there exist a range of unexpected ways of grappling with the most discomfiting and troubling aspects of expanding climate justice, beyond the human and the reaches of human agency. We find rich traditions that question our basic obligations to others that do not center human agency, in which the limits of human agency are regularly checked, queried, and sometimes disciplined.[38] Religious epistemologies that envisage humans as subject to nonegalitarian forces, demands, and obligations—in deference to a god, gods, or other unseen beings—often remain suspect for secular liberal subjects. Yet they stand as testimony and archive of age-old sets of queries through which humans have sought to understand what justice is, while living within societies and worlds where humans are not centered, and where human agency is not the defining ground on which justice can be enacted or sought. As humans today confront a changing climate as something that is now only precariously within our capacity to arrest and alter, we would do well to acknowledge the very long histories of human concerns about precisely these issues.

• • •

In this chapter we have argued for more expansive and nuanced understandings of religion that promise greater resources for the pursuit of climate justice. In mainstream discussions about climate justice, religion is often perceived through the lens of "situation normal," in which religious actors are occasionally invited into secular spaces to offer particular perspectives. In a moment when scholars and activists are searching for more pluriversal approaches to climate justice—especially those that contest the many currents that feed into extractive and exploitative logics across human and nonhuman worlds—a more nuanced view of religion aids in expanding our imaginative horizons and conceptual vocabularies for approaching justice. This requires broadening our ideas about what religion looks like in the first place, moving toward an account of religion as a series of streams, confluences, and flows of ideas and practices that takes many routes. These often transcend modern borders, both those that artificially separate territories on maps and those that install rigid divides between humans and more-than-humans. An expanded conception of religion therefore promises to move beyond such lines of demarcation and provide new concepts for imagining relations and solidarities.

At the same time, religions do not stand outside the world but rather are thoroughly embedded within and shaped in response to it. As such, we do not wish to argue that religion and religious actors are always good or bad for climate justice. Nor do we argue that religious ideas or practices hold the keys to solving climate justice on their own. Instead, we have argued that religious climate justice activities are underway and work both with and against climate justice at a variety of scales. As we collectively seek ways of articulating our place in the world and strive to create more just worlds, we ought not to overlook the long histories of human engagement with fundamental existential questions: To whom and with whom can we build justice and repair? This question is especially pertinent under conditions that seem challenging and perhaps (we might worry) beyond our capacities to reckon with. Attentiveness to religion, we suggest, offers new inroads and helps identify different and willing compatriots in the pursuit of climate justice across a variety of scales, from the local to the planetary.

Notes

1.   While we do not deny that self-identified religious projects are well represented among the many interests and groups who stand opposed to climate justice, we believe that the frequent and common deployment of two stereotypical understandings of "religion" has severely limited climate justice activists' abilities to see,

much less consider, the numerous ways that religion matters to climate justice. Veldman, *The Gospel of Climate Skepticism*; Dochuk, *Anointed with Oil*.

2. Krech, *The Ecological Indian*; Gilio-Whitaker, *As Long as Grass Grows*.
3. Kyyrö et al., "'The Cult of Greta Thunberg.'"
4. Newell et al., "Toward Transformative Climate Justice," 2.
5. Bauman, *Religion and Ecology*; Jenkins et al., "Religion and Climate Change"; McFague, *A New Climate for Theology*; Meziane, *States of the Earth*; Sideris, *Environmental Ethics, Ecological Theology*.
6. Hoskins, "Justice Otherwise"; Khader, *Decolonizing Universalism*.
7. Kothari et al., *Pluriverse*, ix.
8. Newell et al., "Toward Transformative Climate Justice," 2.
9. Taylor-Seymour, "Troubling Climate and Religion."
10. Bornstein, *The Spirit of Development*; Manouchehrifar, "Is Planning 'Secular'?"
11. Schlosberg and Collins, "From Environmental to Climate Justice."
12. Wynter, "Unsettling the Coloniality of Being."
13. Pope Francis, *Laudato Si'*, 10.
14. Jenkins et al., "Religion and Climate Change."
15. Haluza-DeLay, "Religion and Climate Change."
16. Lafont, "Religion and the Public Sphere."
17. Wuthnow and Evans, eds., *The Quiet Hand of God*.
18. MacIntyre, *Whose Justice? Which Rationality?*; Schmitt, *Political Theology*.
19. Meziane, *States of the Earth*; Vasko, "Nature and the Native."
20. Rivera, "Embodied Counterpoetics," 58.
21. Chakrabarty. *The Climate of History in a Planetary Age*, 95, emphasis added.
22. Kyle Powys Whyte, "Our Ancestors' Dystopia Now"; Davis and Todd, "On the Importance of a Date."
23. See also Haraway, *Staying with the Trouble*; Latour, *Facing Gaia: Eight Lectures*; Stengers, *Cosmopolitics I*; Tsing et al., *Arts of Living on a Damaged Planet*.
24. Cruikshank, *Do Glaciers Listen?*; Kimmerer, *Braiding Sweetgrass*.
25. Churchill, *Struggle for the Land*.
26. Gálvez, *Guadalupe in New York*; Beliso-De Jesús, *Electric Santería*; McAlister, *The Kingdom of God Has No Borders*; Mehta, "Bearing the Burden of History."
27. Gade, *Muslim Environmentalisms*. For more discussion of multiscalarity, see chapter 9 in this volume.
28. Gill, "I Am a Messenger"; Palmié, *The Cooking of History*; Tweed, *Crossing and Dwelling*.
29. Lerner, *Sacrifice Zones*.
30. Juskus, "Sacrifice Zones."
31. Hosbey et al., "Global Black Ecologies."
32. Srinivas, "A Lake of Fire, a Runaway Goddess."
33. Chao and Celermajer, "Introduction: Multispecies Justice," 2.
34. Fernando, "Uncanny Ecologies."
35. Berkowitz, "Birds as Dads, Babysitters, and Hats."

36. Roane, "Black Ecologies, Subaquatic Life," 229.
37. Chao, *In the Shadow of the Palms.*
38. Johnson, *Waste and the Wasters.*

BIBLIOGRAPHY

Bauman, Whitney A. *Religion and Ecology: Developing a Planetary Ethic.* New York: Columbia University Press, 2014.

Beliso-De Jesús, Aisha M. *Electric Santería: Racial and Sexual Assemblages of Transnational Religion.* New York: Columbia University Press, 2015.

Berkowitz, Beth A. "Birds as Dads, Babysitters, and Hats: An 'Indistinction' Approach to the Mother Bird Mitzvah in Deuteronomy 22: 6–7." *Worldviews: Global Religions, Culture, and Ecology* 26, no. 1–2 (2021): 79–105.

Bornstein, Erica. *The Spirit of Development: Protestant NGOs, Morality, and Economics in Zimbabwe.* New York: Routledge, 2003.

Chakrabarty, Dipesh. *The Climate of History in a Planetary Age.* Chicago: University of Chicago Press, 2021.

Chao, Sophie. *In the Shadow of the Palms: More-than-Human Becomings in West Papua.* Durham: Duke University Press, 2022.

Chao, Sophie, and Danielle Celermajer. "Introduction: Multispecies Justice." *Cultural Politics* 19, no. 1 (2023): 1–17.

Churchill, Ward. *Struggle for the Land: Native North American Resistance to Genocide, Ecocide, and Colonization.* San Francisco: City Lights, 2002.

Cruikshank, Julie. *Do Glaciers Listen?: Local Knowledge, Colonial Encounters, and Social Imagination.* Vancouver: UCB Press, 2005.

Dalai Lama and Franz Alt. *Our Only Home: A Climate Appeal to the World.* New York: Hanover Square Press, 2020.

Davis, Heather, and Zoe Todd. "On the Importance of a Date; or, Decolonizing the Anthropocene." *ACME* 58, no. 2 (2017): 761–80.

Dochuk, Darren. *Anointed with Oil: How Christianity and Crude Made Modern America.* New York: Basic Books, 2019.

Fernando, Mayanthi. "Uncanny Ecologies: More-than-Natural, More-than-Human, More-than-Secular." *Comparative Studies of South Asia, Africa and the Middle East* 42, no. 3 (2022): 568–83.

Gade, Anna M. *Muslim Environmentalisms: Religious and Social Foundations.* New York: Columbia University Press, 2019.

Gálvez, Alyshia. *Guadalupe in New York: Devotion and the Struggle for Citizenship Rights Among Mexican Immigrants.* New York: NYU Press, 2010.

Gilio-Whitaker, Dina. *As Long as Grass Grows: The Indigenous Fight for Environmental Justice, from Colonization to Standing Rock.* Boston: Beacon Press, 2019.

Gill, Lyndon K. "I Am a Messenger: Spiritual Baptism and the Queer Afterlife of Faith." *Small Axe* 22, no. 1 (2018): 71–84.

Haluza-DeLay, Randolph. "Religion and Climate Change: Varieties in Viewpoints and Practices." *WIREs Climate Change* 5, no. 2 (2014): 261–79.

Haraway, Donna. *Staying with the Trouble: Making Kin in the Chthulucene.* Durham: Duke University Press, 2016.

Hosbey, Justin, Hilda Lloréns, and J. T. Roane. 'Global Black Ecologies.' *Environment and Society* 13, no. 1 (2022): 1–10.

Hoskins, Nicole. "Justice Otherwise." In *Grounding Religion: A Field Guide to the Study of Religion and Ecology,* edited by Whitney A. Bauman, Richard Bohannon, and Kevin J. O'Brien, 221–33. New York: Routledge, 2017.

Jenkins, Willis, Evan Berry, and Luke Beck Kreider. "Religion and Climate Change." *Annual Review of Environment and Resources* 43 (2018): 85–108.

Johnson, Eleanor. *Waste and the Wasters: Poetry and Ecosystemic Thought in Medieval England.* Chicago: University of Chicago Press, 2023.

Juskus, Ryan. "Sacrifice Zones: A Genealogy and Analysis of an Environmental Justice Concept." *Environmental Humanities* 15, no. 1 (2023): 3–24.

Khader, Serene J. *Decolonizing Universalism: A Transnational Feminist Ethic.* Oxford: Oxford University Press, 2018.

Kimmerer, Robin Wall. *Braiding Sweetgrass: Indigenous Wisdom, Scientific Knowledge and the Teachings of Plants.* Minneapolis: Milkweed Editions, 2015.

Kothari, Ariel Salleh, Arturo Escobar, Federico Demaria, and Alberto Acosta. *Pluriverse: A Post-Development Dictionary.* New York: Columbia University Press, 2019.

Krech, Shepherd E. *The Ecological Indian: Myth and History.* New York: Norton, 1999.

Kyyrö, Jere, Tuomas Äystö, and Titus Hjelm. "'The Cult of Greta Thunberg': De-Legitimating Climate Activism with 'Religion.'" *Critical Research on Religion* 11, no. 2 (2023): 133–49.

Lafont, Cristina. "Religion and the Public Sphere: What Are the Deliberative Obligations of Democratic Citizenship?" *Philosophy & Social Criticism* 35, no. 1–2 (2009): 127–60.

Latour, Bruno. *Facing Gaia: Eight Lectures on the New Climatic Regime.* Cambridge: Polity, 2017.

Lerner, Steve. *Sacrifice Zones: The Front Lines of Toxic Chemical Exposure in the United States.* Cambridge: MIT Press, 2010.

MacIntyre, Alasdair. *Whose Justice? Which Rationality?* South Bend: University of Notre Dame Press, 1988.

Manouchehrifar, Babak. "Is Planning 'Secular'? Rethinking Religion, Secularism, and Planning." *Planning Theory & Practice* 19, no. 5 (2018): 653–77.

McAlister, Melani. *The Kingdom of God Has No Borders: A Global History of American Evangelicals.* Oxford: Oxford University Press, 2018.

McFague, Sally. *A New Climate for Theology: God, the World and Global Warming.* Minneapolis: Fortress Press, 2008.

Mehta, Gaurika. "Bearing the Burden of History: Religion and Porous Kinship in the Indo-Caribbean Diaspora." PhD dissertation, Columbia University, 2024.

Meziane, Mohamed Amer. *States of the Earth: An Ecological and Racial History of Secularization.* London: Verso, 2024.

Newell, Peter, Shilpi Srivastava, Lars Otto Naess, Gerardo A. Torres Contreras, and Roz Price. "Toward Transformative Climate Justice: An Emerging Research Agenda." *WIREs Climate Change* 12, no. 6 (2021): 1–17.

Palmié, Stephan. *The Cooking of History: How Not to Study Afro-Cuban Religion*. Chicago: University of Chicago Press, 2013.

Pope Francis. *Laudato Si'*. Papal encyclical. Vatican Press, 2015.

Rivera, Mayra. "Embodied Counterpoetics: Sylvia Wynter on Religion and Race." In *Beyond Man: Race, Coloniality, and Philosophy of Religion*, edited by An Yountae and Eleanor Craig, 57–85. Durham: Duke University Press, 2021.

Roane, J. T. "Black Ecologies, Subaquatic Life, and the Jim Crow Enclosure of the Tidewater." *Journal of Rural Studies* 94 (2022): 227–38.

Schmitt, Carl. *Political Theology: Four Chapters on the Concept of Sovereignty*. Chicago: University of Chicago Press, 2006.

Sideris, Lisa. *Environmental Ethics, Ecological Theology, and Natural Selection: Suffering and Responsibility*. New York: Columbia University Press, 2003.

Srinivas, Tulasi. "A Lake of Fire, a Runaway Goddess, and the Perils of Climate Change in India." *Revealer* (blog), 2021. https://therevealer.org/a-lake-of-fire-a-runaway-goddess-and-the-perils-of-climate-change-in-india/.

Stengers, Isabelle. *Cosmopolitics I*. Minneapolis: University of Minnesota Press, 2010.

Taylor-Seymour, Raffaella. "Troubling Climate and Religion: The Climate Crisis Beyond Disenchantment." *Zygon: Journal of Religion and Science* 59, no. 4 (2024).

Tsing, Anna Lowenhaupt, Heather Swanson, Elaine Gan, and Nils Bubandt. *Arts of Living on a Damaged Planet: Ghosts and Monsters of the Anthropocene*. Minneapolis: University of Minnesota Press, 2017.

Tweed, Thomas. *Crossing and Dwelling: A Theory of Religion*. Cambridge: Harvard University Press, 2006.

Vasko, Timothy Bowers. "Nature and the Native." *Critical Research on Religion* 10, no. 1 (2022): 7–23.

Veldman, Robin Globus. *The Gospel of Climate Skepticism: Why Evangelical Christians Oppose Action on Climate Change*. Berkeley: University of California Press, 2019.

Whyte, Kyle Powys. "Our Ancestors' Dystopia Now: Indigenous Conservation and the Anthropocene." In *The Routledge Companion to the Environmental Humanities*, ed. Ursula K. Heise, Jon Christensen, and Michelle Nieman, 206–15. New York: Routledge, 2017.

Wuthnow, Robert, and John Hyde Evans, eds. *The Quiet Hand of God: Faith-Based Activism and the Public Role of Mainline Protestantism*. Berkeley: University of California Press, 2002.

Wynter, Sylvia. "Unsettling the Coloniality of Being/Power/Truth/Freedom: Towards the Human, After Man, Its Overrepresentation—An Argument." *New Centennial Review* 3, no. 3 (2003): 257–337.

15 Building a Better Model for Flood Protection Planning

PAUL GALLAY

On January 20, 2020, as the final year of his first term began, President Donald J. Trump weighed in on a controversial proposal to build a series of storm barriers in the offshore waterways encircling New York City. These barriers, whose size, complexity, and cost had no precedent, drew a hard no from Trump, who tweeted, "A massive 200 Billion Dollar Sea Wall, built around New York to protect it from rare storms, is a costly, foolish & environmentally unfriendly idea that, when needed, probably won't work anyway."[1] Trump was far from alone in warning of the proposed barriers' adverse impacts on the environment.[2] Similarly, his claim that they probably wouldn't work echoed remarks by the study's own project manager, who told a Manhattan Community Board that we might not even know how well these barriers would work "until the first storm hit."[3]

While Trump's position on the cost, impact, and feasibility of the massive storm barrier proposal may have been well founded, he offered no other plan for the region's increasingly serious flooding problems.[4] Instead, he simply defunded the planning project that had come up with the barrier proposal, known as the New York–New Jersey Harbor and Tributaries Study (HATS), and put the two states on notice that they were on their own when it came to flooding: "Sorry, you'll just have to get your mops & buckets ready!"[5] As the second Trump administration proceeds, HATS remains probably the largest, most complex, and most costly study of its kind in the history of the United States, designed as it is to protect sixteen million people living along 900 miles of coastline in two of the nation's most densely populated states.[6] President Joseph R. Biden restored funding for this critical study in 2021, and, less than two years later, the United States Army Corps of Engineers (Army Corps) released a new HATS proposal omitting the massive barriers that Trump had vetoed, proposing instead a series of smaller, localized in-water barriers and onshore sea walls.[7]

While an improvement on its predecessor, the Army Corps' 2022 HATS proposal remains in limbo, due to concerns voiced by over 2,600 local residents, dozens of community-based organizations, members of congress, and the Corps' own federal, state, and local agency study partners.[8] A central criticism of the 2022 HATS proposal is that it is designed to protect the region only from wind-driven storms, not from stationary rainstorm-driven flooding, which took at least thirty-six lives in New York and New Jersey during Hurricane Ida in 2021, or from sea level rise, which threatens communities during storms, and, increasingly, even on clear, sunny days.[9] Another major critique calls out the Army Corps for failing to deliver on its promise to put frontline communities "at the front and center" of projects like HATS to help create flood risk reduction projects more effectively tailored to local needs.[10]

Fortunately, significant steps have been taken toward the establishment of new and more innovative approaches to flood reduction planning in New York and New Jersey, which could be useful in other coastal regions as well. On January 8, 2024, the two states invoked a never-before-used provision of the US Water Resources Development Act of 2022, which will require HATS planners to address *all* major sources of flooding, including stationary downpours and sea level rise, in addition to storm surge–related flooding, and to give greater consideration to natural and nature-based approaches, instead of relying solely on concrete seawalls and in-water barriers for flood protection, as do earlier HATS proposals.[11]

The manner in which the HATS project team engages with involved communities is also changing. On November 16, 2023, echoing calls from numerous community and academic organizations, New York, New Jersey, and the City of New York, in their formal roles as HATS study partners, demanded that the Army Corps adopt a specific, vastly expanded public engagement plan designed to ensure that communities on the front lines of flood risk are meaningfully engaged in the HATS planning process.[12] Soon afterward, the Corps announced the intention to create a first-of-its-kind Environmental Justice Coordination Committee to promote more genuine dialogue, accountability, and empowerment during the remainder of the HATS project.[13]

In view of the change in presidential administrations in January 2025, it's questionable whether this new, more community-centered, holistic, and nature-based planning framework, which I will call "HATS 2.0," will actually be implemented and deliver the protection from flooding that New York and New Jersey so desperately need.[14] However, should it move forward, not only would it make New York and New Jersey's coastal communities safer and more ecologically healthy, it would also provide a model for flood protection planning in other at-risk communities as well.

Even if it does move forward, HATS 2.0 will only be fully successful if it can transcend traditional concepts of resilience, which are based on the implied expectation that frontline communities must endure continued hardships and repeatedly strive to *bounce back* from flooding, and be based, instead, on the goal of helping such communities steadily *advance* toward more restorative and equitable conditions. Instead of merely following the rebuild and bounce back model, a truly transformative HATS 2.0 would also support other community needs, such as access to open space and recreation, ecosystem restoration, job creation, and community revitalization, especially in communities that face structural disadvantages due to legacies of environmental injustice.[15]

This chapter offers recommendations as to how agencies, community-based organizations, and researchers can help advance the HATS 2.0 vision, as well as suggestions for planners in other communities who seek to ensure that their residents can rely on something better than "mops and buckets" to protect themselves from flooding.

PLANNING FOR FLOODS WITH COMMUNITIES, NOT FOR THEM

The New York–New Jersey HATS project illustrates the challenge of flood risk reduction planning in the face of growing climate-influenced threats that our communities were not built to withstand, such as heightened storm surge, record-breaking downpours, and seas that are in the process of rising by roughly 1–2 feet during the first half of the twenty-first century. At the same time, HATS planners must remain mindful of other, potentially conflicting community priorities, such as providing public access to waterfronts, protecting community character and biodiversity, redressing past inequality, and building social cohesion. For example, enormous proposed flood barriers on the Miami waterfront and along shoreline parks in Brooklyn might stop storm surge, but they are nonstarters for local residents who seek not only to protect their homes from flooding but also to preserve the neighborhood character and other qualities that attracted them to those homes in the first place.[16] The urgent need to address the increasingly complex problem of climate-influenced flooding led the Columbia Climate School to establish the Resilient Coastal Communities Project (RCCP), a partnership between the Climate School and the New York City Environmental Justice Alliance (NYC-EJA), a citywide network linking grassroots organizations from low-income neighborhoods and communities of color in their struggle for environmental and climate justice.[17] Since November 2021, RCCP has worked to develop actionable, fundable, and equitable solutions to flood risks that also deliver complementary benefits, like habitat restoration, job creation, and greater community cohesion, through a combination of iterative engaged scientific research and active support for enhanced community participation in public planning.[18]

In the spring of 2022, RCCP invited representatives of ten local environmental and climate justice organizations to share their past experiences in resilience planning, provide their perspectives on what a truly just and equitable planning process would look like, and explain what resources they would need to participate fully and effectively in future planning processes.[19] The community-based organizations RCCP interviewed in 2022 seek fundamental changes in the way flood risk reduction planning is done, to lift up community knowledge, support community efforts to act as leaders, and center social cohesion and restorative justice. They know that none of this will happen unless the planning process moves away from its current model, in which government agencies prepare resilience plans and then perfunctorily take comments on them from affected communities. The goal of the groups RCCP interviewed is for planning efforts like HATS to reset under a new model of full and early engagement with affected communities throughout the entire study process, in order to harness community expertise and establish shared leadership between government and community.[20]

RCCP's 2022 interviewees expressed a deep willingness to help reform resiliency planning. They offered reasonable, implementable ideas for immediate action to address flood risk and eliminate exclusions and gaps in resiliency planning.[21] They also explained why narratives of specific places are essential to flood risk reduction planning and illustrated the interconnectedness of flooding and environmental justice in communities with underlying problems of inadequate and poor housing, high asthma rates, insufficient educational opportunities, and other structural disadvantages. Finally, they argued forcefully that community coleadership in the planning process is just as essential to effective resilience planning as agency expertise. As Dariella Rodriguez, director of community development at the Point Community Development Corporation, put it during her interview: "We need community members in those conversations . . . if we're not moving at the speed that our people need us to move in, then all the policy in the world, without that community power . . . we're gonna hit a wall."[22] Effective collaboration with communities also requires that planning agencies make full use of any relevant plans created by those communities themselves. As outlined in RCCP's 2022 working paper, "Designing Community-Led Plans to Strengthen Social Cohesion: What Neighborhoods Facing Climate-Driven Flood Risks Want From Resilience Planning," virtually all of the community-based organizations involved in RCCP's 2022 research project have prepared resilience-related plans, reflecting the high level of locally driven resiliency planning in the New York City metropolitan area generally.[23]

Finally, communities see social cohesion as an essential consideration for resilience planners. Community plans created by RCCP interviewees, such as UPROSE's Green Resilient Industrial District and Staten Island Urban Center's Maritime, Education and Recreation Corridor, actively seek to maintain social cohesion and counter gentrification by creating jobs and strengthening community institutions based on principles of mutual support, a circular economy, and eco-industrial/environmental justice.[24]

To summarize, frontline organization leaders want flood risk reduction planners to make full use of the deep store of wisdom that communities possess, rather than simply defaulting to the technical expertise of agency staff. Only by braiding the twin strands of local knowledge and agency expertise, they believe, can we hope to develop fully informed and effective flood protection plans.[25] For now, this sort of coproduced resilience planning remains an unfulfilled but deeply imagined vision for the future, vividly illustrated by the following statement, made during RCCP's interview with leaders of the Williamsburg, Brooklyn-based organization, El Puente: "The deeper context and source of what we might call resiliency is our being able to imagine a future that we ourselves are not just existing but thrive in, and that we ourselves are active leaders in really creating, and re-creating, and continuing to develop."[26]

BUILDING A BETTER FLOOD PROTECTION PLANNING MODEL

Efforts to Reform US Flood Risk Reduction Policy

Community aspirations for more collaborative and better-informed approaches to flood risk reduction planning are increasingly being reflected in federal, state, and local policy. On February 15, 2024, the Army Corps released proposed "Agency Specific Procedures" explicitly directing that environmental justice considerations be incorporated into all phases of the Corps' planning and decision-making process in order to remove barriers to effective community participation, increase community access to benefits, and drive restorative justice. The published overview of these new rules demonstrates a clear intent to center community experience and makes a promise to "listen to the communities and ensure that they are engaged throughout the planning process. The communities themselves will likely help identify concerns and solutions to their water

resources problems and opportunities as well as participate in the identification of any potential effects, mitigation measures, and benefits, including through sharing Indigenous Knowledge, as they deem appropriate."[27]
While these new rules, which were finalized on December 19, 2024, and
became effective on January 17, 2025,[28] are designed to disrupt the Corps'
traditional top-down approach to planning, they have yet to be tested in
practice, so their actual impact remains to be seen. Realizing this, the offi-
cer in charge of the HATS project, New York District Commander Alexander Young, has expressed his hope that HATS will serve as the "the tip
of the spear" for US flood risk reduction planning reform and that it will
help convince other agencies and communities to embrace innovation in
their own planning processes.[29]

Reforms Intended to Address Flood Risk More Comprehensively

Perhaps the most significant disrupting influence on the HATS study is the
recent decision by the states of New York and New Jersey to invoke section 8106 of the Water Resources Development Act of 2022 (WRDA 2022),
which requires the Army Corps to address *all* major flood risks, rather than
just storm surge, as part of the HATS study.[30] Specifically, this means that
the flood protection projects proposed in the next HATS plan must be
designed synergistically in order to "maximize the net benefits from the
reduction of the comprehensive flood risks within the geographic scope of
the study from isolated or compound effects of: *(i) riverine flooding;
(ii) coastal storms; (iii) tidally induced flooding; (iv) rainfall; (v) tides; (vi)
seasonal water levels; (vii) groundwater upwelling; (viii) sea level rise; (ix)
subsidence; or (x) other drivers of flood risk.*"[31] This is the first time that
section 8106 has been invoked, and it imposes daunting responsibilities on
the Army Corps, New York, New Jersey, the City of New York, and other
stakeholders in the HATS study process, who now must combine the work
already done by the HATS project team, which has only addressed storm
surge risk, with a new investigation into the "isolated or compound effects"
of the nine other types of flooding covered by section 8106. Not only are

HATS planners the first to face this responsibility, they face it in connection with perhaps the largest, most expensive study and project of its kind in Corps history.[32]

Fortunately, a wide range of possible flood risk reduction measures is available to the HATS project team. The Army Corps identified over forty different approaches to flood risk reduction, including structural measures like seawalls, berms and surge barriers, nonstructural approaches such as expanded street-level green infrastructure programs and combined sewer overflow reduction strategies, and nature-based solutions like living shorelines, restoring wetlands, aquatic vegetation, and oyster reefs.[33] The key to success will be picking the right combination of these forty-plus interventions for each community in the 900-mile stretch of coastline covered by the HATS study.

The academic community has pledged to support HATS 2.0 with applied research and consultation. Investigators from eight New York– and New Jersey–based research partnerships are partnering with Army Corps and state and local resilience planning officials to organize workshops to share and discuss relevant findings and proposals for further investigation on topics such as the extent of and interaction between varying flood risks, the most productive ways to deploy natural and nature-based flood risk reduction measures, and best practices for centering community expertise in flood risk reduction planning.[34] Such efforts represent a significant opportunity for academic researchers to put their findings into service outside the university setting, gain a deeper understanding of the perspectives and experiences of communities and community-based organizations, and do more to meet the urgent need for better flood protection.

The complexity of the HATS 2.0 project will require planners, community stakeholders, and academic partners to shoulder extraordinary responsibilities in service of a project whose success is anything but guaranteed. In fact, any such success may be only incremental, given the long history of top-down, storm surge–focused flood risk reduction planning in the New York–New Jersey region. As HATS 2.0 aspires to break the mold that has shaped past resilience plans by addressing multiple flood risks through synergistic solutions informed by traditional science, engineering and modeling, and community engagement, it will be closely watched in other regions seeking innovative ways to address their own resilience planning challenges.

Reforms Intended to Address Flood Risk More Justly and Collaboratively

In addition to the challenges of reformulating HATS to address multiple flood risks, this study faces equally significant and important challenges relating to community engagement. The pressure to make the HATS study process more collaborative has been building for years, as community-based organizations and their allies have become increasingly frustrated at planners' lack of responsiveness to their concerns and priorities.[35] As previously stated, even the Army Corps' own state and local government partners have pointedly called on the Corps to do more to center communities in HATS planning.[36]

To put these requests for a more community-centered HATS planning process into perspective, it's helpful to understand the deficiencies in the Corps' public outreach on the project to date. A chief illustration can be found in the public meetings held by the Corps after issuing its most recent HATS plan in September 2022. At the commencement of the first such meeting, on December 15, 2022, Colonel Matthew W. Luzzatto, then the commander of the Army Corps' New York District, promised *meaningful dialogue*, *community empowerment*, and *agency accountability* through the remainder of the planning process.[37] Yet, the ensuing meetings were poorly advertised and sparsely attended, averaging fewer than twenty public attendees, and there has been no organized follow-up or response to the thousands of written public comments sent to the Corps in the two years since the conclusion of these meetings.[38]

Prior to the start of the second Trump administration, the Army Corps seemed to be on the cusp of more meaningful engagement with communities within the HATS study area. On January 24, 2024, New York District Commander Young admitted that "the model of Corps outreach is failing and needs to be remade," and he pledged to establish an Environmental Justice Coordinating Committee (EJCC) to follow through on his predecessor's December 2022 promise to build HATS on dialogue, accountability, and community empowerment.[39] Colonel Young's commitment to the EJCC demonstrates the Army Corps' deepening acknowledgment that, while government experts may have specialized training and experience in resilience planning that community representatives generally do not possess, those government experts often lack the time and training necessary to establish meaningful collaborations with the communities they are responsible for protecting and thus fail to adequately understand or respond to specific community concerns and priorities, let alone take full advantage of what communities know about the specific risks they face and the best options for reducing those risks.[40]

However, a January 20, 2025, executive order stating the second Trump administration's vehement opposition to equity-oriented initiatives strongly suggests that the success of the EJCC will depend on whether the Army Corps' local partners will insist on its full and effective implementation.[41] Fortunately, since the 2024 election, the New York City Department of Environmental Protection and Mayor's Office of Climate and Environmental Justice have signaled their willingness to support and perhaps even lead the EJCC.[42] On paper, the EJCC represents a promise not just to take and give responses to community input, but *to actively use that input to shape HATS planning*. The Army Corps' draft invitation to potential EJCC members reflects this aspiration in its ten stated goals for the committee:

- Ensure that environmental justice considerations are brought to the fore, both within the community engagement around the plan, and within the implementation and distribution of flood protections.
- Ensure that the HATS plan delivers multiple social, environmental, ecological, and economic benefits and incorporates partnerships with other government agencies that can enhance a variety of solutions, including natural and nature-based features.
- Ensure that the HATS plan includes sufficient protection for disadvantaged frontline communities.

- Facilitate dissemination of information generated by the study.
- Foster relationships and build a network of community members and representatives.
- Facilitate feedback from the community on the framework for, as well as the results of, assessing the benefits and impacts of study alternatives to communities with historical marginalization.
- Identify ways to improve participation in the study process for communities who have historically been marginalized.
- Incorporate community feedback into the formulation of alternatives and design of project features.
- Identify and work to close relevant data gaps on the study that are of importance to communities who have historically been marginalized.
- Ensure that analyses are not skewed, disproportionate, or favored to any particular community.[43]

Based on discussions with advocates in other regions, the HATS EJCC would be the Army Corps' most serious effort so far to put communities "at the front and center" of regional coastal flood risk reduction planning instead of merely presenting them with an agenda that has already been set by the time they are brought into the process.[44] The community-based organizations serving on the EJCC would surely do everything they could to help build a HATS 2.0 based on dialogue, accountability, and empowerment, reflecting community needs and priorities. But for this process to be successful, the Corps or its local partners would have to invest the EJCC with sufficient time, funding, and decision-sharing power to enable it to reshape HATS outcomes. Should they fail to do so, the EJCC would merely replicate the corrosive pattern of failed consultation reflected in other coastal flood protection planning exercises in the New York City metropolitan area and perpetuate what Colonel Young admitted to be a "failing" model of community engagement.[45] Stakeholders in the New York–New Jersey area and in other flood-prone parts of the nation alike will be watching the HATS to see whether it can avoid this fate and deliver on the Army Corps' promise to shift flood risk reduction planning to a process based on dialogue, accountability, and empowerment.[46]

CONCLUSION: SOLVING THE WICKED PROBLEM OF FLOOD RISK

There is no simple answer to the question of what success looks like when it comes to coastal flood risk reduction projects like the New York–New Jersey HATS. Because flooding has so many different causes, including storm surge, erosion, subsidence, intense rainfall, and sea-level rise, planners need to consider a wide range of structural, nonstructural, and nature-based features in order to identify the proper mix of solutions to address the different risk patterns and physical, socioeconomic, and demographic factors in each community under study. As a reminder, to make matters even more complex, many of the available solutions to flooding may work at cross-purposes with other important community goals like maintaining waterfront access and views, protecting neighborhood character and views, and safeguarding natural systems and biodiversity. Flood risk is a truly *wicked* problem, in that it poses a serious threat to life, limb, and property, lacks easily implementable solutions, and is highly interrelated with other societal challenges, not just for New York and New Jersey but for countless other coastal communities as well.[47]

Given the complexity of flood protection planning, it is clear that the more thoroughly understood the conditions in local communities, the more likely it will be that effective combinations of flood safety interventions will be found for each community, and the less likely planners are to propose projects that miss their mark or have unintended negative consequences.[48] The Army Corps' promised commitment to community engagement on the HATS study would reduce the risk of such outcomes. On the other hand, while more data and more holistic thinking can certainly improve planning, there is no way to tell how soon the next major storm will hit, so planners must strive to find the best balance between *planning well* and *planning quickly*.[49]

New York and New Jersey's decision to invoke WRDA 2022 Section 8106 to broaden the scope of the HATS study will most likely result in planners having at least another three years to revisit and reconsider the wisdom of centering the HATS study plan on the shoreline barriers and in-water gates that were the backbone of their 2022 proposal.[50] This will allow sufficient time for new data gathering, research, and modeling and may help generate a HATS 2.0 plan relying more on natural and nature-based solutions, green infrastructure, and other, nonstructural solutions, in addition to or instead of some previously identified shoreline and in-water barriers. Ideas for additional HATS-related research will surely arise from the coastal science convening between the Army Corps and researchers from eight local research partnerships referred to earlier, which could help clarify the extent of, and interaction between, the varying flood risks facing the study area; the intersection between flood risk, vulnerability, exposure, and socioeconomic factors; and the role biodiversity and nature-based interventions and solutions can play in improving equity and stability in urban shoreline social systems.

It is also essential that any additional scientific research and modeling undertaken during HATS 2.0 planning be dovetailed with the community-level information and priorities identified through the EJCC. Specific tasks the EJCC might take on to inform HATS 2.0 in this manner could include:

- Creation of a comprehensive, ranked list of localized priorities for addressing climate-related risks, and a methodology for factoring them into the models developed by risk-reduction planners.
- Establishment of standards for more effective participation in planning decisions by underrepresented and/or disadvantaged communities, through which agency planners not only promise dialogue, accountability, and empowerment but make them central organizing principles.
- Identification of reliable and sufficient funding for community leaders and organizations that would not otherwise have sufficient resources to participate in flood risk reduction and planning.
- Support for a stronger tie between flood risk reduction planning and efforts to achieve complementary goals like community cohesion, habitat restoration, and job creation.
- Fostering of intra- and interregional collaboration between community-based and climate justice–based organizations seeking to provide mutual support for one another's efforts to ensure that local resiliency planning centers community goals.

As encouraging as it may be to witness the HATS study's turn toward multihazard flood risk planning and community engagement, the incoming policymakers of the second Trump administration are not the only ones who must support these reforms; responsible agency staff, also, must truly commit to them before they can succeed. If those staff members instead view community consultation as an obligation rather than an opportunity, perhaps doubting the value of collaborating with community members who are not as highly trained as themselves, studies like HATS will remain mired in top-down thinking and fail to consider critical on-the-ground information.[51] However, if agency staff are ready to join community organizations at the table for a planning process based on dialogue, accountability, and empowerment, those organizations say they are ready to come to that table and help design more collaborative and restorative flood protection plans.[52]

In sum, to address the growing risk of climate-related disruption and repair the damage inflicted on frontline communities by systemic under-representation and disadvantage, coastal resilience planning must center local wisdom, address all major sources of flooding synergistically, and rely more heavily on natural and nature-based solutions. New York and New Jersey's invocation of Section 8106 of the Water Resources Development Act of 2022, which will drive a more comprehensive, "multihazard" approach to flood risk reduction, and the establishment of an Environmental Justice Coordinating Committee for the NY–NJ Harbor and Tributaries Study, are laying the groundwork for a more just and restorative flood risk reduction planning model which could provide sixteen million residents of New York and New Jersey (and, potentially, millions more in other coastal communities in red and blue states alike) with better protection from flooding and more equitable, vibrant, connected, and ecologically sound communities.

Notes

1. Papenfuss, "'Get Your Mops & Buckets Ready!'"
2. Fecht, "Should New York Build a Storm Surge Barrier?"
3. In August 2018, HATS project nanager Bryce Wisemiller told the Manhattan Community Board 1 that this barrier plan "would be a monumental engineering challenge . . . you have the concern that all those gates have to work perfectly while

that storm is approaching, and there's really no way to test those systems until there is a storm in place." Pereira, "Anti-flood Plan Surging Ahead."

4. According to projections by the New York City Panel on Climate Change, sea levels in the 2050s are likely to be 11 to 21 inches higher than in 2000. Heavy downpours like Hurricane Ida and enormous storm surges like those seen during Superstorm Sandy will become more frequent, with the greatest impacts falling on communities already most vulnerable due to a history of redlining, disinvestment, and other inequitable land use policies. New York City Panel on Climate Change, "2019 Report, Executive Summary."

5. Papenfuss, "'Get Your Mops & Buckets Ready!'"

6. Joseph Seebode, deputy New York district commander, United States Army Corps of Engineers, conversation with the author, November 15, 2022.

7. In September 2022 the Army Corps identified five possible approaches to flood prevention from which the corps designated "Alternative 3B" as their tentatively selected plan. The flood protection elements in Alternative 3B are in-water storm barriers at the mouths of Gowanus, Newtown, and Flushing Creeks in Brooklyn and Queens; structural shore-based barriers in Jersey City, on the lower west side of Manhattan, and in East Harlem; a combination of shore-based measures; and in-water barriers from the mouth of Jamaica Bay to the Rockaway Peninsula, lower Brooklyn, and two storm-surge barriers on the mouth of the Arthur Kill and Kill van Kull tidal straits. In total, Alternative 3B comprises 2.2 miles of in-water barriers, 50 miles of shoreline-based walls and berms, and various other measures designed to compensate for the environmental damage that these barriers will cause. Alternative 3B is projected to cost $52 billion dollars, protect 63 percent of the HATS study area, and take fourteen years to construct. New York–New Jersey Harbor and Tributary Draft Feasibility Study and Environmental Impact Statement, September 2022, https://www.nan .usace.army.mil/Portals/37/NYNJHATS%20Draft%20Integrated%20Feasibility%20 Report%20 Tier%201%20EIS_3Oct2022.pdf, 189–219.

8. See, for example, the comment letters sent to the United States Army Corps of Engineers regarding the 2022 New York–New Jersey Harbor and Tributaries Study Plan, March 24, 2023; National Oceanic and Atmospheric Administration, March 29, 2023; New York City Environmental Justice Alliance & Columbia Climate School, March 23, 2023; and Bipartisan Coalition of 14 Members of Congress, September 12, 2023.

9. Hurricane Ida's record-setting rainfall in September 2021 caused widespread inland flooding, killing at least thirty-six New York and New Jersey metropolitan area residents, including eleven who drowned in illegally converted, basement apartments in neighborhoods that house predominantly poor and immigrant New Yorkers. Calvan et al., "More than 45 Dead"; NYC Mayor's Office of Climate and Environmental Justice, "Chronic Tidal Flooding," https://climate.cityofnewyork.us /challenges/chronic-tidal-flooding/; Floodnet.org, "Flooding Data for Hamilton Beach, Queens," https://dataviz.floodnet.nyc/viz?v=WZVYQzqyEgsCp27wUUXvW, both accessed June 8, 2024.

10. Shannon, "Environmental Justice Guidance.

11. Snider, "New Provisions Included."

12. Correspondence from New York, New Jersey, and the City of New York to the United States Army Corps of Engineers, November 16, 2023. For example, in May 2022 RCCP met with the Army Corps' HATS planning team to share the findings of its research interviews and press for a fully collaborative HATS process. At that meeting, the Army Corps made a promise to convene a HATS environmental justice working group. New York district commander Colonel Matthew Luzzatto and a dozen of his colleagues then visited Columbia University on November 18, 2022, for a briefing and dialogue with RCCP staff and advisory board members, at which the environmental justice working group was again discussed. RCCP again called for the establishment of the environmental justice working group in its March 23, 2023, comments on the Army Corps tentatively selected HATS action plan. Finally, on December 11, 2023, over twenty months after the Army Corps' first promise of a HATS environmental justice working group, RCCP and twenty-one frontline community organizations, environmental advocacy groups, and other nongovernmental stakeholders wrote to the states of New York and New Jersey to protest the Army Corps' failure to establish the environmental justice working group and appeal to those agencies for their assistance in this regard (multiparty letter to the commissioner of the Department of Environmental Conservation of the State of New York and the commissioner of the Department of Environmental Protection of the State of New Jersey, December 11, 2023).

13. District Commander Alexander M. Young of the US Army Corps New York District committed to establishing the HATS Environmental Justice Coordination Committee during a January 24, 2024, meeting with the New York City Environmental Justice Alliance, El Puente, and researchers from the Columbia Climate School, including the author. Later that day, Colonel Young shared this announcement with officials from New York, New Jersey, and the City of New York, as well as investigators from six academic research partnerships, at a planning meeting for a proposed HATS technical advisory workshop.

14. The likely impact of the second Trump presidency on HATS and other climate adaptation-oriented studies is uncertain. Although policies relating to decarbonization are widely expected to change, Trump's approach to climate change adaptation (or, as some prefer to put it, preparing for *extreme weather*) may reflect greater continuity, given evidence that flooding and other climate change-driven impacts are likely to hit red states harder than blue states. Muro et al., "How the Geography of Climate Damage"; Kahn et al., "Don't Call It Climate Change"; Wilson, "Rising Tides Are Coming."

15. Gallay et al., "Designing Community-Led Plans."

16. Cunningham, "Locals React to Plan"; Dembicki, "Coastal Residents Fear 'Hideous' Seawalls."

17. New York City Environmental Justice Alliance, "Waste Equity," accessed June 8, 2024, https://www.nyc-eja.org/.

18. Partnerships like the one between NYC-EJA and the Columbia Climate School don't happen overnight; this one has roots that stretch back to the 1990s, when the author served as a policymaker and law enforcement official at the New York State Department of Environmental Conservation with responsibilities for reduction of the disproportionate number of solid waste transfer stations in Brooklyn, Queens, and the Bronx, which was also a top priority for NYC-EJA. Shared successes in this work, like the establishment of an official fair share policy for waste facility distribution in 2006, helped build the trust needed to establish RCCP. New York City Environmental Justice Alliance, "Waste Equity"; Calmes and Khurshid, "Fair Share: Design Flaws."

    RCCP's effort to foster new collaborations on flood risk reduction between environmental justice communities, practitioners, and researchers is also in keeping with Columbia University's commitment in 2019 to adopt an institutional "Fourth Purpose," designed to leverage scholarly knowledge to create more rapid and transformational societal and global impact. "Fourth Purpose Task Force Report on Directed Action," Columbia University, December 15, 2020, https://president.columbia.edu/sites/default/files/content /Additional/Fourth%20Purpose%20Task%20Force%20Report.pdf. The author of this chapter, Paul Gallay, who directs the RCCP, is deeply indebted to its core team and advisory board, who have collaboratively shaped the project's research and advocacy work. RCCP website, https://csud.climate.columbia .edu/research-projects/resilient-coastal-communities-project, accessed June 8, 2024.

19. The community leaders interviewed included staff members from El Puente, GOLES, Guardians of Flushing Bay, Ironbound Community Corporation, New Jersey Environmental Justice Alliance, Newtown Creek Alliance, RISE, Staten Island Urban Center, The Point CDC, and UPROSE.

20. Time and again, interviewees told RCCP researchers in 2022 that they put far more energy into coastal resilience planning processes than they get out of such processes, due to the organizing agencies' inability or unwillingness to make them true partners in developing effective resilience solutions. The growing sense of frustration that communities feel in the wake of such unsuccessful planning efforts threatens to undermine the region's ability to prepare for the growing climate-related risks facing our communities and surrounding ecosystems. Gallay et al., "Designing Community-Led Plans," 1–2.

21. The local organizational leaders spoke extensively about such needs as more extensive and effective floodproofing of homes and businesses, better maintenance of stormwater infrastructure, and more effective agency response in flood situations. They also pointed out that studies like HATS tend to focus too much on building barriers and other physical structures, rather than giving due attention to strengthening community partnerships and local response capacity, which has been shown to save lives during climate-related emergencies. Morris et al., "Advancing

Equitable Partnerships," sec. 3—Findings and Discussion; Klinenberg, "Adaptation: How Can Cities Be 'Climate Proofed'?"

22. RCCP interview with Rodriguez, *The Point CDC*, March 15, 2022.

23. Gallay et al., "Designing Community-Led Plans," 2–3; NYC Climate Regional Plan Mapper, Regional Planning Association, November 2022, https://rpa.org/maps /resilience.html.

24. UPROSE's "Green Resilient Industrial District," in particular, provided the blueprint for the offshore wind turbine assemblage plant currently under construction in Sunset Park, Brooklyn, which unites traditional environmental justice concerns relating to health, safety, and equity with the creation of new green manufacturing jobs, job training programs, and other community benefits. Gallucci, "A Brooklyn Neighborhood's Long Fight."

25. Atalay, *Community-Based Archaeology*, 69; Atalay, "Braiding Strands of Wellness."

26. Staff member at El Puente (name withheld on request), March 8, 2022, https://www .elpuente.us/.

27. "Overview of Proposed Rule: Corps of Engineers Agency Specific Procedures to Implement the Principles, Requirements, and Guidelines for Federal Investments in Water Resources," Federal Register, February 15, 2024, section 234.6(c)(1), https:// www.federalregister.gov/documents/2024/02/15/2024-02448/corps-of- engineers-agency-specific-procedures-to-implement-the-principles-requirements-and. See also section 234.7. The procedures also make reforms to the way in which the Corps will calculate the relative value of different flood risk reduction options. While the Corps' current "benefit-cost" scoring system puts economic goals above all others, these new rules give equal weight to economic, environmental, and social factors, thus rebalancing the scales in favor of more socially beneficial or environmentally restorative flood protection investments. In section 234.4(c), the Corps characterized its new benefit cost calculation rule as follows: "Federal investments in water resources have been mostly based on economic performance assessments [focusing] on investments that will improve national economic efficiency. This focus on national economic gains sometimes resulted in an unduly narrow benefit-cost comparison of the monetized and quantified effects. [R]elevant environmental, social and economic effects should all be considered. . . . This more integrated approach would allow decision-makers to view a more complete range of effects of alternative actions and lead to more socially beneficial investments."

28. 33 Code of Federal Regulations Part 234, 2024–29652 (89 FR 103992), https://www .federalregister.gov/documents/2024/12/19/2024-29652/corps-of-engineers -agency-specific-procedures-to-implement-the-principles-requirements-and.

29. Alexander M. Young, New York District commander, United States Army Corps of Engineers, conversation with nonprofit and academic organization members of the Rise 2 Resilience Coalition, March 27, 2024.

30. Water Resources Development Act of 2022, Division H, Title LXXXI of the National Defense Authorization Act for Fiscal Year 2023, Public Law 117–263, 136 *STAT.* 2395

(2023). Also, correspondence from New York, New Jersey, and the City of New York with the United States Army Corps of Engineers, November 16, 2023; correspondence from the states of New York and New Jersey to the assistant secretary for civil affairs and policy, United States Army, January 8, 2024.

31. Cited in Snider, "New Provisions Included."

32. Joseph Seebode, deputy New York District commander, United States Army Corps of Engineers, conversation with the author, November 15, 2022.

33. New York–New Jersey Harbor and Tributary, Draft Feasibility Study and Environmental Impact Statement, September 2022, https://www.nan.usace.army.mil/Portals/37/NYNJHATS%20Draft%20Integrated%20Feasibility%20Report%20Tier%201%20EIS_3Oct2022.pdf, 151–157.

34. The research partnerships are the Center for Climate Systems Research, the Center for Policy Research and the Environment, the Consortium for Climate Risk in the Urban Northeast, the Megalopolitan Coastal Transformation Hub, the National Center for Disaster Preparedness, the New York City Climate Vulnerability, Impact and Adaptation Study, the New York City Panel on Climate Change, and the Resilient Coastal Communities Project.

35. New York City Environmental Justice Alliance and Columbia Climate School, Center for Sustainable Urban Development, March 23, 2023.

36. For example, in March 2023 New York, New Jersey, and the New York City Mayor's Office of Climate and Environmental Justice demanded that the Army Corps "lead, fund, and facilitate" the promised environmental justice working group, which they referred to as the HATS Climate & Environmental Justice Advisory Group, or CEJAG. New York State, New Jersey, and New York City, March 31, 2023, 2. The two states and New York City later expanded on this request in their November 16, 2023, correspondence with the Army Corps, calling on the Corps to "retain consultants with expertise in reaching and educating affected communities, especially environmental justice and disadvantaged communities, to discuss proposed project elements" and "effectively obtain and appropriately act upon community guidance or critique." Correspondence from New York, New Jersey, and the City of New York to the United States Army Corps of Engineers, November 16, 2023, 3.

37. Colonel Luzzatto's comments to this effect can be accessed at the Army Corps' recording of the December 15, 2022, meeting at which they were made, https://www.youtube.com/watch?v=KoJ4_OaOTE4&t=23s, minute 3:20.

38. The shortcomings contributing to this outcome are illustrated in this excerpt from written comments on the HATS Tentatively Selected Plan submitted by RCCP and NYC-EJA: "Many community members found it difficult to attend or provide feedback in town halls because the information wasn't presented in a clear, timely way. NYNJHATS is a complex, dense project that is difficult to understand for most people." To address this barrier, we asked USACE to create and present simple, accessible briefing materials tailored to specific neighborhoods and languages for their town halls, but have yet to see them do this. USACE also does not make it easy for people to attend their town halls, which is reflected in very low

participation rates. . . . The challenge is that USACE is under- resourced and not positioned to substantively engage with community members and facilitate conversations on a neighborhood's flooding spots, mitigation measures, and project impacts. USACE has failed to deliver on promises to better engage EJ communities, including working with local stakeholders in their proposed "Climate and Environmental Justice Working Group." USACE charged the working group with conducting community engagement on behalf of it but has still declined to provide a scope of work for the group, organize it, or fund it. New York City Environmental Justice Alliance and Columbia Climate School, Center for Sustainable Urban Development, March 23, 2023.

39. Colonel Alexander M. Young, January 24, 2024, meeting between US Army Corps staff, RCCP, NYC-EJA, and El Puente.

40. Morris et al., "Advancing Equitable Partnerships," section 3.11.

41. Executive Order of January 20, 2025, "Ending Radical and Wasteful Government DEI Programs and Preferencing," https://www.whitehouse.gov/presidential-actions /2025/01/ending-radical-and-wasteful-government-dei-programs-and -preferencing/.

42. Correspondence from the United States Army Corps of Engineers, December 13, 2024.

43. Draft invitation [unsent] developed by Army Corps staff on May 3, 2024. These goals were formulated during discussions between the Army Corps, NYC-EJA, El Puente, and the Columbia Climate School during the spring of 2024. As of January 31, 2025, the Corps' stated intention in relation to the issues to be addressed by the Environmental Justice Coordination Committee is to "continue community engagement and consultation efforts, including adherence to the Agency Specific Procedures published in the Federal Register (33 CFR Part 234) on December 19, 2024, and which became effective on January 17, 2025." Correspondence from the United States Army Corps of Engineers, January 31, 2025.

As stated earlier, these regulations are designed to incorporate environmental justice considerations into "all phases of the Corps' planning and decision-making process in order to remove barriers to effective community participation, increase community access to benefits, and drive restorative justice." "Overview of Proposed Rule," Section 234.6(c)(1) Former assistant army secretary for civil works Michael Connor indicated that he would consider it worth spending "a year or two" on the Environment and Climate Justice Working Group to assure that the community enfranchisement goals of the Justice40 program can be achieved during the HATS study. Michael Connor, conversation with the author, September 20, 2022. Looking at this from a different angle, the community buy-in the Army Corps could gain through the successful establishment of such a working group could conceivably save a similar or even greater amount of time during the funding and implementation stage of HATS.

44. In January 2024 coastal storm risk reduction advocates from Galveston, Texas, Miami, Florida, Charleston, South Carolina, Norfolk, Virginia, and New York

organized a mutual support and policy development interregional working group, during which issues associated with community engagement in flood protection planning are frequently discussed. None of these advocates is aware of any previous community engagement initiative with the scope and intention of the NY-NJ HATS Environmental Justice Coordinating Committee. State and local partners in the NY-NJ HATS are mindful of such community concerns and so are themselves working to increase and improve community engagement. For example, New York City has agreed to develop more just, inclusive, accountable, and reliable decision-making models through its Climate Knowledge Exchange program (https://climate.cityofnewyork.us/initiatives/climate-knowledge-exchange/, accessed June 9, 2024).

45. New York City is still dealing with the fallout from its controversial 2018 decision to abandon a proposed $700-million plan for an "East Side Coastal Resiliency Project" (ESCR) on the Lower East Side of Manhattan, which had been developed by consensus through a multiyear partnership with local community organizations. Instead, the city unilaterally selected a vastly different and deeply unpopular alternative plan for the ESCR project, prepared in secret while its community partners still believed the city was going to implement the consensus plan. City officials subsequently acknowledged the fundamentally untransparent and unaccountable manner in which they managed this critical final stage of the ESCR project, but the damage had been done, leading to extensive litigation delays and a near-complete loss of trust in the project in the community. Kimmelman, "What Does It Mean to Save A Neighborhood."

46. Flood risk reduction advocates in the communities referred to earlier are following the reforms described in this chapter to determine how they influence the HATS study and have expressed interest in adopting those reforms that prove effective into their own local processes.

47. On the topic of "wicked problems" generally and in relation to climate change specifically, see Rittel and Webber, "Dilemmas in a General Theory of Planning." See also Incropera, *Climate Change*.

48. For example, researchers from seven universities collaborating as the Megalopolitan Coastal Transformation Hub (https://coastalhub.org/, accessed June 9, 2024) warned that the HATS action plan tentatively selected by the Army Corps in 2022, which includes over fifty miles of shoreline and in-water barriers designed to block storm surge, may increase the likelihood that rainfall-driven flooding will accumulate and worsen flooding in the communities on the land side of those barriers. Such concerns are also referred to as seeking to avoid "maladaptation." Letter on HATS from researchers at Rutgers, Dartmouth, Princeton, and other institutions working together as the Megalopolitan Coastal Transformation Hub (MACH) Project, March 1, 2023.

49. Given the challenges described here, it's fortunate that new funding for flood risk reduction projects is provided for in federal legislation such as the Infrastructure

Investment and Jobs Act of 2021, Pub. L. No. 117–58, 135 STAT. 429 (2021), which will pump over $13 billion dollars into such efforts. Tompkins, "5 Ways the Infrastructure Bill Would Improve."As of the time this book went to print, the second Trump administration has placed on hold new investments under the Investment Infrastructure and Jobs Act.

50. This potential time extension was shared during the January 24, 2024, meetings referenced in note 14. Also, US Army Corps of Engineers, "Request for Additional Resources," December 2023, slide presentation.

51. Most of the community interviewees with whom RCCP spoke in 2022 indicated that, under such circumstances, they would rather not be at the table at all, given the myriad other responsibilities they are balancing at any given time. Gallay et al., "Designing Community-Led Plans," 1–2; Morris et al., "Advancing Equitable Partnerships," section 3.2.

52. Gallay et al.; Morris et al.

BIBLIOGRAPHY

Atalay, Sonya L. "Braiding Strands of Wellness." *The Public Historian* 41, no. 1 (February 2019): 78–89. https//doi.org/10.1525/tph.2019.41.1.78.

Atalay, Sonya L. *Community-Based Archaeology: Research with, by, and for Indigenous and Local Communities.* Oakland: University of California Press, 2012.

Bey, G. "Report on the NOAA Office of Education Environmental Literacy Program Community Resilience Education Theory of Change." 2020. https://doi.org/10.25923/MH0G-5Q69.

Calmes, Maggie, and Samar Khurshid. "Fair Share: Design Flaws, Flashpoints & Possible Updates." *Gotham Gazette*, November 21, 2016. https://www.gothamgazette.com/city/6633-fair-share-design- flaws-flashpoints-possible-updates.

Calvan, Bobby Caina, David Porter, and Jennifer Peltz. "More than 45 Dead After Ida's Remnants Blindside Northeast." *Associated Press*, September 2, 2021. https://Apnews.Com/Article/Northeast-Us-New- York-New-Jersey-Weather-60327279197e14b9d17632ea0818f51c.

Checker, Melissa. *The Sustainability Myth: Environmental Gentrification and the Politics of Justice.* New York: New York University Press, 2020.

City of New York. "Chronic Tidal Flooding." Accessed June 8, 2024. https://climate.cityofnewyork.us/challenges/chronic-tidal-flooding/.

City of New York. "State of Climate Knowledge—Workshop Summary Report." April 2022. https://climate.cityofnewyork.us/wp-content/uploads/2022/04/2022_CKE_Report_10.25.22.pdf.

Cunningham, Mary. "Locals React to Plan to Add 15 ft. Tall Seawall to Greenpoint Waterfront." *Greenpointers*, November 3, 2023. https://greenpointers.com/2023/03/03/locals -react-to-plan-to-add- 15-ft-tall-seawall-to-greenpoint-waterfront/.

Dembicki, Geoff. "Coastal Residents Fear 'Hideous' Seawalls Will Block Waterfront Views." *The Guardian*, January 13, 2023. https://www.theguardian.com/us-news/2023/jan/13/us-cities-ugly- seawalls-climate-crisis-miami.

Fecht, Sarah. "Should New York Build a Storm Surge Barrier?" *State of the Planet*, October 24, 2019. https://news.climate.columbia.edu/2019/10/24/new-york-storm-surge-barriers/.

Floodnet. "Flooding Data for Hamilton Beach, Queens." Accessed June 8, 2024. https://dataviz.floodnet.nyc/viz?v=WZVYQzqyEgsCp27wUUXvW.

Gallay, Paul, Victoria Sanders, Annel Hernandez, Jacqueline Klopp, Bernadette Baird-Zars, and Lexi Scanlon. "Working Paper: Designing Community-Led Plans to Strengthen Social Cohesion: What Neighborhoods Facing Climate-Driven Flood Risks Want from Resilience Planning." Columbia Climate School, Resilient Coastal Communities Project, June 27, 2022. https://csud.climate.columbia.edu/sites/default/files/content/Designing%20Community-led%20Plans%20to%20Strengthen%20Social%20Cohesion-%20What%20Neighborhoods%20Facing%20Climate-driven%20Flood%20Risks%20Want%20from%20Resilience%20Planning%20(6-27-22).pdf.

Galluci, Maria. "A Brooklyn Neighborhood's Long Fight for Green Jobs Is Paying Off." *Canary Media*, October 13, 2022. https://www.canarymedia.com/articles/just-transition/power-by-the-people-green-resilient-industrial-district-sunset-park.

Gould, K., and T. Lewis. *Green Gentrification: Urban Sustainability and the Struggle for Environmental Justice*. New York: Routledge, 2017.

Incropera, Frank. *Climate Change: A Wicked Problem*. Cambridge: Cambridge University Press, 2015.

Infrastructure Investment and Jobs Act of 2021. Pub. L. No. 117–58, 135 STAT. 429 (2021).

Kahn, Debra, Bruce Ritchie, Ry Rivard and Mike Lee. "Don't Call It Climate Change. Red States Prepare for 'Extreme Weather,'" *Politico*, November 23, 2021, https://www.politico.com/states/california/story/2021/11/23/adapting-to-climate-is-a-winning-issue-for-politicians-even-in-red-states-1394620.

Kimmelman, Michael. "What Does It Mean to Save a Neighborhood?" *New York Times*, December 2, 2021. https://www.nytimes.com/2021/12/02/us/hurricane-sandy-lower-manhattan-nyc.html.

Klinenberg, Eric. "Adaptation: How Can Cities Be 'Climate Proofed'?" *New Yorker*, December 30, 2012. https://www.newyorker.com/magazine/2013/01/07/adaptation-eric-klinenberg.

Morris, Aya, Bernadette Baird-Zars, Victoria Sanders et al. "Advancing Equitable Partnerships: Frontline Community Visions for Coastal Resiliency Knowledge Co-Production, Social Cohesion, and Environmental Justice." *Geoforum* 154, no. 104051 (August 2024). https://www.sciencedirect.com/science/article/pii/S001671852400112X.

Multiparty advocates letter to the Commissioner of the Department of Environmental Conservation of the State of New York and the Commissioner of the Department of Environmental Protection of the State of New Jersey, December 11, 2023. Non-web source.

Muro, Mark, David G. Victor, and Jacob Whiton. "How the Geography of Climate Damage Could Make the Politics Less Polarizing." *Brookings*, January 29, 2019. https://www.brookings.edu/articles/how-the-geography-of-climate-damage-could-make-the-politics-less-polarizing/.

National Oceanic and Atmospheric Administration, March 29, 2023. Non-web source.

New York City Environmental Justice Alliance. "Waste Equity." Accessed June 8, 2024. https://nyc-eja.org/campaigns/waste-equity/.

New York City Environmental Justice Alliance and Columbia Climate School, Center for Sustainable Urban Development, March 23, 2023. Non-web source.

New York City Panel on Climate Change. "2019 Report, Executive Summary." *Annals of the New York Academy of Sciences*, March 15, 2019. https://nyaspubs.onlinelibrary.wiley.com/doi/10.1111/nyas.14008.

Nir, S. M. "Trapped in Basements and Cars, They Lost Their Lives in Savage Storm." *New York Times*, September 2, 2021. https://www.nytimes.com/2021/09/02/nyregion/nyc-flooding-deaths.html.

Ottinger, Gwen. "Careful Knowing as an Aspect of Environmental Justice." *Environmental Politics* 33, no. 2 (2024): 199–218. https://doi.org/10.1080/09644016.2023.2185971.

Papenfuss, Mary. "'Get Your Mops & Buckets Ready!'—Trump's Infuriating Answer to Rising Seas." *HuffPost*, January 18, 2020. https://www.huffpost.com/entry/mops-and-buckets-trump-climate-change_n_5e23a83dc5b674e44b993aab.

Pereira, Sydney. "Anti-flood Plan Surging Ahead too Fast, Many Activists Say." *The Villager*, August 23, 2018.

Rittel, H. W., and M. M. Webber. "Dilemmas in a General Theory of Planning." *Policy Sciences*, 4, no. 2 (June 1973): 155–69. https://www.cc.gatech.edu/fac/ellendo/rittel/rittel-dilemma.pdf.

Shannon, Jay. "Assistant Secretary of the Army for Civil Works Issues Environmental Justice Guidance to the Army Corps of Engineers." United States Department of the Army, March 22, 2022, section 10. https://www.army.mil/article/254935/assistant_secretary_of_the_army_for_civil_works_issues_environmental_justice_guidance_to_the_army_corps_of_ engineers.

Snider, Natalie. "New Provisions Included in the Water Resources Development Act of 2022." *Environmental Defense Fund*, December 16, 2022. https://www.edf.org/media/new-provisions- included-water-resources-development-act-2022.

Tompkins, Forbes. "5 Ways the Infrastructure Bill Would Improve America's Flood Resilience." *Pew Charitable Trusts*, August 17, 2021. https://www.pewtrusts.org/en/research-and-analysis/articles/2021/08/17/5-ways-the-infrastructure-bill-would-improve-americas-flood-resilience.

Rutgers University, Dartmouth College, Princeton University et al. Megalopolitan Coastal Transformation Hub, March 1, 2023. Non-web source.

State of New York and State of New Jersey, January 8, 2024. Non-web source.

State of New York, State of New Jersey, and City of New York, March 31, 2023. Non-web source.

State of New York, State of New Jersey, and City of New York, November 16, 2023. Non-web source.

United States Army Corps of Engineers, December 13, 2024. Non-web source.

United States Army Corps of Engineers. "Engineers Request for Additional Resources." Slide presentation, December 2023. Non-web source.

United States Army Corps of Engineers. New York–New Jersey Harbor and Tributary—Draft Feasibility Study and Environmental Impact Statement, September 2022. https://www.nan.usace.army.mil/Portals/37/NYNJHATS%20Draft%20Integrated%20Feasibility%20 Report%20Tier%201%20EIS_3Oct2022.pdf.

United States Army Corps of Engineers. "Overview of Proposed Rule: Corps of Engineers Agency Specific Procedures to Implement the Principles, Requirements, and Guidelines for Federal Investments in Water Resources." Federal Register, February 15, 2024, Section 234.6(c)(1). https://www.federalregister.gov/documents/2024/02/15/2024-02448/corps -of-engineers-agency- specific-procedures-to-implement-the-principles-requirements -and.

United States Army Corps of Engineers. Video recording of HATS December 15, 2022, public meeting. https://www.youtube.com/watch?v=KoJ4_OaOTE4&t=23s. Minute 3:20.

United States Congress (bipartisan coalition of 14 members). September 12, 2023. Non-web source.

United States Environmental Protection Agency. "EPA Report Shows Disproportionate Impacts of Climate Change on Socially Vulnerable Populations in the United States." September 2, 2021. https://www.epa.gov/newsreleases/epa-report-shows-disproportionate -impacts-climate-change- socially-vulnerable.

Water Resources Development Act of 2022, Division H. Title LXXXI of the National Defense Authorization Act for Fiscal Year 2023. Public Law 117–263, 136 STAT. 2395 (2023).

Wilson, Miranda. "Rising Tides Are Coming for Lee Zeldin's Hometown," E&E News, December 19, 2024, https://www.eenews.net/articles/rising-tides-are-coming-for-lee -zeldins-hometown/.

Young, Shalanda, Brenda Mallory, and Gina McCarthy. "The Path to Achieving Justice40." Washington, DC: The White House, July 20, 2021. https://www.whitehouse.gov/omb /briefing-room/2021/07/20/the-path-to-achieving-justice40/.

VI

Pursuing Climate Justice in Academia and Beyond

Embracing Complexity and Bridging Disciplines

16

The Elusive Challenge of Climate Justice and a Call to Action

REBECCA S. MARWEGE, NIKHAR GAIKWAD,
AND JOERG M. SCHAEFER

CRACKING THE CODE OF CLIMATE JUSTICE: BRIDGING DISCIPLINES, EMBRACING COMPLEXITY

A specter moves among us—shadowed, indistinct, silent; it hovers at the edge of vision and then is gone, shape-shifting unrecognizably into plain sight.

—Sharon Krause, *Eco-Emancipation*, 2023

Where Sharon Krause describes the specter of environmental domination as shaping "every aspect of our lives while being virtually invisible to us," this book has sought to illuminate the different debates around the concept of climate justice.[1] Similarly to Krause's description of environmental domination, we note that the exact meaning of the concept of climate justice remains elusive and ever-changing, even if climate justice debates have come to touch every aspect of life, and warnings of the catastrophic consequences of further temperature rise are getting ever louder.

In the process of editing this book, we have observed that each individual discipline provides key insights into understanding the challenges and complexities of climate change and their implications for climate justice. While anthropology allows us to rethink who the subject of climate justice ought to be, for example, the study of political theory allows us to make sense of what we define as "justice" in the first place, including by reorienting justice to comprise relational and intergenerational concerns. Sociology augments this discussion with a focus on the underlying social systems that facilitate structural inequality and therefore shifts the focus away from technological interventions towards more foundational societal change as a mechanism to address the climate crisis. Nonetheless, neither of these debates could meaningfully contribute to climate justice debates without a basic understanding of the scientific foundations and processes that are put in motion by a heating climate, which the natural sciences can supply.

To illustrate the need for a multidisciplinary approach, it is useful to recall our discussion of geoengineering and measuring, reporting, and verification (MRV) tools in chapter 1. The natural sciences are important not only for understanding the foundations of how measurement and technologies can help to achieve carbon emissions reduction and adaptation to increasing climate threats, but also for assessing under what circumstances new technologies could have any impact at all, and at what costs and side-effects. Applying an atmospheric science lens, for example, allows us to recognize that even though the global rate of GHG emissions has slowed, it is far from the sharp reduction needed to achieve the goal of 1.5C warming targeted in the Paris Agreement. While the United States, as one of the top global GHG emitters, needed to reduce its per capita GHG emissions by 6.9 percent to reach the target of the Paris Agreement in 2023, it only achieved a reduction of 1.9 percent.[2] These observations could be interpreted to support a case for a significant reduction of current and future emissions, as well as potentially the use of geoengineering technologies to remove legacy emissions from the atmosphere.

By itself, however, this analysis is incomplete. As Mary Witlacil (drawing on the philosopher Marion Hourdequin) notes in chapter 2, along with various other chapters in this volume, especially in the context of geoengineering "there is a tendency to deploy 'expertise imperialism,' in which experts—who are presumed to know best—have the formal power to research or implement potentially risky climate solutions . . . without sufficiently explaining the methods or consequences, or giving affected communities a seat at the table." Invoking a political theory lens thus raises the question of whose needs are really recognized when geoengineering technologies are applied. Would geoengineering technologies lead to the perpetuation of a "business-as-usual" approach, for example, and thus likely sideline the needs of historically marginalized communities? Could this lead to new forms of climate colonialism? Similarly, does the quantification of emissions under MRV underestimate the climate damages that communities have already internalized?

Applying a sociological and positivist political science lens, we can ask what the systems of power at play are and how equitable global agreements that avoid the continuation of existing hegemonies could be achieved. Separately, employing a religious studies lens, we can ask how societies can avoid perpetuating anthropocentric hubris and recognize the "radical cosmological plurality" in the world rather than reentrench a Western focus on technological solutions.[3] Thus, including a diverse range of disciplines allows us to question who gets included in these discourses and which justice claims are at stake. The aim of this book has therefore been to provide an introspection into how different academic disciplines approach the concept of climate justice and highlight both synergies and tensions across them.

Throughout the volume, we observe that there existed agreement on some of the core themes of climate justice, such as the historical legacy of colonialism, tensions between environmental justice and climate justice, the global and the local scale of analysis, as well as the role of knowledge as both a tool of empowerment and potential privilege. One key concern that emerged from the conversations reflected in this book is that climate justice requires us to think of not only what the concept may include and how it may be defined, but also whom it ought to serve and what the best approach to achieving climate justice may look like. Considering this concern, in chapter 13 Sheng Long highlights, for example, how anthropology has long questioned the assumption of the universal liberal subject and describes how Hakka-speaking pomelo farmers in Meixian refer to climate change in terms very different from Western scientific terminologies. This raises the question of how concepts such as climate justice can be decolonized and made commensurable across radically different experiences of climate change.

Additionally, contributors discussed what the right responses should be to climate change. Focusing on climate adaptation measures, Lew Ziska and his team demonstrate in chapter 8 that climate change is already a threat multiplier, exacerbating precarious conditions of agricultural workers in the United States. Similarly, Adam Sobel and Melanie Bieli, in chapter 12, explore how climate scientists can focus on climate adaptation measures in both academia and industry. Meanwhile, Jacqueline Klopp and Festival Boateng emphasize in chapter 4 that climate justice requires rethinking transport systems across the globe to address both the public health and climate crisis. Focusing on climate mitigation measures, they highlight the need for emission reduction in transport systems focused on private vehicles. Róisín Commane also demonstrates in chapter 5 that increasing building efficiency in cities can significantly lower overall GHG emissions. Nonetheless, she points out that focusing on these climate mitigation measures requires accounting for public health risks if a full switch away from fossil fuels, including natural gas, is not achieved. Thus, Commane highlights the need to account for unintended consequences in both climate mitigation and climate adaptation responses. Together, these observations emphasize trade-offs between mitigation and adaptation-focused policy responses as well as the danger of unforeseen consequences.

Corresponding to these observations, this conclusion is our attempt to narrate the process of communication we experienced in editing this volume and to make sense of both the tensions and synergies across the book. We observe that the chapters in this volume underscore the imperative of "critical pluralism," which describes both the need for the acknowledgment of difference as well as an "open, participatory, and intersubjective process of communication" that allows us to achieve a common yet nuanced understanding of climate justice.[4]

However, multidisciplinary conversations are far from easy. As we noted over the years of our working group, engaging with other disciplines, especially across the divides of the natural sciences, social sciences, and humanities, requires translation efforts from all sides to learn and understand key terminology from other disciplines. For example, a humanist in our group was confused by the technical language around measuring GHG emissions, while a natural scientist in the group had a hard time understanding why different types of justice, such as recognition or participation, which seem evident to many humanists and social scientists, matter. From this experience, we learned that it is nontrivial to call for more multidisciplinary engagement, and it is insufficient to engage only in individual cross-disciplinary dialogue. Instead, we realized that we need to reorganize academic structures to provide the logistical support that can facilitate multidisciplinary conversations on hard topics related to climate justice.

We therefore do not intend this conclusion to be an end-all assessment of the term "climate justice." Instead, the individual chapters reiterate the observation that the concept of climate justice is marked by its dynamic character and impossible to condense into one single, static definition. Importantly, it is not only impossible but also potentially dangerous to conceptualize and tackle climate justice through a monodisciplinary approach and make it fit a box that is too narrow to capture its complexity. As the chapters in this volume have shown, the climate crisis affects all areas of life, and addressing it through a narrow lens can lead to pseudo-solutions that fail to comprehensively address the root causes of the climate crisis and its related systemic injustices. This applies both to technical "solutions" such as geoengineering that overlook the social dimension of the crisis and to theoretical abstractions that fail to engage with the climate science and experiences of communities on the ground. A monodisciplinary focus can thus distort the scope of the climate crisis and perpetuate misguided and inefficient responses. Confronting this challenge, we hope that this book provides the first piece of a puzzle to a much larger multidisciplinary and continuous conversation on the climate crisis.

COMMON THEMES AND TENSIONS ACROSS CHAPTERS

In chapter 1 we argued that there are five key analytical dimensions that can help to highlight the different angles of climate justice:

1. The temporal dimension, which argues that climate justice as a term and concept is not static but rather a moving target that is affected

both by changes in our scientific understanding of the impacts of the climate crisis on human societies and by changes in our humanistic conceptualizations of what constitutes climate justice over time.

2. The spatial dimension, which highlights the necessity of accounting for how both the climate crisis and definitions of climate justice are evolving differently across the world and how different scales of analysis, such as the national and local levels, affect how we approach the topic.

3. The agential dimension, which describes the different types of agency that play a role in any response to the climate crisis, such as by individuals, frontline communities, or states.

4. The structural dimension, which indicates the need to consider the broader structures within which agents operate, such as the international system. These structures both provide opportunities and restrict how agents conceptualize and respond to climate justice in their actions.

5. Ontological considerations, which underscore the role that different worldviews play in identifying the causes, impacts, and justice considerations of climate change.

In the following sections, we illustrate how these five dimensions resonate across the main themes that individual chapters discussed, such as debates over trade-offs between environmental and climate justice concerns, historical injustices and the legacy of colonialism, the right scale of analysis to assess the origins and impacts of the climate crisis, different ontological and epistemological considerations, and the question of what "just" responses to the climate crisis should look like.

Environmental Justice, Climate Justice, and Unforeseen Consequences

To start, we note that there existed some confusion around the differences between the concepts of environmental justice and climate justice. While these terms provoke intuitive reactions, as almost everyone has encountered them in their research in one way or another, and they are often used interchangeably, what uniquely distinguishes one from the other seems less clear. Through our conversations, we found that there exists a significant overlap between the two concepts. Jacqueline Klopp and Festival Boateng show in chapter 4, for example, that climate change reinforces preexisting injustices in cities, such as a lack of green space and the placement of highways in underserved and low-income neighborhoods, which leads to greater exposure to air pollution as well as higher temperatures during summer months and augmented risks for flooding.

Nonetheless, it is misleading to conceptualize climate justice as "just" another subset of environmental justice, as its focus on global GHG emissions and the unequal consequences of rising temperatures make it a discrete concept, both in its assessment of the problem and justice issues at stake and in the suggested solutions and its intertemporal characteristics, which raise questions of historical injustices and responsibilities owed to future generations. A classic environmental justice issue such as the spilling of toxins may raise questions of historical environmental racism and the need for both local and global regulation of toxic materials. However, continued global GHG emissions exacerbate these existing environmental issues by causing unprecedented climate harm in the form of melting glaciers, increased frequency of heat waves, and the heating of oceans. The disastrous impact of these events potentiates the urgent need for both a global and a local approach to rapid climate mitigation and climate adaptation in places where the impact of climate change is already being felt, such as by populations in West Africa and South America. Accounting for these differences allows us to more accurately describe the issues at stake and delineate the normative consequences in responding to each of these justice considerations.

Commenting on these concerns, Mary Witlacil notes in chapter 2 how philosophical approaches to climate justice build on a long tradition of environmental justice that emerged to challenge unjust distributions of environmental harms, but that discourses of climate justice add a focus on the aspect that those "who have contributed the least to climate change are poised to suffer the worst consequences." Similarly, in chapter 6 Jennifer Givens and Mufti Ahmed highlight how scholars such as Robert Bullard coined the language of environmental justice, documenting the racial disparities in exposure to toxic waste sites. They argue that the concept of climate justice contributes to these debates with a focus on how the causes and consequences of climate change are inequitably distributed across "race, class, gender, age, sexuality, queer identities, disability status, Indigeneity, ethnicity . . . [and] other social factors, and [how the] intersections between these factors shape climate inequalities." Additionally, Jennifer Wenzel describes in chapter 9 how the concept of energy justice allows us to better understand the relationship of energy poverty to harmful practices of extraction. Thus, all three of these accounts underline the unequal distributions of climate harm on both a global and a more local level.

Adding to these observations, Róisín Commane notes in chapter 5 how efforts to achieve climate adaptation and mitigation by improving heating efficiencies can lead to environmental justice issues if these policies are not accompanied by efforts to improve ventilation within buildings to avoid toxic levels of unventilated indoor air. Similarly, the planting of trees to absorb CO_2 to ease the urban heat islands effect can also lead to the emission of organic compounds that combine with the pollution emissions from traffic to produce ozone and photochemical smog, thus worsening air quality in the short term and potentially reinforcing existing local environmental injustices. These observations, more generally, underline the need to consider unintended and unforeseen consequences in climate mitigation or climate adaptation policies to avoid reinforcing existing or creating new environmental injustices and displacing the costs of climate policies on historically marginalized groups. Illustrating this aspect in his discussion of the New York–New Jersey Harbor and Tributaries Study (HATS) in chapter 15, Paul Gallay demonstrates how "many of the available solutions to flooding may work at cross-purposes with other important community goals like maintaining waterfront access . . . and safeguarding natural systems and biodiversity."

These observations also highlight that while there exists significant overlap between environmental and climate justice, the trade-offs between the two concepts need to be considered in any political response to climate change. Figure 16.1 lays out some of the overlaps and differences that we will discuss further in subsequent sections. Additionally, climate justice is often framed in terms of climate mitigation versus climate adaptation, which raises distinct questions of historical responsibilities and the adequate scale of analysis, which we discuss next.

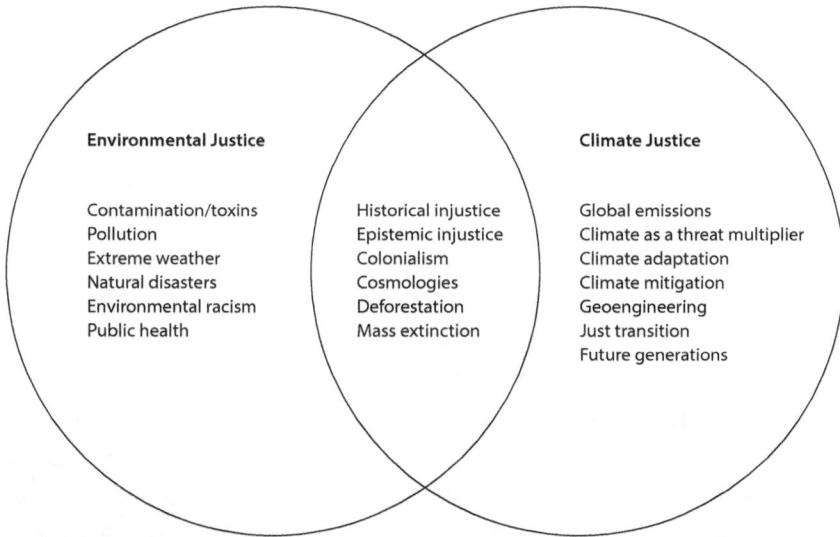

Environmental Justice

Contamination/toxins
Pollution
Extreme weather
Natural disasters
Environmental racism
Public health

Historical injustice
Epistemic injustice
Colonialism
Cosmologies
Deforestation
Mass extinction

Climate Justice

Global emissions
Climate as a threat multiplier
Climate adaptation
Climate mitigation
Geoengineering
Just transition
Future generations

FIG. 16.1

Examples of themes across environmental and climate justice. This figure describes some themes that are often associated with either environmental or climate justice, or both. While environmental justice debates often refer to issues such as contamination and pollution, climate justice debates often focus on the global scale of GHG emissions. Some overlap exists in considerations of historical injustices as a driver of both types of injustices, such as deforestation and mass extinction, which trigger environmental injustice and are exacerbated by climate change. Importantly, the graph does not provide a comprehensive overview but aims to illustrate the differences and overlaps between both concepts.

Just Responses: Mitigation and Adaptation and Navigating Trade-Offs

In addition to raising concerns regarding the just distribution of the costs of climate change, debates around climate justice often include discussions around mitigation and adaptation and the need for just compensation. Implicit in these debates is a concern for the temporal effect of the climate crisis—what are the trade-offs between focusing on climate mitigation as opposed to climate adaptation measures *now*? What is the time horizon of either of these approaches? Who bears the short- and long-term costs of energy transitions?

Some considerations that play into the trade-offs are real or perceived budgetary constraints as well as considerations of timely action on part of economic, civil society, and political actors. While we will consider the role of historical injustices and the legacies of colonialism, we note that just responses to climate change require compensating and protecting those countries and populations who are already experiencing the devastating effects of climate change and have historically contributed the least to climate change, as well as compensating groups that will face economic and social harms from decarbonization efforts.[5] Discussing the continued urgency of climate mitigation efforts, for example, Zara Riaz and Page Fortna describe in chapter 3 how the consequences of failing to act on climate change will have disproportionately higher costs for poorer countries. They demonstrate how these countries have contributed significantly less to global emissions but are already bearing the brunt of the impact of climate change, such as suffering from rising floodwaters and heat waves. Making matters worse, those countries and populations that could achieve the biggest reduction in global GHG emissions, such as the United States, Germany, or China, are also the countries profiting the most from the current economic system and are therefore likely to prefer climate adaptation measures if climate mitigation efforts are seen as too costly.

On the other hand, illustrating the need for climate adaptation measures *now*, Lewis Ziska and his team show in chapter 8 that climate change serves as a threat multiplier, which exacerbates the precarity of agricultural workers who are already facing threats to their health and safety due to their immigration and socioeconomic status. Additionally, Riaz and Fortna introduce the example of Saint Louis, Senegal, the most vulnerable city to climate change in Africa, to illustrate not only how mitigation costs are unfairly distributed globally but also how adaptation efforts can create multiple sets of "losers," such as ethnic groups primarily dependent on fishing for their subsistence that are displaced to inland camps. Similarly, Gallay describes the need to include frontline communities in the planning processes of nature-based protections against climate-related storm surges. Focusing on HATS, Gallay illustrates how these climate adaptation measures need to be not only functional but also sensitive to community needs for job creation and community revitalization. Thus, he argues that HATS "must strive to find the best balance between planning well and planning quickly." Together, such observations emphasize the need for inclusive and actionable climate adaptation measures for communities that already experience climate change as an *existential* crisis.

Implementing mitigation efforts now can help to avoid higher climate adaptation costs in the future. However, adaptation efforts will still be required even with extensive climate mitigation today. Additionally, both of these approaches raise questions of distributive, recognition, and participatory justice, which includes considerations of social status, culture, class, geography, gender, race, and Indigeneity in both climate adaptation and climate mitigation goals. Instead of framing these considerations as zero-sum trade-offs, it is useful to recognize that climate mitigation and adaptation efforts can overlap and reinforce each other, but also that to achieve a just response or transition in either case, community involvement of those affected needs to be centered, and the historical legacies of colonialism need to be considered.

Historical Injustices: Colonialism, Capitalism, and Unjust Power Dynamics

A central complicating factor in addressing climate change justly is that we need to not only rapidly decrease current emissions but also account for legacy emissions that will continue to contribute to global warming even if all emissions were stopped by tomorrow. Both the need for drastic action to achieve rapid decarbonization and the impact of legacy emissions raise questions of historical injustice and the role that colonial structures continue to play in determining who is a major contributor to climate change, who is poised to suffer the most, and who has the strongest capacities (or lack thereof) to respond to climate change. Considering overall GHG emission trends, for example, we can observe that since the beginning of the Industrial Revolution, annual GHG emissions have steadily increased, with a vast majority of these being emitted in the Global North (see figs. 16.2 and 16.3).

Solely focusing on the United States and the European Union, we find that together these jurisdictions contribute almost one-third of global GHG emissions, with emissions from China rapidly growing over the past three decades, which raises questions of what both individual citizens and states in the Global North owe to those most affected by the impact of climate change (see fig. 16.2).

Stressing these interdependencies, Jennifer Wenzel discusses in chapter 9 how a literary approach and an analysis of Amitav Ghosh's book *The Great Derangement* allow us to understand that this twin development was not coincidental. On the contrary, colonialism, empire, and current climate injustices are fundamentally interrelated. As Wenzel argues: "Ghosh not only shows how modern Europe's military and economic dominance was

Combined-Global CO2 Emissions & CO2 Concentrations (1751-2023)

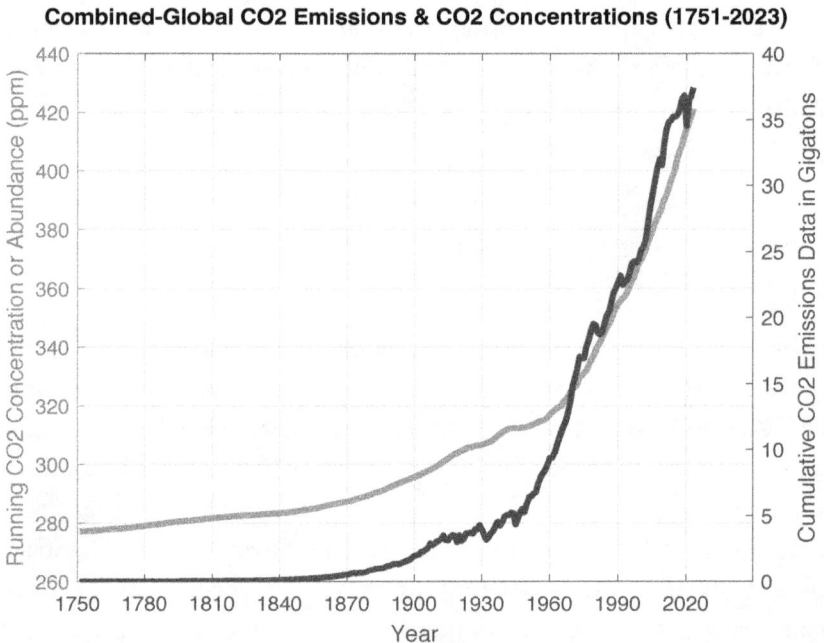

FIG. 16.2

Rising combined global CO_2 emissions and CO_2 concentrations from 1751 to 2023, demonstrating a strong increase in emissions concurrently with the onset of industrialization. Emissions have continued to rise steadily despite the introduction of climate mitigation measures since the 1990s. (NOAA Climate.gov, public domain, 2024).

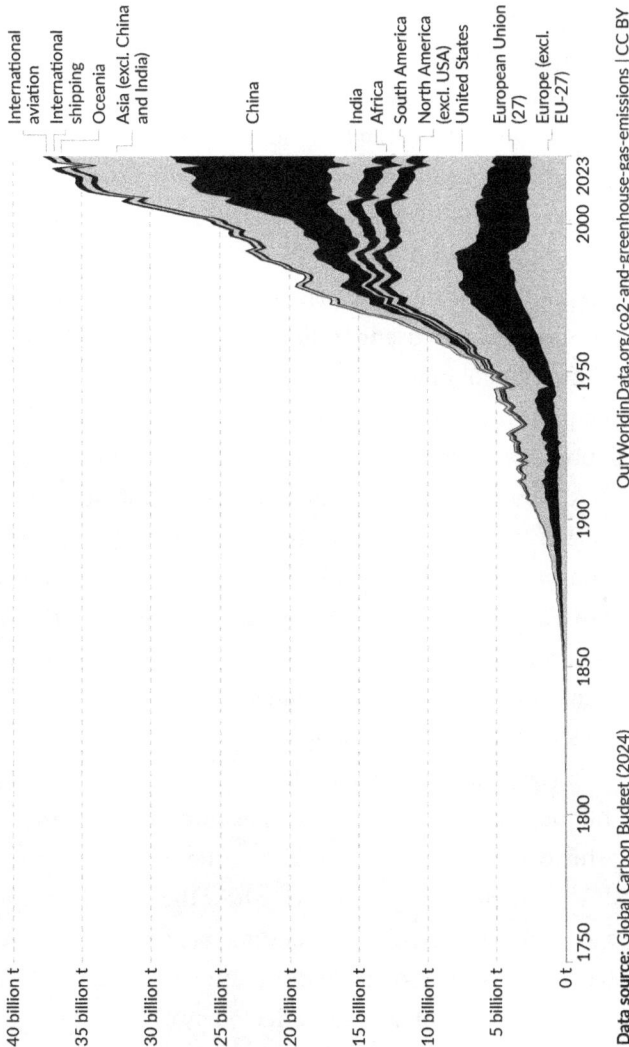

Annual CO₂ emissions by world region

Emissions from fossil fuels and industry[1] are included, but not land-use change emissions. International aviation and shipping are included as separate entities, as they are not included in any country's emissions.

Our World in Data

International aviation
International shipping
Oceania
Asia (excl. China and India)

China

India
Africa
South America
North America (excl. USA)
United States

European Union (27)

Europe (excl. EU-27)

40 billion t

35 billion t

30 billion t

25 billion t

20 billion t

15 billion t

10 billion t

5 billion t

0 t

1750 1800 1850 1900 1950 2000 2023

Data source: Global Carbon Budget (2024)

OurWorldinData.org/co2-and-greenhouse-gas-emissions | CC BY

1. Fossil emissions: Fossil emissions measure the quantity of carbon dioxide (CO₂) emitted from the burning of fossil fuels, and directly from industrial processes such as cement and steel production. Fossil CO₂ includes emissions from coal, oil, gas, flaring, cement, steel, and other industrial processes. Fossil emissions do not include land use change, deforestation, soils, or vegetation.

FIG. 16.3

Annual CO₂ emissions by world region since 1750. Total emissions have surpassed 35 billion tons, with the European Union, United States, China, and Asia (excluding China) constituting the lion's share of total global emissions. (Our World in Data, Creative Commons BY license, 2024.)

enabled by the combustion of fossil fuels; he also highlights how this dominance required the active suppression of a carbon economy in the colonial periphery." This further complicates the notion of economic growth being independent from historical relationships of oppression. Drawing on Livingston, Wenzel instead asserts that "the capacity to understand out-of-control growth as anything other than a historical construct, not something necessary and natural like the air we breathe . . . is the product of a habitual disregard that is mostly a privilege of the rich and powerful."

Similarly, Klopp and Boateng underline how, especially in the Global South, the failure of colonial governments to invest in public transport is today made up for by foreign investments in car-centered infrastructures that ignore preexisting modes of transport such as motorcycle taxis and rickshaws and therefore reinforce unjust embedded structures that ignore the needs of the populations they are meant to serve. Additionally, in chapter 10, Emma Gilheany and Julia Lajus discuss how colonial legacies continue to disrupt the livelihoods of Indigenous communities in the Arctic. For example, the Iñupiat community of Shishmaref, Alaska, continues to suffer from infrastructural dependency as a result of colonization, which has made the use of prior adaptive strategies to the Arctic climate, such as rotating settlements, impossible.

These interdependencies raise the question of how and by whom continuing historical injustices should best be addressed. In chapter 6, Givens and Ahmed, for example, describe how the establishment of a Loss and Damage Fund at the United Nations Framework Convention on Climate Change has aimed to tackle international inequities by providing financial assistance to countries most affected by climate change. Drawing on Falzon and Batur, they note that this mechanism could be used to hold countries in the Global North accountable for their "destructive pasts" and historical contributions to climate change. Nonetheless, political science research has shown that these pledges remain woefully unfulfilled.[6]

Considering both the historical legacies of colonialism and the ambivalent role of the state in having contributed to colonialism and the current climate crisis, as well as the possibility of holding states accountable through international agreements, this tension raises the question of the most appropriate scale of analysis through which to assess and respond to the climate crisis.

Multiscalar Justice: Global, National, Local, and Individual

Addressing debates surrounding the right scale to assess both the origins of climate change and its unequal impacts, the different scales of global, national, local, and individual considerations highlight the entanglement of spatial, structural, and agential dimensions. Some questions were echoed throughout the different chapters in this book: Considering the spatial dimension, what are the trade-offs between a global or local lens of analysis? Does focusing on global connections, for example, overlook local efforts of mitigation or adaptation? Can a structural analysis tie together some of these scales? And finally, what role do individual agents or state actors play in all these responses?

Conceptualizing this overlap of the global and the local, Wenzel demonstrates how transnational forces influence localities, such as in the Niger Delta, North Dakota, and the Mississippi Delta, which are all "profoundly but disparately shaped by (and indispensable to) oil extraction." Wenzel argues that we therefore need a multiscale approach that sees "one place always as juxtaposed and imbricated with another" and exposes how wealth in one place generates poverty and injustice in another. Similarly illustrating this multiscalarity, Alex de Sherbinin and his team note in chapter 7 how climate effects on the local level interact with global migration patterns. In this context, they discuss the passport index, to highlight that passport freedom negatively correlates with climate-caused mobility. This means that those who are most affected by the climate crisis also face the highest institutional barriers to migration.

Recognizing the multiscalarity of the problem of climate change thus requires being attentive to the interlocking causal factors between the different levels of analysis, including the role the state and subnational political actors play in responding to both climate change and its environmental and socioeconomic effects. Presenting a critical analysis of the role governments can play in responding to irregular weather patterns, Long, for example, remarks that "government ecological initiatives are frequently associated with gentrification" and often entrench existing social inequalities. Further examining the role of the state, Riaz and Fortna discuss how even though the state can be a powerful agent to remedy climate injustices through Loss and Damage Funds, for example, climate policies risk displacing costs associated with decarbonization and climate mitigation on poorer countries that are already bearing the brunt of the climatic impacts of GHG emissions. To remedy these shortcomings, Givens and Ahmed draw on sociologist Rikard Warlenius to introduce the concept of *climate debt,* which reflects a restorative climate justice approach that underlines the need to return or decolonize that which has been taken by climate change. Similarly, the concept of a "just transition" underlines the idea that a movement toward a low-carbon economy and climate resiliency also requires a system that is equitable and inclusive and redresses the harms borne by communities on the front lines of decarbonization (for example, compensation for lost livelihoods for individuals and communities in coal country when coal mines and plants are to be shuttered).

Approaching this question from a very different angle, Sobel and Bieli discuss how climate scientists can choose an impactful career, as well as, more generally, what role the market, and oneself as part of it, can play in addressing climate change. The authors make a case that rather than sidelining markets, it is important to engage with them and consider how industries such as the insurance industry can contribute to change by working with financial markets rather than against them. Personal career choices can indeed make important differences, yet it is equally important to work together to change the structural underpinnings of climate injustice. As Klopp and Boateng highlight, the global system of financial capitalism contributes to the emergence of unjust infrastructures that further entrench existing injustices and multiply the harm of climate change impacts. As an example, they discuss how the emergence of highways and personalized transportation systems has displaced popular forms of transport and contributed to a host of environmental and public health problems.

These observations highlight that we need a multiscalar approach, not only to identify the causes and consequences of climate change but also to consider different responsibilities and obligations in climate responses at the individual, local, national, and global levels. However, such an approach also raises questions of the commensurability of the concept of climate justice across governmental, cultural, social, and economic divides. Drawing on philosopher Christine Winter, Witlacil highlights that "meaningfully including Indigenous people in decision-making—when their philosophies and worldviews are incompatible with those in dominant culture and settler colonial states—would require cultural humility and a willingness of settler states to accommodate a plurality of worldviews."[7] Similarly, Long draws on Heather Davis and Zoe Todd to argue that to escape the "pervasive logic of colonialism," governments need to "listen to the voices of those with rich traditions who understand how to coexist with the Earth." This consideration highlights the need to include different ontologies and epistemologies within a climate justice–centered framework—a concern we consider next.

Cosmological Diversity and the Production of Knowledge

Considering the range of perspectives covered in this volume, an unexpected but common point of convergence across different chapters was a focus on how different worldviews and the process of knowledge production shape the way climate injustices are described and addressed. The chapters grappled with the question of how we can avoid reinforcing neocolonial discourses and introduce different worldviews while also recognizing the role scientific expertise can play in empowering local communities to demand stronger action on climate change, as well as compensation for the harms they are already experiencing. In dealing with this question, the chapters underscored the necessity for a mode of engagement that is marked by critical pluralism, which describes both the need for the acknowledgment of difference and an "open, participatory, and intersubjective process of communication" to achieve a common understanding across these differences.[8]

Commenting on epistemological and ontological plurality, for example, Long discusses alternative theories and the absence of scientific language among Hakka-speaking pomelo farmers in southeastern China. Long emphasizes that these farmers relied on existing terminology, such as "Heavenly Years," to describe weather and climate anomalies, while they also updated their vocabulary to interact with state agencies to demand agricultural insurance compensation. Thus, Long suggests that while productive knowledge exchange is important to further research on the complex impacts of climate change, climate-specific knowledge and language can also be a key tool for marginalized groups to demand financial compensation and political recognition from their own state agencies. More generally, Long's observations highlight that an inclusive understanding of climate justice requires not only the inclusion of different worldviews or ontologies but also an awareness of whose stories remain untold in any compensatory process.

Further emphasizing the contribution of Indigenous perspectives, Gilheany and Lajus argue that climate justice requires paying attention to how environmental shifts are felt by people as well as animals living in the circumpolar North. They demonstrate that Inuit populations are already experiencing ecological vulnerability due to both the unpredictability of the weather and its impact on the relationality between human and more-than-human communities. Their observations emphasize the need to open the lens of climate justice to include considerations of interspecies justice and recognize the role colonization has played both in erasing Indigenous presence from the Arctic and in reinforcing an anthropocentric focus of analysis.

Similarly, Raffaella Taylor-Seymour and Courtney Bender highlight in chapter 14 that scholars need to "engage critically with the limits of universalist philosophies rooted in Western traditions" and instead call for a pluriversal understanding of climate justice, which includes "religious 'stuff in the world.'" The authors underline that religions often center the entwinement of humans with other species and the transcendence of material with spiritual concerns and are therefore an example of "radical cosmological plurality." Thus, religions, including Indigenous and animist frameworks, can offer an important resource for new political imaginaries that decenter the human and take into account "ontological diversity [and] relational complexity."

Echoing this aspect and building on philosopher Christine Winter, Witlacil advocates for a relational approach that recognizes that human beings have "obligations and duties . . . to all living beings . . . and nonliving entities," and that the "more-than-human world . . . should be considered subjects of justice." Translating this philosophy into action would require transforming extractive relationships of humans (especially in the West) to the land and actively including guardians of environmental entities in climate policy negotiations.

Together, these chapters discuss the dimensions of cosmological, ontological, and epistemological justice, raising questions of structural injustices and whose agency counts and ought to count in climate justice narratives. In doing so, the chapters underline the importance of opening discourses of climate justice to different worldviews and embracing the collision of these worldviews to foster new prospects for cooperation.

Additionally, these observations underline the need for community-oriented research. We believe it is here that synergies between the different disciplines can be particularly fruitful, given the social sciences' focus on methodologies that engage with communities on the ground in a variety of ways. However, research collaborations with frontline communities also need to be conscious of the costs they impose on these communities and avoid extractive relationships. As Gallay demonstrates in chapter 15 on the New York case study, frontline communities should therefore be involved from the beginning of any major study that will affect their very livelihoods and prospects, and the benefits of such scholarly–community collaborations should clearly flow to those who will be most affected by the research.

A CALL TO ACTION!

The chapters in this volume illustrate the imperative of thinking and communicating across conceptual and disciplinary divides and making processes of knowledge production on climate change inclusive and fair. One of the key insights we gained by editing this book is that researching climate justice–related themes requires us to be aware of and alert to the structural background conditions that cause injustices. Yet we also need to recognize our agency and ability to change such injustices. Focusing on academia, this means pursuing a critical introspection on how existing academic institutions not only prevent cross-disciplinary engagement but also fail to center justice concerns within their own organizational and academic structures.

As Kailani Acosta and Gisela Winckler describe in chapter 11, the lack of representation of historically marginalized communities in the STEM field, including of Black and Indigenous researchers, raises the question of how we can pursue research on topics of justice without also addressing the impacts of systemic racism within our home institutions. Conversely, if we manage to center justice claims within academia, by, for example, increasing the pipeline of underrepresented researchers and giving them a stronger voice, as well as transforming institutional and funding structures to incentivize cross-disciplinary and community-oriented research, academia is more likely to unlock its full potential and provide the foundational research that is necessary to address the climate crisis. This means that climate justice research, pedagogy, and social impact can only be meaningfully pursued if they go hand-in-hand with reforming institutional structures within academia.

A concrete obstacle to cross-disciplinary research is current funding structures that often favor monodisciplinary and natural science–focused research proposals. As one of our contributors pointed out, merely calling for more interdisciplinary engagement is unlikely to yield results if current publication rankings tend to penalize interdisciplinary scholarship. This is particularly the case since, for example, tenure assessments reward publications in the highest-ranking disciplinary journals. Funding structures for new research projects in the sciences often also require monodisciplinary or natural science–focused applications rather than truly cross-disciplinary collaborations. Instead of conceptualizing these as inherent trade-offs, in which funding is shifted from the natural sciences to the social sciences and humanities, we need to promote cross-disciplinary research in the same way that we reward monodisciplinary research to tackle a problem so global and complex as the climate crisis.

Cross-disciplinary synergies will also translate into better learning experiences for students. We already see increased demands for interdisciplinary courses on climate-related topics and the creation of climate-focused schools. Sadly, with the further deterioration of climatic conditions, the research and teaching contributions of these institutions will only become more valuable. If we want our students to understand the complexity of the climate crisis and its natural and social impacts, we also need to provide the pedagogical structures that facilitate teaching these. Thus, we hope that more institutions will follow suit and introduce cross-disciplinary programs that incentivize curricular collaborations across the disciplinary divides. Such programs are not only more equipped to teach climate-related topics but also better prepared to train students for working in the climate sector after graduation.

Academia cannot solve the climate crisis on its own. Throughout the book, contributors have noted how grassroots movements and Indigenous communities have played a key role in both raising attention to the immediacy and unjustly distributed impacts of climate change and demanding political action to prevent further climate destruction. However, academia could potentially play a much more positive role in advancing the foundational research and the multidisciplinary scholarship that could help humanity understand the intricacies of the climate crisis and thus support these grassroots struggles in positive and meaningful ways.

We strongly believe that academia can, must, and will play an important role by advancing cross-disciplinary and justice-focused research and education. This will fortify climate-focused research in the humanities, social sciences, and natural sciences against attacks on academic freedom, which is central to the very obligation and purpose we have as academic researchers. What we need is an all-hands-on-deck approach that moves beyond single disciplines and disciplinary thinking and instead recognizes that researching climate change and centering climate justice will require reforming existing academic structures to reward innovative cross-disciplinarity. This, we believe, is imperative not only for climate justice but also for achieving justice more broadly.

Let's work together to achieve this goal!

Notes

1. Krause, *Eco-Emancipation*, 1–2.
2. King, Gaffney, and Rivera, *Preliminary US Greenhouse Gas Emissions Estimates*. In the present volume, chapter 15, on flood protection, and chapter 13, on anthropology, also discuss the problem of expertise imperialism.
3. See also Taylor-Seymour and Bender's chapter 14 in this volume.
4. Schlosberg, *Environmental Justice and the New Pluralism*, 17.

5. Gaikwad et al., "Creating Climate Coalitions."
6. Gaikwad et al., "Climate Action from Abroad."
7. Winter, *Subjects of Intergenerational Justice*.
8. Schlosberg, *New Pluralism*, 17.

BIBLIOGRAPHY

Gaikwad, Nikhar, Federica Genovese, and Dustin Tingley. "Climate Action from Abroad: Assessing Mass Support for Cross-Border Climate Transfers." *International Organization*. 79, no. 1 (2025): 146–72.

Gaikwad, Nikhar, Federica Genovese, and Dustin Tingley. "Creating Climate Coalitions: Mass Preferences for Compensating Vulnerability in the World's Two Largest Democracies." *American Political Science Review*, 116, no. 4 (2022): 1165–83.

King, Ben, Michael Gaffney, and Alfredo Rivera. *Preliminary US Greenhouse Gas Emissions Estimates for 2023*. New York: Rhodium Group, 2023.

Krause, Sharon. *Eco-Emancipation: An Earthly Politics of Freedom*. Princeton: Princeton University Press, 2023.

Schlosberg, David. *Environmental Justice and the New Pluralism—the Challenge of Difference for Environmentalism*. Oxford: Oxford University Press, 1999.

Winter, Christine J. *Subjects of Intergenerational Justice—Indigenous Philosophy, the Environment and Relationships*. London: Routledge, 2022.

Contributors

KAILANI ACOSTA received her PhD degree in earth and environmental sciences in 2024 from Columbia University Lamont-Doherty Earth Observatory, focusing on biological oceanography in the surface ocean of the Gulf of Mexico. She is currently a Science Program associate at OceanX, working to combine research, science communication, and marine education both onshore and at sea. She is interested in understanding how nutrients vary and how they influence phytoplankton communities and larger-scale ecosystem nutrient cycles. Her research interests include nutrient cycling, microbial ecology, and ecosystem modeling. She is passionate about increasing diversity, equity, and inclusion in STEM.

SUSANA B. ADAMO is a research scientist at the Center for Integrated Earth System Information (CIESIN) of the Columbia Climate School, adjunct associate professor of climate in the Undergraduate Program in Sustainable Development and the Climate and Society Master Program in the Climate School, and coordinator of the Population-Environment Research Network (PERN). Among other projects, she has worked on the development of gridded population databases, vulnerability, climate variability, and climate migration in Latin America; migration, cities, and climate change; and air quality and environmental justice.

MUFTI NADIMUL QUAMAR AHMED is a PhD student in sociology at Utah State University. His research interests include climate change perception, social impacts of climate change and natural disasters and coping and adaptation strategies, environmental dimensions of fertility, and migration. He is from Bangladesh and has completed his bachelor's and master's degrees, majoring in sociology, from Shahjalal University of Science and Technology, Sylhet, Bangladesh. His PhD dissertation focuses on the sociodemographic impacts of climate change and adaptation strategies.

COURTNEY BENDER is Ada Byron Bampton Tremaine Professor of Religion at Columbia University. A sociologist and ethnographer by training, with a PhD degree from Princeton University in 1997, she focuses principally on the production and practice of American religion in the twentieth and twenty-first centuries.

MELANIE BIELI received her PhD in applied mathematics from Columbia University and did a postdoc at Caltech. She worked as a natural catastrophe modeler in the reinsurance industry before moving into quantitative finance and is now a weather analyst at Jane Street. Her academic research focused on tropical cyclones and temperature extremes.

FESTIVAL GODWIN BOATENG is a senior researcher in mobility governance at the Transport Studies Unit (TSU) of the University of Oxford. He has an interest in the everyday mobilities of people, goods, and information within and between cities. His research is particularly focused on the historical, cultural, and institutional political economy dimensions of mobility in cities in the Global South. He is also part of the Partnership for Research on Informal and Shared Mobility (PRISM) Consortium exploring pathways to establish "Living Labs" for transport reforms in Global South Cities.

RÓISÍN COMMANE is an associate professor in the Department of Earth and Environmental Sciences at the Lamont-Doherty Earth Observatory of Columbia University, where her group investigates the sources and sinks of gases in the atmosphere by combining the fields of atmospheric chemistry, composition and transport, terrestrial ecology, and human-related pollution sources. She has a PhD degree in atmospheric chemistry from University of Leeds and was a research associate at Harvard University from 2009 to 2018. She has conducted field research on greenhouse gases and air quality on all continents except Antarctica and currently focuses her research on urban pollution in New York City.

ALEXANDER DE SHERBININ is the director and a senior research scientist at the Center for Integrated Earth System Information (CIESIN), a spatial data and analysis center within the Columbia Climate School specializing in the human aspects of global environmental change. His research interests focus on the human aspects of global environmental change and geospatial data applications, integration, and dissemination. His research and teaching address climate-related mobility; climate vulnerability mapping; urban climate vulnerability; population dynamics and the environment; and environmental sustainability indicators.

PAGE FORTNA is the Harold Brown Professor of US Foreign and Security Policy in the Political Science Department at Columbia University. Her research focuses on terrorism, the durability of peace in the aftermath of both civil and interstate wars, war termination, and, increasingly, the international politics of climate change, including the impact of climate change on power in international relations.

SHEILA FOSTER is a tenured professor of climate at the Columbia Climate School and affiliated professor at Columbia Law School. Previously, she was the Scott K. Ginsberg Professor of Urban Law and Policy and professor of public policy at Georgetown University. She is a leading scholar of environmental and climate justice. Her research spans a broad range of topics, including innovative resource governance regimes, land use policy, and the role of subnational governments and local leaders in addressing cross-border challenges, such as climate change.

NIKHAR GAIKWAD is an assistant professor of political science and a member of the Committee on Global Thought at Columbia University.

PAUL GALLAY is the director of the Columbia Center for Sustainable Urban Development's Resilient Coastal Communities Program, where he and his colleagues seek to foster actionable, equitable solutions to flood risks along with complementary benefits like habitat restoration, job creation, and more empowered communities. Prior to joining Columbia and establishing the Resilient Coastal Communities Program in 2021, he served in legal and policy leadership positions with the New York State Attorney General and Department of Environmental Conservation, Hudson Riverkeeper, and land conservation organizations in New York and Maine.

ELENA GIACOMELLI is a tenure track researcher (RTT) at the University of Bologna, where she conducts her research as part of the Marie Skłodowska-Curie Postdoctoral Global Fellowship project "PANICOCENE. Reframing Climate Change-Induced Mobilities," which aims to investigate and analyze the ways in which media discourses and narratives can lead to stereotypes of climate-induced migration, thereby influencing the perceptions of readers, general public, and ultimately decision-making, and to propose a space for collaboration among activists, artivists, journalists, and academics on the issue.

EMMA GILHEANY is a William Lyon Mackenzie King Postdoctoral Fellow in the Canada Program at Harvard's Weatherhead Center for International Affairs and a Research Affiliate of the Harvard University Native American Program. She is an environmental anthropologist and archaeologist of the contemporary who examines how imperialism and settler colonialism are refused, resisted, and circumvented in the circumpolar North by Nunatsiavummiut. Her current research foregrounds Inuit self-determination at the intersection of Cold War radar bases and climate change.

JENNIFER GIVENS is an associate professor at Utah State University and an environmental and comparative international sociologist. Broadly, she studies coupled human and natural systems. Her research examines environmental and social sustainability across nation-states, and she studies how these relationships change over time. In some of her research she investigates variation in countries' carbon intensity of well-being, which is a way to measure a country's progress toward simultaneous environmental and social sustainability by asking how carbon intensely nation-states produce well-being for residents.

JACQUELINE M. KLOPP is a research scholar at the Columbia Climate School, where she also directs the Center for Sustainable Urban Development. She is an interdisciplinary social scientist with a focus on urban planning, climate change, air pollution, and governance, all with a strong justice lens. She is particularly interested in better understanding the transport services operated and used by underserved communities and how to work in an engaged way to improve these services, access, and equity. She is part of a research consortium looking at equity and transport across the Global South called the Partnership for Research on Informal and Shared Mobility (PRISM).

JULIA LAJUS is a researcher at the University of Helsinki, Finland. In spring 2025 she was a Senior Smithsonian Fellow at the National Museum of Natural History in Washington DC, and in fall 2024 a Senior Fellow at the Institute of Advanced Study in Amsterdam. In 2023 she served as a visiting associate professor at Columbia University, where she taught courses on the history of the Arctic and the history of climate science. Before 2022 she was the head of the Laboratory for Environmental and Technological History and an associate professor at the National Research University Higher School of Economics in St. Petersburg, Russia. From 2011 to 2015 she served as a vice president of the European Society for Environmental History (ESEH). She has published widely on the history of marine and polar environments, related sciences, and resource use, especially fisheries.

SHENG LONG is a postdoctoral research scholar in the Department of Anthropology at Columbia University. Her work focuses on environmental justice in the age of big data: how people quantify farmland, vegetation, and climate, in ways that resonate with urban-rural development disparities, social stratification conflicts, and gender inequalities in statistical engagement. Her project "Numbering Land: Ethical Measures of Geography and Subjectivity in Agrarian Reforms" is an ethnography of environmental data in national reforms and everyday agriculture.

REBECCA MARWEGE is an assistant professor of environmental politics at the American University of Paris.

SARA PAN-ALGARRA is a Venezuelan researcher focusing on the relationship between climate mobility, disasters, and girls' education in the Honduran Sula Valley. She has a multidisciplinary background combining migration studies, international education, policymaking, and human rights. She has conducted research both on climate mobility and on the Venezuelan migration crisis and is currently a Doctoral Fellow at Teachers College, Columbia University.

ZARA RIAZ is a postdoctoral fellow at the London School of Economics. Her research primarily focuses on migration and political behavior in Western Africa and Western Europe.

JOERG SCHAEFER is Lamont Research Professor in the Lamont-Doherty Earth Observatory and adjunct professor of earth and environmental sciences at Columbia University.

ADAM H. SOBEL is a professor of applied physics and applied mathematics and of Earth and environmental sciences at Columbia University. He does basic research on atmospheric and climate dynamics, especially in the tropics, and applied research on the risks to human society from extreme weather events and climate change. He also writes on climate and related issues in the mainstream media, produces a podcast, and collaborates with social scientists and humanists to think critically about the role of climate science in society.

JEFFREY L. SHAMAN is professor of environmental health sciences at the Columbia University Mailman School of Public Health and professor of climate at the Columbia Climate School. He studies the survival, transmission, and ecology of infectious agents, including the effects of meteorological and hydrological conditions on these processes. His work to date has primarily focused on mosquito-borne and respiratory pathogens. He uses mathematical and statistical models to describe, understand, and forecast the transmission dynamics of these disease systems, and to investigate the broader effects of climate and weather on human health.

RAFFAELLA TAYLOR-SEYMOUR is an anthropologist of religion and an assistant professor of religious studies at Columbia University. She received a PhD degree from the University of Chicago in 2022. Her work examines religious transformations over the past two centuries in the context of struggles over gender, sexuality, and the environment in Zimbabwe.

EMILY WEAVER is a staff research associate in the Department of Environmental Health Sciences at the Columbia University Mailman School of Public Health, where she studies the intersection of climate justice and reproductive health.

JENNIFER WENZEL is jointly appointed as a professor in the Department of English and Comparative Literature and the Department of Middle Eastern, South Asian, and African Studies at Columbia University, where she teaches courses on postcolonial literature and theory and environmental and energy humanities. Her most recent monograph, The Disposition of Nature: Environmental Crisis and World Literature, was published by Fordham in 2020; with Imre Szeman and Patricia Yaeger, she co-edited Fueling Culture: 101 Words for Energy and Environment (2017). She is a member of the Petrocultures Research Group and the After Oil Collective. Her current research examines the fossil-fueled imagination in literature, visual culture, and public life.

GISELA WINCKLER is a professor of climate at the Columbia Climate School. She was trained as an oceanographer and environmental physicist and holds a (PhD degree from Heidelberg University. Her research focuses on the history and the causes of climate variability in the past, present, and future. Using climate archives such as deep-sea sediments and polar ice cores, she investigates the interplay of climate change, the carbon cycle, and aerosols. She teaches about the climate system across a wide range of audiences and has been a faculty facilitator for a Seminar on Race, Climate Change, and Environmental Justice. She is interested in connecting science, art, journalism, design, and climate justice pathways toward climate action.

MARY WITLACIL is an assistant professor of political science in the Department of the Humanities, Arts, and Social Sciences at South Dakota School of Mines and Technology. She is an environmental political theorist who draws on contemporary political thought, the Frankfurt School, and critical theory. Her research involves the relationship between climate change, progress, affect, and pessimism to consider how we cope with the climate crisis, climate injustice, and environmental injustice.

DAVID WRATHALL is an associate professor of Natural Hazards at Oregon State University's College of Earth, Ocean and Atmospheric Sciences, and a lead author on the Intergovernmental Panel on Climate Change (IPCC) AR6 on chapter 8: Poverty, Livelihoods, and Sustainable Development. He also coordinated the group of IPCC authors contributing to the cross-chapter treatment of migration, displacement, and mobility for the Sixth Assessment. He studies the ways that climate change impacts are threatening people's livelihoods and rendering parts of the planet uninhabitable, and thus driving new patterns of human migration.

LEWIS H. ZISKA is an associate professor in the environmental health sciences at the Mailman School of Public Health at Columbia University. He also serves as the climate and health certificate lead. He has worked primarily on documenting the impact of climate change and rising carbon dioxide levels on crop selection for CO_2 responsiveness to improve production; climate and agronomic pests, including chemical management; and climate, plant biology, and public health impacts on food security, with a focus on nutrition and pesticide use.

AMI ZOTA is associate professor of environmental health sciences and Community Engagement Core codirector at the Columbia Center for Environmental Health and Justice in Northern Manhattan. Her research focuses on understanding social and structural determinants of environmental exposures and their consequent impacts on women's health outcomes across the life course. She was among the first to frame the disproportionate burden of toxic chemical exposures from beauty and personal care products among women of color as an environmental justice concern. She codeveloped an intersectional framework called "the environmental injustice of beauty," which links systems of power and oppression, such as racism, sexism, and classism, to Eurocentric beauty norms, racialized beauty practices, and adverse environmental health outcomes.

Index

abyssal line, 155

ACA. *See* Affordable Care Act

academia, 3–6, 9–11, 22, 29–38, 256, 295, 318, 333–35; academic freedom and, 246–47, 335. *See also specific disciplines*

accountability, 20, 73, 259, 268, 290, 296–98, 307n44

Accra, Ghana, 69

acid rain, 111–12, 197

Acosta, Kailani, 26, 36–37, 233, 334

activism, climate, 3–4, 22, 27–28, 35, 39n18, 46, 141–42; in Copenhagen, 261–62; distributive justice and, 48; in the Niger Delta, 198; religion and, 272–74, 276, 283n1; of Thunberg, 24, 127, 131, 142

Adamo, Susana, 36

adaptation, climate, 109–10, 187, 213–14, 216, 303n14, 316, 321–22; abilities and, 67–68; albedo engineering as, 24; capabilities approach and, 57; China on, 259; costs of, 32, 325; insurance industry and, 139; mitigation and, 20, 31–32, 35, 48–49, 59, 242, 318, 323–25;

urban greening, 118; voluntary buyout programs as, 10

affluence, 3, 8, 136–37

Affordable Care Act (ACA), US, 186

Africa: Cameroon, 256–57; Egypt, 206; Ghana, 69, 162–63; Niger, 163; Niger Delta, 196–97, 199; Nigeria, 78, 198; Senegal, 324; South Africa, 76; South Sudan, 69

agential dimension, of climate justice, 23–27, 32–33, 155–56, 164, 245–47, 282, 320

Agreement to Prevent Unregulated High Sea Fisheries in the Arctic Ocean, 219

agriculture, 69–72, 115, 159–60, 162, 164–65; agricultural insurance, 262–63, 265–66, 331–32; Black Belt region, 26, 36–37, 226, 228–32, *229*, *230*; health of migrant workers in, 36, 167–68, 179–87, 324; plantations and, 159, 229–30, *230*; pomelo, 33–34, 37, 255–59, 262–68, 317–18, 331–32

Ahmed, Mufti Nadimul Quamar, 35, 69–70, 321–22, 328, 330

air conditioning, 68, 116
Air Force, US, 216
air pollution, 3–4, 90–91, 94–95, 97, 101, 111–20, 321
air quality, 4, 8, 109–20, 322
Alabama, 180, 229–31
Alaska, 114, 209, 216, 218, 328
albedo engineering, 24
American Anthropological Association, 255
American Clean Energy and Security Act, US, 74–75
American Rescue Plan, US, 75
ammonia, 115
Anglican Church, 276
Animal Liberation Front, 27–28
animal rights, 281
Anthropocene, 200, 209, 213, 264, 266
anthropocentrism, 46, 59, 317, 332
anthropogenic climate change, 6, 20–21, 26, 69, 181, 186
anthropology, 215, 217, 219, 255–56, 260–61, 264–65, 316–18; erasure and, 210–11; limits of, 208; political science and, 80; urban studies and, 92
anticolonial approaches, 28, 211
anti-immigration laws, 180
Anyamba, T., 96
Aotearoa (New Zealand), 55, 58
Arctic, 24, 208, 264; Indigenous communities in the, 36, 209–12, 214–17, 328, 332; oceanic nonhuman mobility in, 217–19
Arctic amplification, 209
Arizona, 53
Army Corps of Engineers, United States: EJCC pledged by, 290, 296–301, 303n13, 307nn43–44; HATS, 34, 37–38, 288–301, 301n3, 302n7, 303nn12–14, 304n21, 306n36, 306n38, 308n44, 308n46, 308n48, 322, 324–25
arson, 28
Ashton-Jones, Nick, 197

assimilation, 215
asthma, 4
asylum, 159, 169, 171n62
Atlantic Meridional Overturning Circulation (current), 12
atmosphere, 35, 57, 109–10, 118, 240
atmospheric science, 32, 35, 236, 316
autocentric urban mobility systems, 91
automobility, 90–91, 95–97, 100–102

Bacigalupi, Paolo, 203–205
Bali Action Plan, Conference of the Parties 13, 19–20
banana plantations, 159
Bangladesh, 69–70, 135, 203–4
Baton Rouge, Louisiana, 184
Batur, Pinar, 140, 328
Baucom, I., 166
Behe, C., 211–12
Bender, Courtney, 37, 81, 263, 332
Benjaminsen, Tor, 51
benzene, 117
Bering Sea, 218
Bering Strait, 217
Berlin, Germany, 69
Berlin West African Conference (1884–85), 171n53
"Between Mammon and Mother Earth" (Meikle, Wilson, Jafry), 39n18
Bhavnani, Kum-Kum, 29
Biden, Joseph R., 7, 289
Bieli, Melanie, 37, 238–39, 318, 330
biodiversity, 291, 299–300, 322
biofuels, 113
biology, 226–33
Black, Richard, 164
Black Belt region, US, 26, 36–37, 226, 228–32, 229, 230
black carbon, 91, 111
Black communities, 280–82; climate change impacting, 232–33; environmental racism impacting, 3–7, 226, 229–32; flood risks and, 71; pollution

impacting, 3–5, 94, 275–76; transport and, 92

Black Rights/White Wrongs (Mills), 49

blue economy, 218

Boas, Franz, 210–11

Boateng, Festival, 25–26, 35, 318, 321, 328, 330

Bolet, Diane, 76

Bolivia, 28

borders, 156, 159, 163, 170n42, 279–80, 283

Botswana, 201–2

Brandt line, climate vulnerability and, 69

Brazil, 28, 70–71, 171n53

Brazilian Rubber Tappers campaign, 28

"Bridging the Gulf" (exhibition), 233

Briggs, Charles, 259–60

Britain, 162, 199, 201, 203

Brock, Andrea, 39n26

Brooklyn, New York, 233, 291, 293, 305n24

Brulle, Robert J., 139

Bryant, Bunyan, 6

Bullard, Robert, 6, 129–30, 321

Bullard Center, 11

Burtynsky, Edward, 203–4

bus depots, 4

"business-as-usual," 36, 199–200; geoengineering and, 33, 201–2, 317

Cadena, Marisol de la, 255, 260–61

California, 180, 182–83, 255

Camden, New Jersey, 3, 6–7

Cameroon, 256–57

Canada, 69, 216; Indigenous communities in, 215, 261

Canary Islands, 166–67

Caney, Simon, 49

capabilities approach, 55–57

capitalism, 58, 211, 214, 237–38, 244–45, 248–49, 330; colonialism and, 36; extractive violence of, 29; fossil, 29, 199–203; global, 33, 89, 199; justice and, 58; racial, 8, 89; sociology on, 137, 140

carbon capture, 243

carbon dioxide (CO_2) emissions, 18, 57, 99, 109, *327*

carbon economy, 36, 139, 200–201, 326, 328, 330

carbon emissions, 81, 141, 316

carbon-free fuels, 114, 116

carbon intensity, 74–75, 77–78, 137; infrastructure and, 25–26, 93; of well-being, 134–35

carbon taxes, 49, 79

car-centered infrastructure, 25, 90–97, 100, 328, 330

carcinogens, 3, 98, 116–17

Caribbean, 171n53

cars, 90–96, 98, 111

"Case for Letting Anthropology Burn, The" (Jobson), 255

catalytic converters, 111, 113

Catholicism, 276

Celermajer, Danielle, 281

Census, US, 184, *230*

Center on Sustainable Investment, Columbia University, 244

Central America, 154, 156, 158–62, 167, 170n42, 180, 182, 186

Chakrabarty, Dipesh, 213, 278, 280–81

Chao, Sophie, 281, 282

Chavis, Benjamin, 5

Chester, Pennsylvania, 3–4, 6–7

Chester Residents Concerned for Quality of Life (CRCQL), 4

child labor, 161, 179–80, 183

China, 70–71, 211; GHG emissions by, 325–26, *326*, *327*; pomelo farming in, 33–34, 37, 255–59, 262–68, 317–18, 331–32

Chipko Andolan movement, 28

Chittagong, Bangladesh, 203–4

Christianity, 272, 274, 277

chronic kidney disease (CKD), 182

Ciplet, David, 139

circumpolar North. *See* Arctic

cities, 4, 32, 88–102, 114, 118–19, 245, 318; energy efficiency of buildings in, 115–17, 120. *See also specific cities*
citizenship, 158, 180, 279
civil rights, 7, 93, 231
civil wars, 159
CKD. *See* chronic kidney disease
Clark, Brett, 137
class, socioeconomic, 48, 49–50, 202–3, 324
Clean Air Act (1956), United Kingdom, 111–12
climate action, 11, 32, 67, 72–76, 78–79, 100, 239, 245
climate catastrophe, 66, 69–70, 75, 79, 136
climate change, 130–31, 140–41, 160, 162, 227, 237, 240, 315–18; agriculture and, 181–83; anthropogenic, 6, 20–21, 26, 69, 181, 186; Black communities impacted by, 231–32; capabilities approach and, 57; climate reductionism and, 208; costs of, 16, 71, 323; distributive problems of, 72–79; Late Cretaceous period, 26, 36–37, 226, 228, *229*; lived experiences of, 211; scientific language and, 256–57, 259–61; as a threat multiplier, 186, 318, 324; violence and, 70
climate crisis, 45–47, 82n18, 102, 275, 277, 318, 334–35; colonialism and, 14, 328; geoengineering and, 18; social drivers of, 126–38; urban, 90–92
climate debt, 132–33, 330
climate delay, 139
climate denial, 139, 201
climate disasters, 161, 232–33, 263–64
Climate & Environmental Justice Advisory Group, HATS, 306n36
climate ethnography, 211
climate finance, 99, 140
Climate Futures (Bhavnani), 29
climate inaction, 35, 67–68, 70–72, 78, 80, 139

climate inequalities, 130, 132, 142, 328; racial disparities and, 129, 321–22
climate injustice, 26, 36–37, 232, 237–38, 267; in the Arctic, 209, 217; colonialism and, 326; corrective justice and, 14; identities and, 81; misrecognition and, 51; political science and, 32; sociology on, 33, 137, 140; structural underpinnings of, 330; transport systems and, 91, 99–100; types of, 116; violence and, 70
climate justice, 13–14, *14*, *15*, 22–25, 28, 31–38, 130–32. *See also specific topics*
Climate of Injustice (Roberts, Park), 131
climate reductionism, 33, 208–9, 213, 215–16, 220
climate science, 14, 25, 236–49, 319
Clinton, Bill, 7
cloud brightening, 19, 112
cloud seeding, 263
CO_2. *See* carbon dioxide (CO_2) emissions
coal, 74, 111–12; mining, 76, 209, 330
coccolithophores, 228
Cochabamba, Bolivia, 28
Coen, Deborah, 249n8
cogeneration power plants, 3
Cold War, 215–16, 218–19
collective action, 68, 73–74
collectivity, 24, 33, 58
colonialism, 14, 26–27, 82n18, 140, 228, 266, 317, 320; Arctic, 209–16, 220; automobility and, 95–96; borders and, 279; capitalism and, 36; in Central America, 158–59; decision-making and, 55, 216; European, 199–200; GHG emissions and, 324–26, *326*, *327*, 328; Indigenous people impacted by, 26, 214, 328; migration and, 154–56; progress narratives and, 204; public transport and, 97; religion and, 272, 277, 281; settler, 23, 25, 55, 58, 210, 214, 330–31; slavery and, 226, 229–30, *230*, 272n53; violence and, 88–89, 166, 200; West Africa impacted by, 162, 166

Columbia University, 244, 250n13; Columbia Climate School, 21–22, 206n38, 291–93, 303nn12–13, 304n18, 304n20, 309n51

Commane, Róisín, 35, 318, 322

commercial fishing, 219

commercial logging, 28

Commission for Racial Justice, UCC, 5

communication, 241, 263, 281, 318–19, 331, 333; Briggs on, 259–60; cross-species, 264–65; Latour on, 261. *See also* language

communism, 167

communities of color, 3–4, 7, 48, 94, 231–33, *233*, 291

comparative politics, 67

compensation, 25, 76, 185; agricultural insurance as, 262–63

Comprehensive Environmental Justice Enforcement Strategy, US Justice Department, 231

Conference of the Parties 13 (2007), 19–20

Connor, Michael, 307n43

consumption/degradation paradox, 134

consumption dynamics, 88–89, 95–96, 137–38, 201–3

Conway, Erik M., 237–38

cooling, power demand for, 115–17

Copenhagen, Denmark, 261–62

Corporate Watch Group, 28

corporations, 74, 77, 137; agriculture and, 159, 181

corrective justice, 14

corruption, 98, 163, 168

cosmology, 33, 59, 277, 331–33

costs, 68, 78–80, 120; adaptation, 32, 325; of climate change, 16, 71, 323; climate inaction related, 71–72; decarbonization, 74; mitigation, 32, 74, 324

cotton, 199, 226, 228–30, *230*

COVID-19 pandemic, 116, 119, 161, 170, 183

Crate, Susan A., 211

CRCQL. *See* Chester Residents Concerned for Quality of Life

critical pluralism, 10, 22, 30, 318, 331

critical race theory, 14, 16

cross-disciplinarity, 7, 34, 128, 333–35

crude oil, 197–98

cultural preservation, 56

custodial relationships, with land, 58–59

cyclical temporality, 57–58

Dakar, Senegal, 167

Dakota Access Pipeline, 50–51, 145n67

Dalai Lama, 276

Daniel, R. A., 211–12

Davis, Heather, 266–67, 331

Dawson, Ashley, 29

death rates, 69, 181

deaths, 111, 120; of agricultural workers, 184; air pollution related, 110; of environmental activists, 28; Hurricane Ida related, 302n9; of Indigenous people, 210, 228; migrant, 163; social, 164; transport related, 94–95

debt, 164, 202; climate, 35, 132–33; peonage, 162

decarbonization, 48, 66, 131, 201, 243–44, 303n14, 324–25, 329–30; air quality and, 110; costs, 74; geoengineering and, 19; job losses associated with, 75; justice implications of, 66, 198; transport systems and, 99

decision-making, 34, 51–53, 100, 102, 157, 214, 247; Army Corps, 293, 298, 307n43; climate policy, 54, 66; colonialism and, 55, 216; migration, 163

decolonization, 25, 132, 201, 277, 317–18, 330

Deep South Center for Environmental Justice, 11

deforestation, 28

degli Uberti, Stephano, 165

dehumanization, 29, 181

delay, climate, 139

dematerialization, 134

democracies, industrialized, 68, 79
democratization, 201
demographics, 3–5, 165, 226, *230*, 232, *233*
Demuth, Bathsheba, 217
denial, climate, 139, 201
Department of Agriculture, US, 181
Department of Environmental Conservation, New York State, 304n18
Department of Environmental Protection, New York City, 297
Department of Labor, US, 184
dependency theory, 154–55
desertification, 202
"Designing Community-Led Plans to Strengthen Social Cohesion" (RCCP), 292
Detroit, Michigan, 199
developed countries, 28–29, 68–71, 76, 79, 133–34, 137
developing countries, 28–29, 68–71, 76, 87, 133–34, 137
Dibee, Joseph, 28
dictatorships, 159, 167
diesel, 90, 111, 113, 120
difference, politics of, 49–50
dignity, human, 55–57, 59, 129, 158–59, 198
disadvantaged communities, 66
discrimination: land use and, 8, 89; racial, 3–7, 14, 16, 26–27
discursive misrecognition, 51–52
disinvestment, 6, 8–9, 302n4
displacement, of people, 33, 135, 153, 210, 216, 245, 279; climate disasters and, 161; forced, 158, 214–15, 217; gentrification and, 9; of Indigenous people, 88–89, 228; transport and, 94
Disposition of Nature, The (Wenzel), 195–96, 199
dispossession, 155, 161, 167
distributive injustice, 50, 53, 78
distributive justice, 14, 47–51, 67, 75, 129
distributive politics, 6, 67–80
Drake, H. F., 242

drought, 154, 202, 255–56, 258–59, 262–63, 267
drug trafficking, 154, 159, 168
Dumping in Dixie (Bullard), 6
Dunlap, Alexander, 39n26

EA. *See* Effective Altruism
Earth Beings (Cadena), 255, 260–61
Earth First, 27–28
Earth Institute, Columbia University, 21
Earth Liberation Front, 28
East Side Coastal Resiliency Project (ESCR), New York City, 308n45
Eco-Emancipation (Krause), 315
ecologically unequal exchange, 35, 133–34
economic development, 133
economic growth, 29, 69, 78, 136–38, 162–63, 218, 328
economic inequalities, 21, 57, 70, 131, 154
eco-racism, 46
education, 55, 69, 246, 248; climate mobility and, 161–62; climate science, 236–40; geoscience, 232–33, *233*
Edwards, Guy, 139
Effective Altruism (EA) movement, 239, 247–48
Egypt, 206
EJAtlas. *See* Environmental Justice Atlas
EJCC. *See* Environmental Justice Coordination Committee, HATS
electric vehicles (EVs), 120
electrification, 101, 116
El Salvador, 159
emissions, 7, 68, 70, 78, 115, 134; carbon dioxide, 18, 57, 99, 109, *327*; coal, 111–12; global, 16, 68, 73, 134, 324, *327*; legacy, 19, 316, 325; militarization and, 137–38; Paris Agreement and, 28–29; per capita, 93; reduction of, 20, 28–29, 132–33, 318; smoke, 182–83, 185; transport, 25–26, 89–95, 98–99, 112–13; urban greening and, 118–19; visas and, 158. *See also* greenhouse gas emissions

empirical evidence, 20, 70, 82n17, 133–35, 137

energy efficiency, 8, 115–17, 120

energy justice, 31, 33, 36, 196–97, 322

energy poverty, 196, 200, 322

energy transition, 198

Enlightenment epoch, 201, 213

entitlement theory, 155

entrepreneurship, Indigenous, 97

environmental anthropology, 211

environmental benefits, 48, 53

environmental harms, 27, 46, 51, 128–29, 133–34, 280, 321

environmental health sciences, 31, 33, 36

environmental injustice, 128–29, 140, 231, 290, 322, *323*; environmental benefits and, 48; environmental racism and, 27; transport systems and, 92–100

environmentalism, 129–30, 239–40, 257

environmental justice, 3–4, 7, 27, 46–47, 198–99; climate justice and, 35–36, 109–20, 130, 231, 317, 320–23, *323*; Indigenous, 58; misrecognition and, 50; participatory parity and, 53; social justice and, 129–30

Environmental Justice Atlas (EJAtlas), 130

Environmental Justice Coordination Committee (EJCC), HATS, 290, 296–301, 303n13, 307nn43–44

environmental justice movement, 4–5, 46, 131, 140, 275–76

environmental law, 7

environmental load displacement, 35, 133–35

environmental personhood, 55

Environmental Protection Agency (EPA), US, 7, 48, 111, 117

environmental racism, 3–8, 27, 226, 229–30, *230*, 321

environmental sociology, 128, 131, 136

EPA. *See* Environmental Protection Agency

epistemic injustice, 46, 51

epistemological justice, 18, 333

epistemology, 51, 58, 208, 282, 331–33

erasure, 51, 210–11

ESCR. *See* East Side Coastal Resiliency Project

essential workers, 179, 186–87

ethnography, 211, 216, 256

etymology, 206

EU. *See* European Union

Europe, 69, 75, 96, 111; international migration to, 154, 156, 163–65, 168–69, 171n57, 171n62. *See also specific countries*

European Union (EU), 112–14, 163; GHG emissions by, 325–26, *326*, *327*, 328

EVs. *See* electric vehicles

exclusion, 6, 18, 89, 94, 100–101, 226, 232; accessibility and, 53; structural, 26

expertise, 227, 292, 293, 306n36, 331; climate science, 237, 243; imperialism, 52, 316–17

externalities, 202

extractivism, 29, 158, 162, 164–66, 168, 229, 267; dependency theory and, 154–55; parachute science and, 227

extreme weather, 240, 259, 262, 268, 303n14; agricultural labor and, 180, 185; fisheries and, 20

Falzon, Danielle, 140, 328

Farrell, Justin, 139

Federal Bureau of Investigation (FBI), US, 28

federal government, US, 7, 10, 94, 305n27, 308n49

feminism, 51

Fernando, Mayanthi, 281

fertilizers, 115

fetal development, pregnant farmworkers and, 180–81

First National People of Color Environmental Leadership Summit (1991), 27

fisheries, 20, 218–19

fishing, 79, 166–167

flood risk, 8, 70–71, 302n4, 302n9, 304n18, 305n27, 308n49, 321; Arctic, 216; in Central America, 154, 161; New York–New Jersey HATS on, 34, 37–38, 288–301, 301n3, 302n7, 303nn12–14, 304n21, 306n36, 306n38, 307nn43–44, 322, 324–25, 333; transport and, 95, 99
Florida, 9, 185, 291
food scarcity, 179
food security, 153, 160, 164, 181, 183, 187
forced displacement, 158, 214–15, 217
forced migration, 153, 157–58, 171n53
forecasting, weather, 258
foreign aid, 76
foreign investment, 162–63, 165, 328
Fortna, Page, 35, 324, 329–30
Fossil Capital (Malm), 36, 198–201
fossil capitalism, 29, 199–203
fossil fuels, 24, 59, 112–16, 213, 318, 326; air quality and, 110; capitalism and, 29, 199–203; carbon taxes on, 49; combustion of, 110, 119–20, 200, 327–28; Corporate Watch Group on, 28; costs of transitioning from, 78; decarbonization and, 74; energy poverty and, 196; internal combustion engines and, 90; natural gas as, 50, 110–11, 197–98, 206n2, 218; urban greening and, 119
Foster, John Bellamy, 140
Foster, Sheila, 27, 89
France, 166, 171n57
Francis (Pope), 276
Fraser, Nancy, 50, 52
freedom, 55, 57; academic, 246–47, 335
French colonialism, 162
Fridays for Future school strikes, 131
FRLD. *See* Fund for Responding to Loss and Damage
frontline communities, 11, 24, 47, 141–42, 301, 320, 333; flood risk and, 289–90, 293, 297, 303n12, 324
Fullilove, Mindy, 94

FUME. *See* Future of Migration to Europe
Fund for Responding to Loss and Damage (FRLD), UNCCC, 75
Future of Migration to Europe (FUME) project, 165

Gallay, Paul, 37–38, 322, 324–25, 333
gangs, 159, 161–63
gas flaring, 197–98
gasoline, 90
Gaza, 138
Gazmararian, Alexander F., 76
GBV. *See* gender-based violence
GDP per capita, *17*, 137
gender, 49, 72; climate vulnerability and, 130; education and, 161–62; migrant labor and, 181
gender-based violence (GBV), 161
genetic selection, agriculture and, 184
gentrification, climate, 9–10, 263, 293, 329
geoengineering, 18–19, 52, 111–12, 316–19
geology, 213, 226–33, *229*
geoscience, 227, 232–33, *233*
Germany, 69, 171n53
Ghana, 69, 162–63
GHG. *See* greenhouse gas emissions
Ghosh, Amitav, 199–201, 204, 249n8, 326, 328
Giacomelli, Elena, 36, 167
Gilheany, Emma, 36, 264, 328, 332
Givens, Jennifer, 45, 69–70, 321–22, 328, 330
global capitalism, 33, 89, 199
global climate negotiations, 131–32
global dimming, 111–12
global equity, 134
globalization, 95–96, 257
Global Network for Popular Transport, 99–100
Global North, 25, 68–71, 72, 74, 81, 133, 168, 203; cities in, 88–90; GHG emissions by, 16–17, *17*, 23–24, 36, 325,

326, *327*, 328; transport systems in, 92–93

global political economy, 131–34, 138

Global South, 68, 133, 135–36, 139, 197, 200, 203–4, 328; financing pledges and, 76; Global North and, 16–17, *17*, 23–25, 36, 69–71; mitigation costs for, 74; transition costs for, 78; transport systems in, 92–93, 96–99

global surface air temperatures, 111

Global Tipping Points report, 12

global warming, 182, 197–98, 237, 256–57, 259–60, 265–66

Gonzalez, Carmen, 155

Gramsci, Antonio, 139, 145n67

Great Derangement, The (Ghosh), 199–200, 326

green energy policies, 74–76

Greenhouse Gangsters vs. Climate Justice report (Corporate Watch Group), 28

greenhouse gas emissions (GHG), 46, 99, 113, 197, 201, 237–38, 321, 330; air quality and, 109–20; by China, 325–26, *326*, *327*; efficiency in cities and, 115–17, 120, 318; by EU, 325–26, *326*, *327*, 328; gas flaring, 198; geoengineering and, 18; global rate of, 316; historical injustices in, 325–26, *326*, *327*, 328; indirect, 115; by industrializing countries, 70–71; measurement of, 318–19; mitigation efforts and, 241, 324; from New York City, 118–19; Passport Index and, 156; triple injustice and, 54; US, 16–17, *17*, 156, 316, 325–26, *326*, *327*

Greenland, 12, 219

Greenpeace, 27

green space, lack of, 321

green tobacco sickness (GTS), 183

greenwashing, 244, 248–49

grief, ecological, 332

ground-level ozone, 90

GTS. *See* green tobacco sickness

Guatemala, 159

Gulf Coast, US, 203–4

Guyana, 266

H-2A visa program, 184

Haber-Bosch process, 115

Hadden, Jennifer, 261–62

Hakka (language), 37, 255–59, 262–68, 317–18, 332–33

Harlem, New York, 4, 6–7

harm multiplier, 160

Harrison, Jill Lindsey, 140

HATS. *See* New York–New Jersey Harbor and Tributaries Study

Health and Human Services, US, 185

health care, 55, 185–86

heat islands effect, 71, 118–19, 322

heat stress, 180–82, 184–87

heat waves, 8, 69

Henderson, G., 242

High Court, Nigeria, 198

high-density urban developments, 245

high-sulfur bunker fuel, 112

highways, 92–94, 97, 321, 330

Hill, Patricia, 145n67

historical climate responsibility, 16–17, *17*, 22, 25, 67–68, 158, 212, 322

historical injustices, 320–21, 325–26, *326*, *327*, 328

historically marginalized communities, 9, 317, 322, 334

historical responsibility, 17, 22, 25, 67–68, 158, 212, 322

Holland, Breena, 54, 56–57

Honduras, 159–60, 161–62

Honneth, Axel, 50

Hopedale, Nunatsiavut, 216–17

horizontality, 54

House of Representatives, US, 185

How Cities Work (Marshall), 89–90

Hulme, Michael, 213

human-animal relationships, 130

humanism, 45, 318–20

humanities, 18, 34, 38, 246, 318–19. *See also specific disciplines*
human rights, 5, 29, 97, 101, 159, 256
Huq, Saleemul, 135–36
Hurricane Ida (2021), 289, 302n4, 302n9
Hurricane Iota, 161
Hurricane Mitch (1998), 159
Hurricane Sandy (2012), 240, 302n4
hydrogen, 114–15

identity, 14–16, 31, 50, 52–53, 75, 81
Ijaw Youth Movement, 198
illegibility, 277
immigration court systems, 169
immigration status, 324
imperialism, 210, 266; expertise, 52, 316–17; religion and, 272, 279
income, 49, 70, 156, 160
India, 28, 70–71, 75, 200
Indigeneity, 28, 61n46, 261, 280–81, 325, 330–31, 333, 335; anthropology and, 260; Arctic, 36, 209–10, 214–17, 328, 332; in Canada, 215, 261; Central American, 159; climate justice and, 46–47; colonialism impacting, 26, 214, 328; cultural preservation and, 56; displacement impacting, 88–89, 228; Indigenous entrepreneurship, 97; Indigenous knowledge systems and, 80, 208, 211–17, 219, 294; Indigenous philosophy and, 16, 55–57, 59; Indigenous rights and, 5, 27, 50, 265; intergenerational climate justice and, 58–59; marginalization and, 256; #NoDAPL movement and, 50–51; ontology and, 30; participatory justice and, 53–55; religion and, 272, 282; sociology and, 129; temporality and, 278; United Nations Declaration on the Rights of Indigenous People, 50; in the US, 26, 79, 130, 140, 145n67. *See also specific Indigenous groups*
indoor air quality, 8, 116–17, 120

industrialization, 12, 26, 68–69, 77, 154–55, 197, 199–200, 229
Industrial Revolution, 325–26, *326*, *327*, 328
"informal" transport, 26, 97–99
informed consent, 52–54
infrastructure, 8, 24, 71, 109–10, 186, 215, 231; dependency on, 216, 328; transport, 90–97, 101, 328, 330; water, 265–66. *See also* New York–New Jersey Harbor and Tributaries Study
institutions, 22, 53, 77–78, 96. *See also specific institutions*
insurance industry, 139, 239–41, 245, 330; agriculture insurance and, 262–63, 265–66, 331–32
interdisciplinarity, 7, 21, 39n18, 67–68, 80–81, 101–2, 273, 334
intergenerational climate justice, 46–47, 58–59, 67, 316
Intergovernmental Panel on Climate Change (IPCC), 7–8, 19, 101, 141, 157–58, 242
internal climate migration, 153
internal combustion engines, 90
international agreements, 24–25, 29, 218–19, 328
international inequalities, 131, 140, 328
International Maritime Organization, 112
International Organization for Migration, 156
international relations, 67
International Transport Forum, 99
internet, 161–62
intersectionality, 9, 27, 129–30, 145n67, 160
interspecies justice, 217, 332
intersubjectivity, 22, 318, 331
Inuit Circumpolar Council, 209, 212
Inuit people, 210–12, 215–17, 332
Iñupiat community, 216, 328
invasive species, 218
investments, 25, 75, 97, 118, 244; climate risk impacting, 248; federal, 305n27, 308n49; foreign, 162–63, 165, 328;

land-use, 89; transport related, 93, 96–98, 100–101
invisibility, 156, 280, 315
Io Capitano (film), 163
IPCC. *See* Intergovernmental Panel on Climate Change
irregular migration, 159, 163, 167
Islam, 154, 165–66
isoprene, 119
Israel, 138

Jafry, Tahseen, 39n18
job loss, 75
Jobson, Ryan, 255
Johnson, Lyndon, 18
Joint Fisheries Commission, 219
Jorgenson, Andrew W., 137
justice, 102, 127, 273, 277, 281; corrective, 14; distributive, 14, 47–51, 67, 75; energy, 31, 33, 36, 196–97, 322; epistemological, 18, 333; intergenerational climate, 46–47, 58–59, 67, 316; interspecies, 217, 332; and the more-than-human world, 18, 33, 47, 51, 58; multiscalar, 329–31; multispecies, 33, 130, 140, 208, 214–20, 264–66; participatory, 51–55, 67, 80; procedural, 80, 129, 157, 169; recognitional, 50–52, 129, 158–59; redistributive, 52, 76, 132–33, 169; relation theory of, 47, 58–59, 316; restorative, 132, 158; social, 130; types of, 32, 35, 45–57, 129, 316, 319. *See also* climate justice; environmental justice
Justice Department, US, 231
"Just Transition Agreement," (PSOE), 76
"just transition," climate, 13, 22, 76, 116, 139–40, 198, 330

Kaplan, Robert, 204
Keane, Webb, 259
Kenya, 96
Kirsch, Stuart, 265

Klopp, Jacqueline, 25–26, 35, 318, 321, 328, 330
knowledge, 29–30, 34, 158, 257, 261, 266, 317, 331–33; coproduction of, 209, 211–12, 219; Indigenous, 80, 208, 211–12, 219, 294; traditional environmental, 211, 214, 267–68. *See also* expertise
Kohn, Eduardo, 264
Kono, Daniel Yuichi, 74–75
Krause, Sharon, 315
Kyoto Protocol, EU, 113

labor movements, 139
Lajus, Julia, 36, 264, 328, 332
Lamb, William F., 139
land use, 3–6, 90, 92, 232, 302n4; discriminatory, 8, 89; distributive justice and, 48
language, 33, 186, 206, 209; Hakka, 37, 255–59, 262–68, 317–18, 332–33; scientific, 37, 80, 211, 256–57, 259–61, 266–68, 317–18, 331; translation and, 256–57, 260–61, 264–65
Late Cretaceous coastline, 26, 36–37, 226, 228, *229*
Latour, Bruno, 261
law enforcement, 169
legacy emissions, 19, 316, 325
LeQuesne, Theo, 140, 145
less developed countries, 133
liberalism, 58
Libya, 163
linear temporality, 57
literary studies, 195–206
Livingston, Julie, 201–3, 328
loan forgiveness, 132
logging, commercial, 28
London Fog event (1952), 111–12
Long, Sheng, 37, 211, 317, 329, 331–32
long-durée, 212–13
longitudinal analysis, 212
Loss and Damage Fund, UNFCC, 28–29, 140, 169, 328–30

Louisiana, 184
lower-income countries, 154, 227
low-income communities, 7–8, 91, 94–95, 98, 101, 231, 321; participatory parity impacting, 53–54; pollution impacting, 5, 48
Lowndes County, Alabama, 231
Luzzatto, Matthew W., 296, 303n12

machine-learning, 139
Malawi, 69
Malm, Andreas, 35, 198–202
Māori people, 58
marginalized communities, 10, 25–26, 53–54, 71, 237, 263, 280, 331–32; anthropology and, 256; climate actions and, 78–79; distributive justice and, 48; First National People of Color Environmental Leadership Summit addressing, 27; historically, 9, 317, 322, 334; sociology addressing, 32–33; transport and, 94
marine cloud brightening, 112
Marino, Elizabeth, 216
Marshall, Alex, 89–90
Martínez-Alier, Joan, 130
Marx, Karl, 204
mass manufacturing, 92–93
mass transit, 100–101
Mayfield, Zulene, 3–4
measuring, reporting, and verification (MRV) tools, 18–21, 316–19
Megalopolitan Coastal Transformation Hub, 308n48
Meikle, Mandy, 39n18
Meixian, China, 211, 255–59, 262–67, 317–18, 331–32
Mendes, Chico, 28
menstrual poverty, 161
mental health, 162
meta-capabilities, 56–57
meteorology, 236
methane, 109–10, 115

Mexico, 69, 159, 170n42
Miami, Florida, 9, 291
microplastics pollution, 91–92
migration, 31, 33, 89, 217–19, 329; agricultural work and, 36, 167–68, 179–87, 324; borders and, 156, 159, 163, 170n42, 279–80, 283; forced, 153, 157–58, 171n53; international, 153–69, 170n37, 171n57, 172n62; seasonal, 167–68, 184–86
Mildenberger, Matto, 77
militarization, 137–38, 154, 159, 162, 201
Mills, Charles, 49
minibus systems, 26, 98
minoritized communities, 49, 53–54
misrecognition, 50–53
missionaries, 215–16
Mississippi, 229–30
Mississippi Delta, 199
mitigation, climate, 79, 109, 113, 118, 241, 245, 321–22; adaptation and, 20, 31–32, 35, 48–49, 59, 242, 318, 323–25; costs of, 67–68, 72, 74; discursive misrecognition and, 51–52; gentrification and, 9; MRV tools and, 20; role of states in, 329–30
mobility, climate, 10, 36, 89–91, 153–69, 157; politics of, 80–81, 216–19
Mohai, Paul, 6
monkeywrenching, 27
monodisciplinarity, 18–19, 21, 34, 319, 334
more-than-human world, 130, 214–20, 261, 264–65, 279–83, 332–33; agency and, 24; capabilities approach to, 56; justice and, 18, 33, 47, 51, 58; representation and, 55
mortality rates, 69
MRV. See measuring, reporting, and verification
multidisciplinarity, 9, 127, 132–33, 138, 141–42, 318–19, 333–35; climate justice and, 12–14, 14, 15, 16–22; climate mobility and, 153; need for, 32, 316

multimodal transport systems, 100–101
multiscalar, 11, 197, 199, 206, 278–80, 329–31
multiscalar justice, 329–31
multispecies justice, 33, 130, 140, 208,
    214–20, 264–66
Mumford, Lewis, 93–94

NAACP. *See* National Association for the
    Advancement of Colored People
Nadasdy, Paul, 261
Nairobi, Kenya, 96
narcotics, 154, 168
National Association for the Advance-
    ment of Colored People (NAACP), 93
National Environmental Protection Act
    (NEPA), US, 53
nation-states, 11, 82n18, 131, 133, 135,
    278–81; sovereignty of, 162, 215
natural gas, 50, 110–11, 114–15, 117, 197–98,
    206n2, 218
natural sciences, 14, 17–18, 23, 208–9, 316,
    318–19, 334
nature-based climate solutions (NbCS),
    118–19
Nature Climate Change (journal), 14
Navajo Nation, 53, 61n46
NbCS. *See* nature-based climate solutions
neocolonialism, 331
neoliberalism, 29
NEPA. *See* National Environmental
    Protection Act
net zero pledges, 244
New Jersey, 3, 6–7
New Orleans, 9, 93, 229–30
New York, 4, 6–7, 98, 110, 112, 117–19;
    Brooklyn, 233, 291, 293, 305n24. *See
    also specific departments, offices*
New York City Environmental Justice
    Alliance (NYC-EJA), 291–92, 304n18
New York City Panel on Climate Change
    (NPCC), 7–9, 302n4
New York–New Jersey Harbor and
    Tributaries Study (HATS), Army

Corps, 34, 37–38, 301n3, 304n21, 322,
    324–25; Climate & Environmental
    Justice Advisory Group, 306n36;
    EJCC, 290, 296–301, 303n13, 307nn43–
    44; Megalopolitan Coastal Transfor-
    mation Hub on, 308n48; possible costs
    related to, 288–89, 302n7, 306n27;
    RCCP and, 291–93, 303n12, 304n18,
    304n20, 306n38, 309n51; role of
    reform in, 293–98, 301, 305n27,
    308n46; Trump impacting, 288–89,
    296–97, 301, 303n14, 308n49
New Zealand (Aotearoa), 55, 58
Nicaragua, 159
Niger, 163
Niger Delta, 196–97, 199
Nigeria, 78, 198
nitrogen oxide, 110–11, 113–15, 117, 119
#NoDAPL movement, 50–51
noise pollution, 94
nonhuman species, 55–56, 129–30, 210,
    214–20, 281–83. *See also* more-than-
    human world
North Carolina, 5, 27, 117, 275–76
North Dakota, 197, 199
North River Wastewater Treatment
    Plant, 4
Norway, 77, 218–19
NPCC. *See* New York City Panel on
    Climate Change
nuclear testing, 27
numerical simulation, 236, 240
Nussbaum, Martha, 55
nutritional shortages, 179
NYC-EJA. *See* New York City Environ-
    mental Justice Alliance

oak trees, 118–19
Occupational Health and Safety
    Administration (OSHA), US, 184–85
oceanography, 26, 33
Office of Climate and Environmental
    Justice, New York City, 297, 306n36

oil extraction, 199, 329
oil spills, 196
Oklahoma, 180
ontology, 23, 25–27, 30, 33, 264, 281, 320, 331–33
Oreskes, Naomi, 237–38
OSHA. *See* Occupational Health and Safety Administration
Ottinger, Gwen, 51
Our Only Home (Dalai Lama), 276
ozone, 110, 113, 119

Page, Edward, 57
Pan-Algarra, Sara, 36, 161
Panama Canal, 159
parachute science, 227
paratransit, 97–98
Paris Agreement (2015), 28, 243, 316
Parks, Bradley, 131–32
Parks Department, New York City, 118–19
participatory justice, 51–55, 67, 80
participatory parity, 53–55
particulate matter (PM), 90, 110–11, 113–15, 117, 182
Passport Index, 156, *157*, 329
PCBs. *See* polychlorinated biphenyls
Pellow, David Naguib, 129, 140
Pennsylvania, 3–4
personhood, environmental, 55, 58
Peru, 260–61
pesticides, chemical, 180, 182, 186–87
petrol, 113
petroleum, 28
philosophical approaches, to climate justice, 14, 35, 321, 331–33
photosynthesis, 118
physical science, 13, 31–33
phytoplankton, 26, 228
piece work, agricultural, 185
place-based custodial relationships, 58–59
plantations, 159; slavery and, 229–30, *230*
pluralism, critical, 10, 22, 30, 318, 331

PM. *See* particulate matter
Point Community Development Corporation, 292
Polanyi, Karl, 139
policy, climate, 46, 72–75, 77–80; air quality and, 111–14, 119–20; decision-making and, 54, 66
political repression, 196
political science, 31–32, 35, 66–68, 72–75, 77, 80–81, 92, 317
political theory, 31–32, 45–47, 316–17
pollution, 48, 51, 77, 216–17, 279; air, 3–4, 90–91, 94–95, 97, 101, 111–20, 321; Black communities impacted by, 3–5, 94, 275–76; PM, 90, 110–11, 113–15, 117, 182; transit, 90–91, 94–95, 97–99, 101. *See also* emissions; greenhouse gas emissions
polychlorinated biphenyls (PCBs), 27, 117
pomelo farming, Chinese, 33–34, 37, 255–59, 262–68, 317–18, 331–32
positivism, 67, 81n4, 317
postcolonial identities, 75
poverty, 153, 159, 162, 168, 180, 263, 329; energy, 196, 200, 322; infrastructure and, 71; menstrual, 161; rising temperatures and, 69; urban greening and, 118–19
power dynamics, 139, 164, 208, 292, 325–26, *326*, *327*, 328; asymmetrical, 214; in climate policymaking, 54; environmental anthropology and, 211; expertise imperialism in, 316–17; fossil capitalism, 201; migration, 154–55; political science on, 66–67; race and identity in, 14, 16; redistribution of, 67; research, 52
pregnancy, 161, 180–81
preventive medical care, 185–86
private sector, 33, 37, 237–38, 243–45, 247
procedural justice, 80, 129, 157, 169
propane, 117
property values, 4, 10

PSOE. *See* Spanish Socialist Party
public communication, 241
public goods, 237–38
public health, 91, 96, 100, 179, 182, 186–87,
231; air pollution and, 94, 97; climate
mitigation and, 318
public opinion, 242
public sector, 165
public transit, 4, 91, 97
El Puente (organization), 293, 303n13
Purucker, David, 140

qualitative methodologies, 75, 76, 128,
136, 141, 166
quantitative finance, 239
quantitative methodologies, 20, 128,
135–36

*Race and the Incidence of Environmental
Hazards* (Bryant, Mohai), 6
racial capitalism, 8, 89
racial discrimination, 3–7, 14, 16, 26–27
racial disparities, 3–10, 129, 321–22
racialization, 275–76
racial segregation, 94
racism, 94, 210; colonialism and, 140;
eco-racism, 46; environmental, 3–8, 27,
226, 229–32, *230*, 321; systemic, 232, 334
radon, 116–17
Rawls, John, 47, 49
Raymond-Yakoubian, J., 211–12
RCCP. *See* Resilient Coastal Communities Project
recognitional justice, 50–52, 129, 158–59
redistributive justice, 52, 76, 132–33, 169
redlining, 6, 8, 302n4
redress mechanisms, climate policy,
74–75
reductionism, climate, 33, 208–9, 213,
215–16, 220
reform, 68, 72, 77–78, 293–98, 301, 305n27,
308n46; activism compared to, 27–28;
system change compared to, 133, 140

refugees, 157–58, 163
regenerative agriculture, 184
reinsurance industry, 239
relational theory, of justice, 47, 58–59, 316
religion, 34, 37, 81, 215–16, 272–83, 283n1,
317, 332. *See also specific religions*
religious studies, 273–74, 277–78, 281, 317
renewable energy, 79, 113–15, 243
representation, 52, 54–55, 67, 77–78, 195;
of historically marginalized communities, 334; political, 80
resettlement, 157–58, 210
resilience, climate, 13, 89, 109–10, 130, 136
Resilient Coastal Communities Project
(RCCP), Columbia Climate School,
291–93, 303n12, 304n18, 304n20,
306n38, 309n51
resource extraction, 165, 168, 229, 272,
276, 279–80
respiratory illnesses, 3–4, 98
responsibility, climate, 14, 18, 20, 46, 49,
57, 132, 134–36; historical, 16–17, *17*, 22,
25, 67–68, 158, 212, 322
restorative justice, 132, 158
reverse anthropology, 265
Riaz, Zara, 35, 324, 329–30
Ribot, Jesse, 164
rickshaws, 26, 96–98, 328
rights, 56, 163, 231; animal, 281; civil, 7, 93,
231; to education, 161; human, 5, 29, 97,
101, 159, 256; Indigenous, 5, 27, 50, 265
rising temperatures, 19, 23, 28–29, 68–69,
71, 154, 182, 321
rituals, 37, 258, 263, 265, 275–76, 278
Rivera, Mayra, 277
Rivet and Sons LLC, 184
Roane, J. T., 281–82
Roberts, J. Timmons, 131–32, 138
Rodriguez, Dariella, 292
Rodriguez, Iokine, 29
root shock, 94, 97
Royal Canadian Air Force, 216
Royal Canadian Mounted Police, 215

Ruiz Soto, A., 160
rural communities, 53, 118; in China, 211, 216, 231, 256, 258–59, 263, 266; disasters displacing, 161; migrants leaving, 153, 165; suburbanization of, 94
Russia, 69, 138, 166, 197, 218–19

Said, Edward, 199
Saint Louis, Senegal, 79, 324
Salako, Ishag, 78
sanitation, right to, 231
scales, 14, 89, 195–206, 278–83, 320; global, 35–36, 97, 131, *323*; international, 132–33; local, 131, 317; multiscalar, 11, 197, 199, 206, 278–80, 329–31
SCCA. *See* South Camden Citizens in Action
Schlosberg, David, 56, 60n11
Science Advances (journal), 38n1
scientific expertise, 331
scientific language, 37, 80, 211, 256–57, 259–61, 266–68, 317–18, 331
sea ice, 36, 209, 214, 216–17, 219
sea-level rise, 47, 68, 70–71, 79, 135, 166, 299
seasonal migration, 167–68, 184–86
seawalls, 8, 24, 79, 289, 295
secularism, 37, 211, 274–78, 282
Securities and Exchange Commission, US, 243
securitization, 82n18
sedentarization, 215–16
segregation, racial, 94
self-determination, 198, 217
self-devouring growth, 201–3
Self-Devouring Growth (Livingston), 201–2
Sen, Amartya, 55, 155–56
Senegal, 79, 162–67, 324
settler colonialism, 23, 25, 55, 58, 210, 214, 330–31
sewage treatment plants, 3
sexual harassment, 181
shale gas, 206n2

Shaman, Jeffrey, 36
Sheller, Mimi, 156
Sherbinin, Alex de, 36, 329
Ship Breaker (Bacigalupi), 203–5
Shishmaref, Alaska, 216, 328
"sick-building syndrome," 116
SIDS. *See* small island developing states
Simpson, Audra, 210–11
slavery, 26, 36–37, 171n53, 226, 229–32, *230*
small island developing states (SIDS), 70–71
smart cities, 91–92
smartphones, 167, 258
smoke emissions, 182–83, 185
smugglers, human, 163, 167
Sobel, Adam, 37, 239–41, 318, 330
social cohesion, 69–70, 128, 135, 141, 291–93
social death, 164
social justice, 130
social media, 167, 263, 288–89
social sciences, 6–7, 16–18, 34, 38, 128, 212–13, 246, 318–19
social welfare, 155
societal goods, 47–48
socioeconomic status, 3–8, 48, 49–50, 324
sociology, 31–33, 35, 81, 127, 135, 139–42, 316–17; climate debt and, 132–33; climate justice and, 130–32; ecologically unequal exchange and, 133–34; environmental justice and, 129–30; environmental sociology and, 128, 131, 136; social drivers of the climate crisis and, 136–38
solar energy, 245
solar engineering, 18
solar radiation management (SRM), 18, 52
solidarity, 59, 280, 283
South Africa, 76
South America, 321
South Camden Citizens in Action (SCCA), 3
South Sudan, 69

Soviet Union, 171n57, 216
Spanish Socialist Party (PSOE), 76
spatial dimension, of climate justice,
    23–27, 33, 320, 329–31
special interest groups, 72–73, 77–78
speculative fiction, 203–5
spirituality, 37, 81, 272–73, 275, 280, 332
SRM. *See* solar radiation management
Standing Rock Sioux Nation, 50–51
states, 140, 274, 278–81, 329–32; repression
    by, 27–28. *See also* nation-states
statistical analysis, 236
steam power, 200
steel manufacturing, 114
Stensrud, Vaughn, 265–66
Stockholm conference (1972), 131
storm barriers, 288–91, 301n3, 302n7
Storm Surge (Sobel), 240
structural dimension, of climate justice,
    24–27, 134, 139–40, 154–55, 157, 320, 330
structural disadvantages, 290, 292
structural exclusion, 6
structural inequality, 53, 81, 128, 316
structural injustices, 17, 158–60, 290, 292,
    316, 333
subjectivity, 52, 56
Subjects of Intergenerational Justice
    (Winter), 30
subnational levels, climate inaction at,
    67–68, 71–72
subsistence farmers, 181
suburbanization, 90, 94–95
suffering, 46, 99–100, 127
sugar, 229
sugarcane workers, 182
sulfur dioxide, 111
sulfur gas, 111–12
Superfund sites, 3
Superstorm Sandy (2012), 240, 302n4
survival, 27, 136–37
sustainability, 35, 57, 101, 136–37
system change, 133, 140, 142
systemic racism, 232, 334

Taylor, Charles, 50
Taylor-Seymour, Raffaella, 37, 81, 263, 332
technocracy, 275
technologies, 59, 138, 167, 206n2, 213;
    adaptation, 68; cloud seeding, 263;
    geoengineering, 18–19, 316–17; smart
    cities, 91–92; smartphone, 258;
    transport, 90–93
temperature rise, 28–29, 315
temporal dimension, of climate justice,
    23, 26–27, 33, 59, 220, 319–20, 323;
    Chakrabarty on, 278; cyclical time
    and, 57; historical injustices and, 321,
    324; historical responsibility and, 17,
    22, 25, 67–68, 158, 212, 322; scales of,
    198–99, 204–6
temporary protected status (TPS), 159,
    170n37
territorial sovereignty, Indigenous, 51
textile mills, British, 199
threat multiplier, climate change as, 186,
    318, 324
Thunberg, Greta, 24, 127, 131, 142
Tingley, Dustin, 76
tobacco industry, 183
Todd, Zoe, 209, 214, 266–67, 331
toxicity, 48, 90–91, 116–17, 202, 216–17; of
    coal emissions, 111–12; of fossil fuel
    combustion, 110; racial disparities in
    exposure to, 129, 321–22; regulation of,
    321; of waste facilities, 3–5
"Toxic Wastes and Race in the United
    States" (UCC), 5
TPS. *See* temporary protected status
trade-offs, 245, 329, 334; between
    environmental and climate justice,
    320, 322; between mitigation and
    adaptation, 318, 323–25
trade relations, 133–34, 154, 163, 165
Traditional Environmental Knowledge,
    211, 214, 267–68
trafficked Indigenous people, 210
Trail of Tears, 228

Transatlantic Slave Trade, 229
transition risks, 198, 244
transitions, energy, 54, 77, 79, 198, 323
translation, language and, 256–57,
    260–61, 264–65
transport systems, 53, 243, 245, 318;
    cities and, 4, 88–102; emissions and,
    25–26, 89–95, 98–99, 112–13;
    environmental injustices and, 92–100;
    in the Global South, 25, 328;
    infrastructure and, 330; infrastruc-
    ture for, 90–97, 101, 328, 330;
    investments and, 93, 96–98, 100–101;
    safety in, 184
trauma, 94, 161, 211–12, 214
triple injustice, 54, 66
tropical meteorology, 240
Tropical Storm Eta, 161
Trump, Donald J., 7, 184, 240, 288–89,
    296–97, 301, 303n14, 308n49
Turpo, Nazario, 260–61

UCC. See United Church of Christ
Ukraine, 138
unaccompanied minors, migrants as,
    159, 168
underdevelopment, 69, 200
underinvestment, 95
underrepresented minority communities,
    53, 232
underserved communities, 91, 94–95,
    97–98, 100, 321
undocumented migrants, 186; children as,
    179–80
unemployment benefits, 74–75
unequal distribution: of climate change
    costs, 16, 71; of climate harm, 322; of
    global change risks, 183–84; of societal
    goods, 48
UNFCC. See United Nations Framework
    Convention on Climate Change
unforeseen consequences, 24, 318, 320–23,
    323

United Church of Christ (UCC), 5
United Fruit Company, 159
United Nations Declaration on the Rights
    of Indigenous People, 50
United Nations Framework Convention
    on Climate Change (UNFCC), 28–29,
    75, 140, 158, 169, 328–30
United States, 69, 129–30, 166, 170n42,
    185, 206n2, 281–82; Affordable Care
    Act, 186; Air Force, 216; American
    Clean Energy and Security Act,
    74–75; American Rescue Plan, 75;
    Black Belt region, 26, 36–37, 226,
    228–32, 229, 230; car-centered
    infrastructure, 90–96, 328, 330;
    Census, 184, 230; climate justice in,
    92–95, 275–76; climate policies in,
    77–78; coal power plants in, 111;
    Department of Agriculture, 181;
    Department of Labor, 184; environ-
    mental justice movement in, 275–76;
    EPA, 7, 48, 111, 117; FBI, 28; federal
    government, 7, 10, 94, 305n27, 308n49;
    GHG emission by, 16–17, 17, 156, 316,
    325–26, 326, 327; global emissions
    and, 134; hydrogen in, 114; Indig-
    enous communities in, 26, 79, 130,
    140, 145n67; international migration
    to, 159–60, 163, 167–69, 170n37; Inuit
    resettlement by, 215; job loss in, 75;
    Justice Department, 231; migrant
    agricultural workers in, 36, 167–68,
    179–87, 324; NbCS in, 118–19; NEPA,
    53; OSHA, 184–85; redlining in, 6, 8,
    302n4; Securities and Exchange
    Commission, 243; Water Resources
    Development Act, 289, 294, 299, 301.
    See also Army Corps of Engineers;
    specific states
universalism, 145n67, 273–75, 277, 279
UPROSE (organization), 292–93,
    305n24
uranium, 53, 116

urban development, 35, 70, 93, 245. *See
also* cities
urban greening, 118–19
urbanization, 88–89, 97–98
urban-rural divides, 263
urban sprawl, 202
urban studies, 25–26, 32, 88–92, 101–2

Vaughn, Sarah, 266
ventilation, 116–17, 120, 322
vertical flow, of exports, 133–34
Vietnam, 70
violence, 29, 155, 159, 163, 210; climate
change and, 70; colonial, 88–89, 166,
200; of displacement, 88–89; gang, 162;
GBV, 161
visa policies, *157*, 158, 184
Volkswagen, 113
volunteerism, 245
voters, 72–76, 78, 83n31, 245
vulnerability, climate, 9–10, 154, 237, 264,
266–67, 302n4, 324; agency and, 32–33;
of agricultural workers, 36, 167–68,
179–87, 324; Brandt line and, 69;
capabilities impacted by, 57; climate
finance and, 140; decision-making and,
51; forced relocation and, 215;
gendered, 130; intersectionality and,
160; intersectoral, 160; marginalized
communities and, 53–54, 71; policies
and, 74; transport planning and, 95

wages, 180
walkability, 91, 96–97, 100
war, 70, 138, 159
Warlenius, Rikard, 132, 330
"war on terror," 162
Warren County, North Carolina, 5, 27
water, 114–15; access to potable, 50, 181–82;
contaminated, 101; right to, 231
Water Resources Development Act (2022),
US, 289, 294, 299, 301
WE ACT (organization), 4, 11

wealth redistribution, 201
Weaver, Emily, 36
well-being, human, 55, 81, 132–36, 138,
231, 273–74; OSHA and, 184; psychological, 52
Wenzel, Jennifer, 36, 322, 326, 328–29
West Africa, 154, 156, 162–68, 171n53, 321
West Papua, New Guinea, 282
white flight, 95
whiteness, 49
wildfires, 113, 182–83, 185–87, 255
Wilson, Jake, 39n18
Winckler, Gisela, 26, 36–37, 333–34
wind energy, 245
Winter, Christine, 30, 54–55, 58–59,
330–33
Wisemiller, Bryce, 301n3
Wit, Sara de, 256–57
Witlacil, Mary, 35, 66, 316–17, 321, 330–33
Wolin, Sheldon, 45
wood burning, 112–14
work authorization, migrants and, 180
workplace standards, 184–85
World Bank, 26, 96, 133, 260–61
World People's Conference on Climate
Change and the Rights of Mother
Earth, 28
world systems theory, 133, 154–55
World Trade Organization, 133
Wrathall, David, 36

yi (Chinese concept), 264
Young, Alexander, 294, 296–97, 303n13
Young, Iris Marion, 49
youth bulge, migration and, 165
youth climate movement, 24, 131
Yua, E., 211
Yukon, Canada, 161

zero emissions, 78
Ziska, Lewis, 36, 318, 321
zoning practices, 6, 10, 48
Zota, Ami, 36

GPSR Authorized Representative: Easy Access System Europe, Mustamäe tee
50, 10621 Tallinn, Estonia, gpsr.requests@easproject.com

www.ingramcontent.com/pod-product-compliance
Lightning Source LLC
Chambersburg PA
CBHW022133020426
42334CB00015B/874